Energy and the Living Cell

An Introduction to Bioenergetics

Wayne M. Becker

University of Wisconsin
Madison, Wisconsin

J. B. Lippincott Company

Philadelphia   New York   San Jose   Toronto

Library of Congress Cataloging in Publication Data

Becker, Wayne M
  Energy and the living cell.

  Includes bibliographical references and index.
  1. Bioenergetics. 2. Cell physiology. I. Title.
QH510.B4   574.1'9121   77-1103

1 3 5 7 9 8 6 4 2

Book produced by Ken Burke & Associates

Text and cover designer: Michael Rogandino
Illustrator: Carl Brown
Copyeditor: Don Yoder
Proofreader: Marjorie Hughes
Makeup: Douglas Luna
Compositor: Typesetting Services, Ltd.
Printer: Halliday Lithograph

To the students of Biocore 303—past, present, and future—who have made the writing of this book both necessary and pleasurable

Nach Energie drängt,
An Energie hängt
  Doch alles.

                    . . . with apologies to Goethe

Preface

The book you have before you began as a series of notes and handouts on topics in bioenergetics and cellular energy metabolism prepared for use in Biocore 303, a course in cellular biology at the University of Wisconsin—Madison. At the suggestion of students and colleagues, I have endeavored to collect and organize these materials into text form in the hope that they might prove useful in other quarters as well. This book should be particularly suitable for undergraduate courses in cellular biology and biochemistry and perhaps also as a self-teaching aid for advanced students.

The format and the readership may have changed somewhat in the publication process, but the intention remains unaltered—to sketch in as lucid a manner possible the essential concepts underlying the flow of energy in biological systems. This is a fundamental topic in biology, touching as it does on virtually every aspect of what a cell is and can do. The book begins with a general discussion of the overall macroscopic flow of energy and mass in the biosphere (Chapter 1) and continues with the fundamental principles of bioenergetics (Chapter 2) and enzyme kinetics (Chapter 3) that determine in what direction and at what rate the various cellular reactions and processes occur. The stage is then set for a discussion of energy metabolism in chemotrophs (Chapters 4, 5, and 6) and phototrophs (Chapter 7). This discussion is followed by a consideration of the major uses of energy in biological systems, including the work of biosynthesis (Chapter 8), uptake and concentration (Chapter 9), and mechanical movement (Chapter 10). Each chapter concludes with a set of problems designed to test your understanding of the content of that chapter, as well as a reference list intended as a guide for further reading. Two appendixes are also included. Appendix A provides numbers, formulas, and definitions related to the subject matter and useful in working the problems; Appendix B provides answers to all the problem sets.

A note is probably in order concerning the apparent preoccupation of this book with chemistry. If that is the charge, I readily plead guilty, since the role of chemistry as a foundation for modern biology is one of the major tenets of my teaching, and it would be surprising indeed if this conviction did not pervade my writing as well. Trying to appreciate cellular biology without a knowledge of chemistry would be like trying to appreciate Goethe without a knowledge of German—most of the meaning would obviously come through anyhow, but much of the beauty and depth of understanding would be lost in translation. I commend both to you: Goethe in German and cellular biology in its chemical framework. Nonetheless I recognize that the undergraduate student of cellular biology is often still in the midst of acquiring competence in chemistry. Accordingly I have presumed here, as I do in my teaching, a background equivalent to no more than a year (but a solid one!) of general chemistry at the college level, along with a mathematical competence that includes the use of exponents and logarithms but without recourse to much calculus. Prior or concurrent organic chemistry is obviously of great benefit but is not absolutely essential, since I have tried to introduce and explain the necessary concepts from that discipline as they are encountered. For students with a special interest in organic chemistry, I have included detailed discussions of selected topics in the footnotes of Chapters 4 and 5 and to a lesser extent in later chapters as well. Although the text deals primarily with the energetics of living cells, it includes—inevitably, perhaps—many fundamental principles of biochemistry that I hope will be regarded as a serendipitous side effect.

A word should also be said about the role of problem solving as a study aid. My own conviction is that one learns science not just by reading or hearing about it but by working with it, and this conviction is likewise on display here. For those who share my enthusiasm for problem solving, I have provided a set of problems and questions at the conclusion of each chapter. The problems are arranged in order of increasing difficulty, and those marked with a double dagger (‡) are intended only for the undaunted few who delight in special challenge. Detailed answers for all questions appear in Appendix B, but I recommend that they be consulted only for final confirmation or, when all else fails, as a last resort.

I want to acknowledge the contributions of all those who have made this book possible (as well as necessary!). First and foremost I am indebted to the many students in Biocore 303 whose words of encouragement catalyzed the writing and collecting of these chapters and whose thoughtful comments and willing criticisms have contributed much to whatever level of lucidity the text may be judged to have. A special debt of gratitude is owed to David Spiegel for his cogent initial draft of what has now become Chapter 2. Reviewers whose words of appraisal and counsel are hereby gratefully acknowledged include Marcia Allen, Gary Borisy, Peter Hinkle, Lee Peachey, John Schaeffer, and William Sistrom. General editorial advice was provided by Arthur Giese and Lewis Kleinsmith. A word of thanks is also in order to Mary Wheeler for expert secretarial assistance and unfailing cooperation in early drafts of these materials and to Mrs. J. J. Donaldson for her competence in preparation of the final typed manuscript. I am also indebted to Mary Purucker for her proofreading aid and to Patricia Becker and Velda Barnes for their assistance with the index. The generous suppliers of micrographs are credited in the figure legends. Finally, I am grateful beyond measure to my wife Pat and my daughters Lisa and Heather for their patience, understanding, and encouragement, without which this book would never have been written.

Though I have enjoyed the counsel, assistance, and encouragement of many, the final responsibility for what you read here rests with me, and you may confidently attribute to me any errors or shortcomings you may encounter. Indeed I would welcome any comments or criticisms concerning this book, as well as any suggestions as to how it can better help the reader to understand cellular energy metabolism, for that is the single goal to which all these efforts are dedicated.

W. M. Becker
Madison, Wisconsin
April, 1977

ILLUSTRATION CREDITS

5-8a   Courtesy of K. R. Porter, Univ. of Colorado.

5-8b,c   Courtesy of E. Racker, Cornell Univ.

7-4   Courtesy of M. D. Hatch, C.S.I.R.O., Canberra, Australia. *In* C. L. Black and R. H. Burris, eds., *Carbon Dioxide Metabolism and Productivity in Plants.* University Park Press, Baltimore (1976).

7-7   After F. P. Zscheile and C. L. Comar. Botan. Gaz. *102*:463 (1941).

7-8   After F. T. Haxo and L. R. Blinks. J. Gen. Physiol. *33*:389 (1950). By permission of The Rockefeller University Press.

7-9   Courtesy of D. L. Ringo, Menlo Park, Calif. From D. L. Ringo (1967). J. Cell Biol. *33*:543. By permission of The Rockefeller University Press.

7-10a,b   Courtesy of W. P. Wergin and E. H. Newcomb, Univ. of Wisconsin at Madison.

7-10c   Courtesy of D. Branton, Harvard Univ.

7-11   Courtesy of S. W. Watson, Woods Hole Oceanographic Inst.

7-12   After J. T. O. Kirk. Ann. Rev. Biochem. *40*:161 (1971).

8-9b   After J. L. Sussman and S. H. Kim. Science *192*:853 (1976). Copyright 1976 by the American Association for the Advancement of Science.

9-5   After S. J. Singer and G. L. Nicolson. Science *175*:720 (1972). Copyright 1972 by the American Association for the Advancement of Science.

9-11, 9-12   After E. Epstein, D. W. Rains, and O. E. Elzam. Proc. Nat. Acad. Sci. *49*:684 (1963).

10-1a   Courtesy of F. A. Pepe, Univ. of Pennsylvania.

10-4   Courtesy of D. L. Taylor, Harvard Univ.

10-5a   Courtesy of J. J. Paulin and A. Steiner, Univ. of Georgia.

10-5b   Courtesy of H. S. Wessenberg, San Francisco State Univ.

10-5c   Courtesy of J. J. Paulin.

10-5d   Courtesy of J. Hren, Oregon Regional Primate Research Center.

10-6   Courtesy of W. L. Dentler, Univ. of Kansas.

10-7a, c   After H. C. Berg. Sci. Am. *233*:36 (Aug. 1975). Copyright © 1975 by Scientific American, Inc. All rights reserved.

10-7b   Courtesy of J. Adler, Univ. of Wisconsin at Madison.

Contents

PART 1

Fundamentals of Cellular Energy Flow

The Flow of Energy and Matter in the Biosphere

1

One of the fundamental properties of all living organisms is their ability to grow and reproduce. At the cellular level, these essential activities are manifested in the tendency of cells to become larger and to divide, giving rise to daughter cells. Cellular growth and division depend in turn upon the ability of cells to take up from their environment both the *raw materials* out of which cellular mass can be fashioned and also the copious quantities of *energy* needed to organize, rearrange, and maintain cellular structure. Thus a major and fundamental topic in cellular biology deals with this basic requirement for the raw materials and energy that are requisite to the growth and maintenance of individual cells, of whole organisms, and, ultimately, of the biosphere itself. And it is to the resulting flow of both energy and matter through living systems that this text addresses itself.

The Need for Raw Materials

All living cells require certain indispensable inorganic and organic compounds, because these are the very building blocks from which life itself is constructed. Although inorganic elements such as calcium, magnesium, potassium, and iron are absolutely essential to living cells and their importance to the nutritional well-being of plants, animals, and microorganisms should not be minimized, the present discussion will concentrate almost exclusively upon the organic (carbon-containing) molecules central to both cellular structure and energy metabolism. In chemical terms, it is clear that all living matter is essentially a collection of organic molecules—a complicated collection, to be sure, but at the same time an increasingly well-understood collection. Thus all forms of life require carbon as an essential material from which to fashion the proteins, nucleic acids, carbohydrates, and other molecules of which living cells consist.

It is true, of course, that this universal requirement for carbon is satisfied in quite different ways by different organisms. The nutritional requirements of African violets, soil microorganisms, and humans are, for example, obviously dissimilar, as are also the means by which these needs are met. Basically, however, all organisms fall into two major classes, depending upon the chemical form of carbon they require from the environment. Those organisms capable of using carbon dioxide as their sole carbon source for the synthesis of all needed organic molecules are called *autotrophs*. A brief lesson in Greek will inform us that the term *troph* has to do with feeding, and *auto*, of course, means "self." Thus the autotroph is a "self-feeder," or, more explicitly, an organism that can satisfy all its organic carbon needs if provided with inorganic CO_2. Only the green plants, the blue-green algae, and some bacteria are autotrophs. All the rest of us depend for our carbon (and, as we shall see later, usually for our energy needs as well) upon the preformed organic food molecules we obtain from the autotrophs. We are called *heterotrophs*, from the Greek prefix *hetero*, meaning "other." Heterotrophs are therefore "other-feeders," organisms that must feed on others to meet their carbon needs. But whether taken up from the environment by heterotrophs or manufactured internally by autotrophs, organic molecules are required by all cells as the building blocks for the manufacture of new cell parts and the repair of old ones and, as we shall discover, as fuel for cellular energy metabolism.

The Need for Energy   In addition to the need for raw materials from which new biomass can be fabricated, all living systems require an ongoing supply of energy. Before discussing why cells need energy, however, we might first consider the actual meaning of energy. The usual definition of energy, of course, is the capacity to do work. But, as we shall see in Chapter 2, work is usually defined in terms of energy changes, so a less circular and more useful definition might be this: Energy is the ability to cause specific changes. Since life is characterized first and foremost by change, this definition underscores the utter dependence of all forms of life upon the availability of continuous and copious quantities of energy.

If we now inquire into the cellular need for energy in terms of this definition, we should ask what kinds of changes cells must effect or what sorts of work cells must do. To answer this, we can think of three major classes of cellular activities that involve change and require energy. These are discussed in detail in Part 3; for the present, a simple enumeration will suffice. In terms of the kinds of changes effected, these categories are as follows:

1. *Changes in location or orientation (mechanical work)* : Mechanical work involves a physical change in the location or orientation of an organism, a cell, or a part thereof. Examples include the contraction of a muscle, the movement of a sperm cell up the oviduct, the opening and closing of a flower, the migration of chromosomes toward the opposite poles of the mitotic spindle, and the movement of a ribosome along a strand of messenger RNA.

2. *Changes in concentration (osmotic and electrical work)* : Concentration work involves the movement of cellular components (most commonly minerals, organic nutrients, and waste products) across a membrane against a concentration gradient. This type of work is less conspicuous than mechanical work, but it is of great importance to the cell since it is essential to the establishment of the localized concentrations of specific molecules and ions upon which most essential life processes depend. All cells, for example, must be able to accumulate essential raw materials from their environment and concentrate them within the cell. Equally important products or by-products of cellular functioning must be removed from the cell, and this, too, often involves active pumping against a concentration gradient and therefore requires energy. Examples of concentration work (also referred to, though less satisfactorily, as *osmotic work*) include the uptake of the sugar glucose by the cells of your body, the pumping of sodium out of a marine microorganism, and the movement of nitrate from the soil into the cells of a plant root. A subcase of concentration work is *electrical work*, in which the establishment of a differential concentration of charged ions across a membrane is used to build up an electrical charge. Although perhaps the most dramatic example of this is the large potential developed by the electric eel, electrical work is a common phenomenon and is in fact the mechanism of excitation and of conduction of impulses in nerve and muscle cells.

3. *Changes in chemical bonds (biosynthetic work)* : Synthetic work involves the formation of the energy-rich chemical bonds necessary to fabricate and assemble the complex organic molecules of which cells are composed. Since most of these are energy-rich, oxidizable molecules, it should be clear that energy must be

expended to form such molecules from the simpler and less energy-rich molecules the cell characteristically obtains from its environment. It is important to note that biosynthetic work occurs not only during periods of active growth of an organism, when there is an obvious accumulation of new cellular material, but also in nongrowing mature organisms, in which existing structures are continuously being repaired and replaced. The continuous expenditure of energy to elaborate and maintain ordered structures out of less ordered raw materials is, in fact, one of the most characteristic properties of living organisms.

Sources of Energy We have already noted that, based on the way they meet carbon needs, all organisms fall into two major classes, the autotrophs and the heterotrophs. Organisms (and cells) can be similarly classified in terms of the way they satisfy energy requirements, and again two groups emerge. One group depends for its energy supplies directly upon the radiant energy of the sun. These are the *phototrophs* (literally, "light-feeders"), those marvelous green forms of life that can use solar energy to form energy-rich, oxidizable organic (or, in the case of some microorganisms, inorganic) compounds whose chemical bonds effectively store, transfer, and, upon oxidation, release the energy originally derived from the sun. The second group of organisms in this energy-based classification consists of those of us who meet our energy needs by oxidizing the energy-rich compounds we obtain from the phototrophs. We are called *chemotrophs* (literally, "chemical-feeders"), since we require the intake of oxidizable chemical compounds. Thus both the writer and the reader of these lines, along with all animals and most microorganisms, are chemotrophs; we are parasites living off the energy that has been packaged for us by the phototrophs into the chemical bonds of the food molecules we consume. A world composed only of chemotrophs would last only as long as food supplies held out. Then all of us would die for lack of energy on a planet that is daily flooded with solar energy but in a form inaccessible to our energy-requiring processes.

At this point you may be somewhat puzzled by the dual classification of organisms according to carbon source (autotrophs versus heterotrophs) and according to energy source (phototrophs versus chemotrophs). Indeed you may question the need for two sets of categories, since it may occur to you that the common phototrophs you are likely to think of (the green plants, notably) are also autotrophs and the common chemotrophs that come to mind (the higher animals, most likely) are also heterotrophs. And in fact, as Table 1-1

TABLE 1-1 Classification of Organisms According to Carbon and Energy Sources

Classification According to Carbon Source	Classification According to Energy Source	
	Phototrophic (sunlight)	Chemotrophic (chemical compounds)
Autotrophic (CO_2)	All green plants All algae Photosynthetic bacteria	Hydrogen bacteria Sulfur bacteria Iron bacteria Denitrifying bacteria
Heterotrophic (organic compounds)	Nonsulfur purple bacteria	All higher animals Most microorganisms

indicates, the great majority of organisms are either autotrophic and phototrophic (shaded box, upper left) or heterotrophic and chemotrophic (shaded box, lower right). Nevertheless, the other two categories should not simply be written off as esoteric exceptions. Even though they contain only a relatively few species of bacteria, some of these species play very important and unique roles in the overall economy of the biosphere, especially of the soil. Note especially that, in addition to the chemotrophs which meet their energy needs by oxidizing organic compounds (and which are almost always accorded the greatest attention in introductory texts, including this one), there are also chemotrophic bacterial species which obtain energy exclusively by the oxidation of inorganic substances such as H_2, S, Fe^{++}, or NH_3, and these organisms play unique and essential roles in the metabolism of these inorganic elements in the biosphere. Nonetheless, it remains true that the majority of organisms are either both autotrophic and phototrophic or both heterotrophic and chemotrophic.

Since we are concerned in this text primarily with energy metabolism, we shall classify all organisms as either phototrophic or chemotrophic, bearing in mind that some phototrophic microorganisms are heterotrophs with respect to carbon source and some chemotrophic bacteria are autotrophic in their ability to use CO_2. It is also important to note that all organisms that can utilize solar energy, and are therefore classified in Table 1-1 as phototrophs, also are capable of functioning as chemotrophs—and do so every night in fact. Furthermore most multicellular phototrophic organisms (higher plants, notably) are, at the cellular level, really a mixture of phototrophic and chemotrophic cells. A plant root cell, for example, is in most cases obligatorily chemotrophic, even though it is part of an obviously phototrophic organism.

Energy and Matter in Flux

With this perspective, we can look briefly at the flow of energy through the biosphere and see where the energy enters, what eventually happens to it, and how the phototrophs and chemotrophs fit into place. As shown in Figure 1-1, the ultimate source of energy for the entire biosphere is the sun. Solar energy is trapped by the phototrophs and used to convert carbon dioxide into more complex (and more reduced) cellular materials in the process of *photosynthesis*. (Remember, though, that some phototrophic bacteria cannot fix carbon dioxide and therefore do not fit the stereotype that otherwise applies to the phototrophs.) The chemotrophs, on the other hand, cannot use solar energy directly but obtain their energy from chemical compounds (which are often, but not always, organic molecules). The energy needs of the chemotrophs can be met either *anaerobically* (that is, in the absence of oxygen) by a variety of *fermentation* processes or *aerobically* (in the presence of oxygen) by the oxidation of chemical compounds in the process of *respiration*. The chemotrophs, then, require chemical energy in the form of bond energies of fermentable or oxidizable food molecules obtained, directly or indirectly, from the phototrophs.

Both the phototrophs and the chemotrophs use energy to perform various kinds of work or, in other words, to effect various specific changes. But since no chemical or physical process can occur with 100 percent efficiency, every chemical or physical change results obligatorily in an incomplete conversion of energy from one form to another. As a result some energy always becomes dissipated into the environment and is thus unavailable to do work. This *entropy loss* is

Figure 1-1 The role of the phototrophs and chemotrophs in the flow of energy and matter through the biosphere.

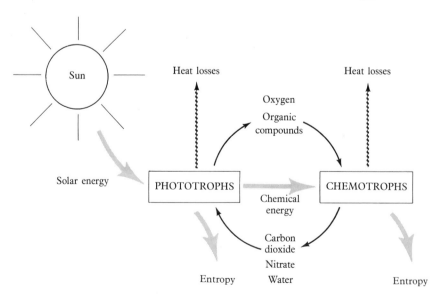

discussed in the next chapter; for the present, simply note that just as the ultimate source of all energy in the biosphere is the sun, so the ultimate fate of all energy in the biosphere is to become randomized in the universe as increased entropy or randomness.

On a cosmic scale, then, there is a continuous, massive flow of energy from a high-energy source (the nuclear fusion reactions occurring in the sun) to a low-energy sink (the entropy of outer space). We in the biosphere are the transient custodians of a small portion of that energy, but it is precisely that small but critical fraction of energy and its flow through living systems that is of concern to us here. For it is the flux of energy through living matter that provides cells with the capacity to do work and to effect the changes necessary for the maintenance of those structures and functions that are collectively defined as life. And the price that is paid for this energy flow is the inevitable increase in the entropy or randomness of the universe that accompanies every energy transaction.

Concurrent with the flow of energy in the biosphere is a flow of matter or mass, since energy in living cells is almost inevitably stored and transferred as chemical bond energies of organic molecules or organic mass. Thus energy enters the biosphere as pure energy (photons of light from the sun) and leaves the biosphere ultimately as pure energy (heat losses and increases in entropy). But all the while it is in flux through the biosphere, energy exists primarily as chemical bond energies such that the flow of energy in the biosphere is obligatorily coupled to a correspondingly immense flow of matter. This relationship is shown in Figure 1-1, which illustrates both the unidirectional flow of energy from phototrophs to chemotrophs and the accompanying cyclic flow of matter. During respiration, aerobic chemotrophs take in organic nutrients from their surroundings and oxidize them to carbon dioxide and water. These substances are in turn used by phototrophic cells to make new organic molecules, with a return of oxygen to the environment. At the same time, there is a constant cycling of available nitrogen since the phototrophs also obtain nitrogen from the environment in inorganic form (usually from the soil as nitrate), convert it into ammonia, and use it in the synthesis of amino acids, proteins, and nucleic acids. Eventually these molecules, like other components of phototrophic cells, may be consumed by chemotrophs; the nitrogen is then converted back into ammonia and, via soil microorganisms, back to nitrate. Carbon dioxide, oxygen, nitrogen, and water thus cycle continuously between the phototrophic and chemotrophic worlds, such that the two great groups of organisms live in symbiotic relationship with each other, with both matter and energy as components of that symbiosis.

When we deal, as we have been here, with the overall flux of energy and matter through living organisms, we find cellular biology interfacing with ecology, since the ecologist is very much concerned with overall energy cycles, with the roles of various species in the cycles, and with environmental factors that affect these fluxes. We cannot stay at this interface very long, however, since ultimately the flux of energy and matter, which we have thus far been considering on a macroscopic scale, must be expressed and explained on a molecular scale in terms of energy transactions and the chemical reactions that go on within cells. We shall leave the macroscopic cycles to the ecologist, then, and turn our attention instead to the processes that occur within individual cells of bacteria, plants, and animals to account for those cycles.

Important to our discussion will be the thermodynamic principles governing energy transactions (Chapter 2), the enzymes that catalyze all such transactions in biological systems (Chapter 3), the utilization of organic food molecules as sources of chemical energy (Chapters 4, 5, and 6), the synthesis of these molecules by photosynthetic autotrophs (Chapter 7), and the kinds of biological work that make cellular energy flow so vital (Chapters 8, 9, and 10).

Practice Problems

1-1. Solar Energy: Although we are concerned at present with an energy crisis, we actually live on a planet that is flooded continuously with an extravagant amount of energy in the form of solar radiation. Every day, year in and year out, solar energy arrives at the upper surface of the earth's atmosphere at the rate of 1.94 cal/cm^2-min (the *solar energy constant*).

(a) Assuming the cross-sectional area of the earth to be about 1.28×10^{18} cm^2, what is the total annual amount of incoming energy?

(b) A substantial portion of that energy, particularly in the wavelength ranges below 3000 Å and above 8000 Å, never reaches the earth's surface. Can you suggest what might happen to it?

(c) Of the radiation that reaches the earth's surface, only a small proportion is actually trapped photosynthetically by the phototrophs (you get to calculate the actual value in Problem 1-2c). Why do you think the efficiency of utilization is so low?

1-2. Photosynthetic Energy Transduction: Photographs of the earth taken from outer space show our planet to have a blue-green color, a pleasant distinction owed to the thin but prominent layer of vegetation covering every bit of the earth's surface that is the least bit hospitable to the maintenance of life. The amount of energy trapped and carbon converted into organic form by this network of photosynthetic energy transducers is mind-boggling—about 5×10^{16} g of carbon per year over the entire earth's surface.

(a) Assuming the average organic molecule in the cell to have about the same proportion of carbon as that found in the simple sugar glucose ($C_6H_{12}O_6$), how many grams of organic matter are produced annually by the carbon-fixing phototrophs?

(b) Assuming all the organic matter in part (a) to be glucose (or any molecule with an energy content equivalent to that of glucose), how much energy is represented by that quantity of organic matter? Assume that glucose has an energy content (free energy of combustion) of 3.8 kcal per gram.

(c) Refer to the answer for Problem 1-1a. What is the average efficiency with which the radiant energy incident upon the upper atmosphere is trapped photosynthetically on the earth's surface?

(d) What proportion of the net annual phototrophic production of organic matter calculated in part (a) do you think is consumed by the chemotrophs each year?

1-3. Meeting Carbon and Energy Needs: All organisms are either autotrophic or heterotrophic with respect to carbon requirements and either phototrophic or chemotrophic in meeting energy needs. Classify each of the following organisms according to both criteria:

(a) A corn plant growing in a field in Iowa.

(b) A soil bacterium that is found in the same cornfield, has no chlorophyll, converts carbon dioxide into organic matter, and oxidizes nitrite (NO_2^-) to nitrate (NO_3^-).

(c) A cow leaning over the fence to eat the corn plant.

(d) A protozoan that lives in the rumen (part of the stomach) of the cow and ferments the sugars of the corn leaf into a mixture of organic acids.

‡1-4. Origin of the Phototrophs: It is usually argued that primitive organisms capable of photosynthesis appeared earlier during evolution than those capable of aerobic respiration. Assuming that the earth originally had a reducing atmosphere (probably consisting mainly of methane and carbon dioxide, with little or no free oxygen present), can you suggest an explanation for the postulated early advent of photosynthetic organisms?

References

The following sources are good general references for the whole area of cellular energy metabolism and are therefore suitable for most of the topics covered in this book. In addition, each succeeding chapter contains a list of references especially pertinent to the subject matter of that chapter.

Avers, C. J., *Cell Biology* (New York: Van Nostrand, 1976).

Baker, J. J. W., and G. E. Allen, *Matter, Energy and Life: An Introduction for Biology Students* (Reading, Mass.: Addison-Wesley, 1965).

Dyson, R. D., *Cell Biology—A Molecular Approach* (Boston: Allyn & Bacon, 1974).

Gates, D. M., "The Flow of Energy in the Biosphere," *Sci. Amer.* (Sept. 1971), *224*:88.

Goldsby, R. A., *Cells and Energy* (New York: Macmillan, 1969).

Krebs, H. A., and H. L. Kornberg, *Energy Transformations in Living Matter* (Berlin: Springer Verlag, 1957).

Lehninger, A. L., *Biochemistry,* 2nd ed. (New York: Worth, 1975).

Lehninger, A. L., *Bioenergetics,* 2nd ed. (Menlo Park, Ca.: Benjamin, 1971).

Stryer, L., *Biochemistry* (San Francisco: Freeman, 1975).

Wood, W. B., et al., *Biochemistry—A Problems Approach* (Menlo Park, Ca.: Benjamin, 1974).

Energy in Biological Processes

2

Upon first encounter, a chapter dealing with fundamental considerations of energy and thermodynamic principles may seem somewhat out of place in a book for biologists and biochemists, since the topics belong so obviously to the domain of physics and physical chemistry. The fact is, of course, that the living systems we are attempting to understand are an integral part of the physical universe and as such are governed by the basic principles of thermodynamics just as surely as are the systems that traditionally occupy the attention of physicists and chemists. In the final analysis, every process and every reaction necessary to sustain the activities of a living cell is an energy transaction, and each is governed by the same physical principles that dictate for every other process and reaction whether, in what direction, and to what extent it can occur. If we are to understand cellular activities, then, we must come to grips with the fundamental principles of energy flow. This brings us to the properties of energy, to the laws of thermodynamics, and, eventually, to the very useful concept of free energy. These, therefore, are the topics of this chapter, and a convenient place to begin it all is by defining what we mean by energy and systems.

Energy and Systems

Energy is often defined as the ability to do work. But, as pointed out in Chapter 1, this is really somewhat of a circular definition, since work in turn is understood only in terms of energy changes. A more useful definition might be that energy is the ability to cause change.

This emphasizes the fact that, without energy, all processes would be at a standstill—including the processes of living organisms, for life is characterized first and foremost by constant change.

As you are probably already aware, energy can exist in a variety of forms, many of them of interest to biologists. Think, for example, of a distorted chemical bond, a pressurized fluid, an electron excited to an outer orbital, or a mass suspended in a gravitational field. Or consider the movement of molecules or atoms, the concentration of atoms, molecules, or ions within a membrane, or electromagnetic radiation such as light and X rays. These represent a variety of phenomena, yet all are manifestations of energy and all are governed by the same basic principles.

Similarly the amount of energy present in a given form can also be expressed in a variety of units, such as joules, calories, or electron volts. All these units are interconvertible, however, and for our purposes it is convenient to express energy uniformly in terms of the *calorie*, which is the amount of energy required to warm a gram of water $1°C$ (rigorously, from 14.5 to 15.5°C) at a pressure of 1 atmosphere. The calorie is equal to about 4.18 joules; it should not be confused with the Calorie, or kilocalorie, which is commonly used to express the energy content of foods and is equal to 1000 calories.

Energy is distributed throughout the universe. Usually, however, one is not interested in the whole universe, but rather only in a small portion of it, as, for example, a beaker of chemicals, a cell, or an ice cube. By convention, the restricted portion of the universe one wishes to consider is called the *system* and the rest of the universe is called the *surroundings*. A system may have a natural boundary such as a beaker, a membrane, or the surface of a solid; or the boundary between the system and its surroundings may be an imaginary one used only for convenience of discussion, such as the imaginary boundary around a mole of glucose molecules in a solution.

A system may exchange energy with its surroundings by two general processes—heat and work. *Heat* is energy transfer from one place to another as a result of a temperature difference between the two places. The spontaneous transfer is always from the hotter place to the colder, and as a consequence the temperature of the hotter place falls while that of the colder place rises. Heat is considered separate from work because it has some unique properties that need not concern us here. *Work* is the transfer of energy from one place or form to another place or form by any process other than heat flow. For example, work is performed when a muscle cell expends chemical energy to lift a weight, when a spinach leaf uses light energy to synthesize sugar, or when an electric eel depletes ion concentrations to deliver a shock. Similarly work is done on a cell when it absorbs a glucose molecule, because chemical energy has been transferred to the cell from its surroundings.

Although energy can be converted from one form to another, it can never be created or destroyed, a principle referred to as the *law of conservation of energy* or the *first law of thermodynamics*. Applied to the universe as a whole (or to any isolated system that does not exchange energy with its surroundings), this principle of energy conservation means that the total amount of energy present in all forms must be the same before and after any process occurs; in other words, a process can only convert energy from one form to another. Applied to an open (unisolated) system such as a cell interacting with its surroundings, conservation of energy means that the total amount of energy entering the system (as heat, light, chemical compounds, or mechanical distortion, for example) over a period of time must be equal to the energy leaving the system during the same period plus any increase in energy stored within the system:

Conservation of Energy

$$\text{Energy in} = \text{energy out} + \text{energy stored} \tag{2-1}$$

The total energy stored within a system in all forms is called the *internal energy* of the system and is represented by the symbol E. In practice the absolute amount of energy in a system can never be determined, since energy is only evident when it does work or flows as heat (that is, only when it undergoes change). It is, however, possible to calculate the *change* in internal energy, ΔE, that occurs as a result of interactions of a system with its surroundings, provided we know how much energy enters and leaves the system during the reaction. Thus if a system undergoes some reaction that changes its total internal energy content from E_1 to E_2, ΔE is *defined* as the difference between E_1 and E_2, but it is *calculated* by keeping track of the energy entering and leaving the system during the reaction:[*]

Internal Energy

$$\Delta E = E_2 - E_1 = \text{energy in} - \text{energy out} \tag{2-2}$$

---

[*] In fact, since energy can enter and leave a system only as heat or work, the first law of thermodynamics can also be expressed in the form $\Delta E = q - w$, where q is the net heat entering the system during the time period in question and w is the net work done by the system on its surroundings. (The negative sign in this equation often confuses the novice who thinks the two forms of energy should be summed. It is, however, necessary because q is expressed as energy *entering* the system and w as energy *leaving* the system. When considered in the historical context of heat engines, the signs of q and w are very practical and the equation becomes much less arbitrary.) Expressed in terms of q and w, the equation for ΔE is in the form usually encountered in physics and chemistry, but we shall make no further use of it in this form.

If the system under consideration is a muscle cell undergoing contraction, the change in internal energy (negative in this case since the system is expending energy) will equal the mechanical work performed during the contraction plus any heat evolved by the cell in the process. Similarly, if the system consists of glucose, oxygen, and appropriate enzyme molecules to catalyze the oxidation of the glucose to carbon dioxide and water (Equation 2-3), the decrease in internal energy of the system represents energy transferred from the chemical bonds of the glucose molecules to the surroundings. Conversely the formation of glucose and oxygen from carbon dioxide and water (Equation 2-4), as is accomplished during photosynthesis in green plants, requires a transfer of energy from the surroundings to the system. We can in fact quantitate these energy changes by saying that the oxidation of glucose to CO_2 and H_2O (Equation 2-3) occurs under physiological conditions with a change in internal energy ΔE of -673 kcal per mole of glucose oxidized. Under the same conditions, the change in internal energy for the synthesis of glucose (Equation 2-4) would be $+673$ kcal per mole of glucose.

$$C_6H_{12}O_6 + 6O_2 \rightarrow 6CO_2 + 6H_2O + \text{energy} \qquad (2\text{-}3)$$
$$\text{Glucose}$$

$$6CO_2 + 6H_2O + \text{energy} \rightarrow C_6H_{12}O_6 + 6O_2 \qquad (2\text{-}4)$$
$$\text{Glucose}$$

These ΔE values serve also to illustrate two points that will prove useful later. The first is that values for ΔE (and, more generally, for the other thermodynamic parameters we shall encounter as well) are always expressed in terms of *energy units* (calories or kilocalories, usually) *per mole of reactant consumed or product formed.* Secondly, note that once the value of ΔE (or, again, of any other thermodynamic parameter) is known for a reaction proceeding in a given direction (for example, the value of -673 kcal/mole for Equation 2-3), the corresponding value for the same overall reaction in the reverse direction (Equation 2-4 in this case) has the same numerical value but the opposite sign.

The State of a System

A system is said to be in a specific *state* if each of its variable properties (such as temperature, pressure, and volume) is held constant at a specified value. In such a situation, the internal energy of the system, while still not measurable, has a unique value. Consequently, if a system changes from one state to another as a result of some interaction with its surroundings, the change in its internal energy is completely determined by the initial and final states of the system, as already implied by Equation 2-2. This means that the net amount of energy transferred to or absorbed from the surroundings in going from the initial state to the final state is independent of both the mechanism by which the change occurs and the intermediate states through which the system passes. This fact enables one to determine the energy change resulting from a sequence of reactions from a knowledge of the initial and final states only; the actual route is irrelevant.

If one or more of the system variables can be held constant during a process, the changes that can occur in the system will be limited and the problem of keeping track of them will be simplified. Fortunately this is the case with most biological reactions, which occur in dilute solutions within cells that are at approximately the same temperature and pressure. These environmental conditions, as well as the cell volume, are slow to change in comparison with the speed of biological reactions. Therefore three of the most important system variables— temperature, pressure, and volume—are essentially constant for biological reactions.

The principle of energy conservation is useful in that it enables one to determine how much energy would be liberated or absorbed if a given process were to occur in a system, but it provides no indication whatever as to whether the process will in fact occur under specified conditions. The first law of thermodynamics, in other words, provides us with a balance sheet for calculating energy changes in the event that a particular reaction or process takes place, but it provides no indication of whether the reaction or process of interest is in fact possible. Yet we have an intuitive feeling, in certain cases at least, that some reactions or processes are clearly possible whereas others are not. You are somehow very sure that if you were to tear out this page and set a match to it, it would burn; the oxidation of cellulose to carbon dioxide and water is, in other words, a possible reaction or, in the terminology of thermodynamics, a *spontaneous* process. Furthermore there is quite clearly a directionality to the process, for you are doubtless equally convinced that the reverse process will not occur (i.e., it is thermodynamically nonspontaneous). You are somehow quite certain that if you were to stand around clutching the charred remains of the page, the page would not spontaneously reassemble in your hand. You have, in other words, a feeling for both the potential and the directionality of cellulose oxidation.

You can probably think of other processes, such as the melting of an ice cube or the dissolving of sugar in water, for which you can

**Spontaneous
Reactions**

make such thermodynamic predictions with equal confidence. When we move from the world of familiar physical processes to the realm of chemical reactions, however, we quickly find that we cannot depend upon prior experience to guide us in our predictions. It is then that we realize clearly how valuable it would be to have some reliable means of actually determining whether a given physical or chemical change can occur under specific conditions.

A great deal of attention has been devoted to the search for a parameter (that is, a measurable property) that could be used to predict whether a given process will occur spontaneously under a specified set of conditions. Such parameters were at first defined in mathematical terms and were only later given physical interpretations. The details need not concern us greatly, but an awareness of the general concepts is important since they determine what kinds of changes can occur in biological systems and what conditions will prevail at equilibrium. To simplify the discussion, we shall restrict our attention to biological systems, where, as already noted, temperature, pressure, and volume may be assumed constant.

Bound Energy and Extractable Energy

Of the total internal energy E of a system, a portion can be extracted by various processes and transferred to the surroundings. This available portion of the total energy is called the *extractable energy* (or the *Helmholtz energy*) of the system. The remainder of the internal energy, which cannot be extracted from the system without lowering its temperature, is called the *bound energy* of the system. The bound energy, in other words, is that portion of the internal energy which is trapped within the system by virtue of the fixed system temperature. Together, then, the bound and extractable forms make up the total internal energy of the system:

$$E = \text{total internal energy} = \text{extractable energy} + \text{bound energy} \quad (2\text{-}5)$$

Entropy

The amount of energy bound within a system is, as implied above, dependent upon the final temperature of the system; the only way more energy could be extracted by a given process would be to allow the temperature of the system to fall during the process. And since biological systems operate at a constant system temperature (or at least at a temperature that is slow to change in comparison with the speed of biological reactions), temperature is clearly one of the factors determining the amount of bound energy. In fact the amount of bound energy in a system at constant temperature is the product of that temperature T (in degrees Kelvin) and a mathematical parameter called *entropy*, represented by the symbol S. Thus the amount of bound energy in a biological system is calculated as the product of temperature and entropy:

$$\text{Bound energy} = \text{temperature} \times \text{entropy} = TS \qquad (2\text{-}6)$$

This allows Equation 2-5 to be rewritten as

$$E = \text{extractable energy} + TS \qquad (2\text{-}7)$$

Since we shall eventually be interested mainly in *changes* in energy, it is useful to note that, because the temperature is held constant, we can write the following expression for the change in internal energy ΔE:

$$\begin{aligned} \Delta E &= \Delta(\text{extractable energy}) + \Delta(\text{bound energy}) \\ &= \Delta(\text{extractable energy}) + T\Delta S \end{aligned} \qquad (2\text{-}8)$$

Although entropy is basically a mathematically derived function, a physical approximation is possible by regarding entropy as a measure of randomness or disorder. This means that, for any system, the *change* in entropy ΔS, in which we are really interested, represents a change in the degree of randomness or disorder of the components of that system. If, for example, you place a drop of ink in a beaker of water and watch the ink diffuse through the water, you are observing an increase in the randomness of the location of individual molecules and hence in the entropy of the system.

We have not yet seen how any of this can help predict what changes will occur in a cell. As it turns out, however, there is a very important link between spontaneous events and entropy change: whenever a process occurs in nature, the amount of bound energy in the universe (that is, the entropy of the universe) invariably increases. In other words, the value of $\Delta S_{\text{universe}}$ is positive for every real process or reaction. A hypothetical process that by its occurrence would lead to a decrease in the entropy of the universe has never been observed to take place.[*] This principle that no process can occur which would decrease the entropy of the universe is called the *second law of thermodynamics*.

Considering entropy as a measure of randomness, we can rephrase the second law to state that the universe becomes more random with

Entropy Change as a Measure of Reaction Spontaneousness

---

[*] The special case where the entropy of the universe remains constant during a reaction represents the equilibrium condition for the reaction—the point at which products are being formed at the same rate as they are being used to regenerate reactants. When a process is at its equilibrium point, it is said to be *thermodynamically reversible*.

every reaction that occurs and that reactions which would make it less random are never observed. As an illustration, consider the diffusing ink drop again. As you place the ink in the water, you fully expect that it will diffuse spontaneously throughout the water, as individual molecules move from a region of high concentration (the droplet) to one of lower concentration (the surrounding water), thereby becoming more randomly distributed in the beaker. You can also predict with equal confidence that the ink, once diffused in the water, will not spontaneously reassemble as a concentrated droplet somewhere in the beaker. You are, in effect, basing these expectations on prior experience with the second law and on the tendency toward increased entropy that it represents.

It should be kept in mind, however, that these formulations of the second law pertain to the universe as a whole and, despite our example, do not in general apply to smaller systems. While every real process must without exception be accompanied by an increase in the entropy of the universe, the entropy of a given *system* may increase, decrease, or stay the same as a result of a specific reaction or process. For example, the oxidation of glucose to carbon dioxide and water (Equation 2-3) under standard conditions (25°C, 1 atmosphere pressure, and pH 7) is accompanied by an *increase* in system entropy ($\Delta S_{system} = +43.6$ cal/mole-degree) and is spontaneous. The freezing of water at $-0.1°C$ is also a spontaneous event, even though it involves a *decrease* in system entropy ($\Delta S_{system} = -0.5$ cal/mole-degree). Thus while the change in entropy of the universe is a valid measure of the spontaneousness of a process, the change in entropy of the system, like the change in internal energy of the system, is not.

To understand how the entropy of the universe can increase during a process while the entropy of the system decreases, it is only necessary to realize that the decrease in entropy of the system can be accompanied by an equal or even greater increase in the entropy of the surroundings. In this way $\Delta S_{universe}$ (equal to $\Delta S_{system} + \Delta S_{surroundings}$) can be positive (and the reaction therefore possible) even though ΔS_{system} is negative.

This is, in fact, the situation normally seen within living cells, which by their growth and reproduction produce local increases in order. On the basis of the second law, such local increases in order (decreases in entropy) must be offset by an even greater decrease in the order (increase in entropy) of the surroundings. Of course, if the system being considered is prevented from interacting with its surroundings in any way, processes occurring within the system will not affect the surroundings. Under this condition, the processes will inevitably produce an increase in the system entropy, since $\Delta S_{surroundings} = 0$. In practice, however, it is not possible to isolate a system this completely, and it certainly is not the case for biological systems such as cells, which interact with their surroundings in a variety of ways, including the acquisition of nutrients and the elimination of wastes. Since such interactions always take place, it is entirely possible for local increases in order to occur within the cell, for they are offset by decreases in order in the surroundings. This also means, however, that the second law, stated in terms of entropy of the universe, is of limited value in predicting the spontaneousness of biological processes, for it requires that one keep track of changes occurring not only within the system but also in its surroundings. It would be far more convenient if a parameter could be found that would enable prediction of the spontaneousness of reactions from a consideration of the system alone.

Fortunately such a parameter has been found to exist, and it lies in **Free Energy** the extractable component of the internal energy (Equation 2-7). The new function is called the *Gibbs free energy*, represented by the symbol G and named after W. Gibbs, who first developed this concept. Because of its predictive value, the Gibbs free-energy function is one of the most useful functions in biochemistry. In fact it is used so often that it is customary to refer to it simply as the free energy of the system, a convention we shall adopt also.

Like many other thermodynamic functions, free energy is defined only in terms of a mathematical relationship. But the important point for us is that, for biological systems at constant pressure, volume, and temperature, the free energy of a system differs from the extractable (Helmholtz) energy only by an additive constant, such that the *changes* in them when a biological system undergoes a reaction are identical.[*] This in turn means that, for biological systems, Equation 2-8 can be modified to express the change in internal energy as follows:

$$\Delta E = \Delta G + T\Delta S \qquad (2\text{-}9)$$

or

$$\Delta G = \Delta E - T\Delta S \qquad (2\text{-}10)$$

where ΔE represents the change in the internal energy of the system,

---

[*] The Helmholtz and Gibbs functions are in fact related by the equation $G = A + PV$, where G is the Gibbs free energy of the system, A is the extractable (Helmholtz) energy, and P and V are the pressure and volume of the system. The statement that internal energy equals extractable energy plus bound energy (Equation 2-5) can be written $E = A + TS$. Combining these two equations gives the following relationship between E and G:

$$E = G - PV + TS \qquad \text{or} \qquad \Delta E = \Delta G - \Delta(PV) + \Delta(TS)$$

This relationship holds for any system and is the form most likely to be encountered in introductory thermodynamics texts. For the special case of biological (or other) systems where P, V, and T are constant, this equation reduces to $\Delta E = \Delta G + T\Delta S$, as shown in Equation 2-9.

T is the (constant) system temperature, ΔS is the change in system entropy, and ΔG represents the change in the free energy of the system.

Free-Energy Change as a Measure of Reaction Spontaneousness

The free-energy function is an extremely powerful one. As we shall see later, ΔG for a postulated reaction may be readily calculated from the equilibrium constant for the reaction and from measurable system variables such as chemical concentrations. Once determined, ΔG provides exactly what we are looking for: a measure of the spontaneousness of a reaction based solely upon the properties of the system in which the reaction is occurring. This is so because under conditions of constant temperature and pressure, ΔG for the system is related in a simple way to ΔS for the universe, which we recognize already as a valid criterion of spontaneousness. The relationship may be derived mathematically, but we shall content ourselves with the following summary:

If a spontaneous reaction occurs within a biological system and the system neither does work on its surroundings nor has work done on it, the entropy of the universe increases and the free energy of the system decreases. For a reaction occurring at its equilibrium point in such a system, both ΔS of the universe and ΔG of the system are zero.

Thus the free-energy change of the system is just as valid as the entropy change of the universe in predicting whether a reaction will occur spontaneously in a biological system. Any process that results in a decrease in free energy of the system (that is, for which ΔG is negative) is spontaneous and is termed *exergonic*. Exergonic means "energy-yielding," but note carefully that the reference is specifically to the free-energy change and not to the total internal energy, which may increase, decrease, or remain constant during a spontaneous process. Conversely any process that would result in an increase in the free energy of the system (that is, for which ΔG is positive) is called *endergonic* ("energy-requiring") and cannot proceed under the conditions used to calculate ΔG. If, however, the conditions are altered (as, for example, by increasing the concentration of the reactants), it is possible to convert an endergonic reaction into an exergonic one and thus render it spontaneous.

A Note on the Meaning of "Spontaneous"

Before proceeding further with a consideration of ΔG as a measure of thermodynamic spontaneity, it might be well to qualify carefully our use of the word "spontaneous." The term is used in thermo-dynamics to describe reactions that have a negative ΔG and therefore possess the potential to proceed in the direction indicated, but we have to note carefully that such reactions need not (and, indeed, in bio-logical systems often do not) proceed at a measurable rate. This is because, as a thermodynamic parameter, ΔG can really only tell us whether a reaction or process is thermodynamically possible; i.e., whether it has the potential for occurring. Whether a thermodynami-cally feasible (exergonic) reaction will in fact proceed depends not only upon a favorable (i.e., negative) ΔG but also upon the availability of a mechanism or path to get from the initial state to the final state and frequently upon an initial input of activation energy as well (such as occurs when a match is used to ignite a piece of paper). Thus thermo-dynamic spontaneousness (negative ΔG) is a necessary, but not sufficient, criterion for determining whether a reaction will actually occur. We shall have an opportunity to explore the question of reaction rate further in Chapter 3, when we encounter enzymes as cellular catalysts. For the present discussion, it is adequate to note that by designating a reaction or process as spontaneous, we simply mean that it is a thermodynamically possible event that will liberate free energy if and when it occurs; time and route, however, are of no thermodynamic significance.

The Standard State

Because it is a thermodynamic parameter, the change in the free energy of a system as a result of a reaction depends only upon the initial and final states of the system. Consequently, while ΔG is independent of the actual path of the reaction, it does depend upon the conditions under which the reaction occurs. A reaction characterized by a large decrease in free energy under one set of conditions may have a much smaller (but still negative) ΔG or may even go in the opposite direction under a different set of conditions. The melting of ice, for example, is critically dependent upon temperature, since it proceeds spontane-ously above $0°C$ but goes in the opposite direction (freezing) below that temperature. It is therefore important to identify the conditions under which ΔG is measured.

As a convenience for listing or discussing energy changes for different chemical reactions, certain arbitrary conditions have been selected as defining the *standard state* of a system. For systems con-sisting of dilute aqueous solutions, these are usually a standard temperature of $25°C$ ($298°K$), a pH of 7.0 (neutrality), and all substances present in their most stable forms at a standard concentration of 1 mole/liter (or 1 atmosphere pressure for gases). *

Energy changes for reactions are usually reported in standardized form as the change that would occur if the reaction were run under standard conditions. More precisely, the standard change refers to the conversion of a mole of a specified reactant to products, or the formation of a mole of a specified product from the reactants, under the condition where the concentrations of all substances, the temperature, pressure, and pH are maintained at their standard values. (The maintenance of concentrations at $1\ M$ implies that reactants be added as they are being used up and that products be removed as they are being formed.)

The free-energy change calculated under these conditions is called the *standard free-energy change* and is designated by a superscript zero as ΔG^0. (This is sometimes written with a prime as well, as $\Delta G^{0\prime}$, to emphasize the fact that the hydrogen-ion concentration is not $1\ M$ but rather is determined by the standard pH.) Similar symbols are used for the standard entropy change and the standard internal-energy change, so that for a biological reaction occurring under standard conditions Equations 2-9 and 2-10 can be rewritten as follows:

$$\Delta E^0 = \Delta G^0 + T\Delta S^0 \qquad (2\text{-}9\text{a})$$

$$\Delta G^0 = \Delta E^0 - T\Delta S^0 \qquad (2\text{-}10\text{a})$$

In such standard equations T must be the standard temperature, $298°K$, unless otherwise specified.

Free-Energy Changes under Nonstandard Conditions

The standard free-energy change is a convenient but arbitrary point of reference for a reaction. Usually the system one wishes to consider is not in its standard state, and consequently the actual ΔG for the reaction will differ from the ΔG^0 for that reaction. It is therefore necessary to correct ΔG^0 to account for the differences between the actual state of the system and its standard state. Separate corrections must be made for each system variable that differs from its standard value.

For biological systems, the largest deviation from standard con-

---

*Water and hydrogen ions are exceptions to the general rule that all reactants and products be present at a concentration of 1 mole/liter in the standard state. The concentration of water in a dilute aqueous solution is approximately $55.5\ M$ and does not change significantly during the course of reactions, even when water is a reactant or product. Similarly a pH of 7.0 means that hydrogen ions (and, therefore, hydroxyl ions as well) are present at a concentration of $10^{-7}\ M$. Accordingly the specified concentration of $1\ M$ does not apply to water, nor to H^+ or OH^- when the pH is specified.

ditions is generally in the concentrations of the reactants and products. Whenever a reactant or product is present at a concentration other than 1 mole/liter, a correction must be made to ΔG^0 in order to get the actual free-energy change for the reaction under the prevailing conditions. This correction takes the form of Equation 2-11:

$$\Delta G = \Delta G^0 + RT \ln Q = \Delta G^0 + 2.303RT \log_{10} Q \qquad (2\text{-}11)$$

where R is the universal gas constant (1.987 cal/mole-degree), T is the temperature in degrees Kelvin ($^\circ$C + 273), and Q is the ratio of product to reactant concentrations.

To illustrate how Q is calculated, consider the general reaction of Equation 2-12 and notice the expression for Q as indicated in Equation 2-13, where $[A]$, $[B]$, $[C]$, and $[D]$ are the actual concentrations of reactants and products prevailing under the conditions for which we wish to know ΔG.*

$$aA + bB \rightleftharpoons cC + dD \qquad (2\text{-}12)$$

$$Q = \frac{[C]^c[D]^d}{[A]^a[B]^b} \qquad (2\text{-}13)$$

This expression for Q is similar to that for the equilibrium constant K, with which you may already be familiar. We shall in fact encounter the equilibrium constant shortly, but here it is more important to note the crucial difference between Q and K than to dwell on similarities: K is always calculated from *equilibrium* concentrations of reactants and products, whereas Q is determined from the actual concentrations prevailing at the time the calculation is made. Except for the special case of a reaction at equilibrium, then, Q will always have a numerical value different from that of K.

The correction for concentration (Equation 2-11) has an interesting consequence. We have already noted that when a reaction is at its

ΔG^0 and the Equilibrium Constant

---

*For reactions occurring in a dilute aqueous solution, the concentration of water does not change significantly even if water molecules are formed or consumed in the reaction. It is therefore customary to ignore the water when calculating free-energy changes for reactions involving water. This simplification does not lead to difficulty provided that it is used consistently. A similar situation exists with respect to the hydrogen-ion concentration. Most biological systems contain buffers that keep the pH approximately constant, even though the reaction may involve the consumption or generation of free hydrogen ions. Accordingly it is common practice to ignore the H^+ also when calculating free-energy changes for systems in which the pH is fixed. When the symbol $\Delta G^{0\prime}$ is used, therefore, it may be assumed that hydrogen ions need not be included in the calculation of Q. (Actually neither water nor hydrogen ions are really "ignored" by these conventions; the effect of their fixed concentration is instead incorporated into the value of ΔG^0 directly.)

equilibrium point, $\Delta G = 0$. (This must be so, since if ΔG were negative either for the forward or for the reverse direction of the reaction, the reaction could proceed spontaneously in that direction and thus would not be at equilibrium.) In addition, it should be clear from the preceding section that a reaction at equilibrium represents the special case in which the value of Q is in fact equal to that of the equilibrium constant. In other words, when a reaction is at equilibrium under standard conditions of temperature, pressure, and pH, the Q for the reaction will be the standard equilibrium constant K^0. For Equation 2-12, therefore, we can write

$$K^0 = \frac{[C]^c_{\text{equil}}[D]^d_{\text{equil}}}{[A]^a_{\text{equil}}[B]^b_{\text{equil}}} \qquad (2\text{-}14)$$

At equilibrium, then, Equation 2-11 can be expressed in a way that relates ΔG^0, a parameter we would like to be able to calculate, to K^0, a value we can determine experimentally:

$$\Delta G = 0 = \Delta G^0 + 2.303RT \log_{10} K^0 \qquad (2\text{-}15)$$

or $\qquad\qquad \Delta G^0 = -2.303RT \log_{10} K^0 \qquad (2\text{-}16)$

Equation 2-16 is extremely useful, for it provides a means of calculating the standard free-energy change for a reaction directly from its equilibrium constant at standard temperature, pressure, and pH. Once ΔG^0 has been calculated, Equation 2-11 can be used to compute the actual ΔG for the reaction when reactants and products are present at nonstandard, nonequilibrium concentrations. Because of the great usefulness of Equations 2-11 and 2-16, they are reiterated here along with the conditions necessary to use them properly:

1. *For the general reaction :*

$$aA + bB \rightleftharpoons cC + dD$$

The ratio Q is defined as the ratio of prevailing product concentrations to prevailing reactant concentrations:

$$Q = \frac{[C]^c_{\text{prevailing}}[D]^d_{\text{prevailing}}}{[A]^a_{\text{prevailing}}[B]^b_{\text{prevailing}}}$$

Note that the concentrations used to calculate Q are the prevailing concentrations, which are in general *not* equilibrium conditions. However, for the special case where the reaction is in fact at equilibrium, the ratio Q becomes the equilibrium constant K^0:

$$Q_{\text{(at equilibrium)}} = \frac{[C]^c_{\text{equil}}[D]^d_{\text{equil}}}{[A]^a_{\text{equil}}[B]^b_{\text{equil}}} = K^0$$

2. *To calculate the standard free-energy change,* use the relationship:

$$\Delta G^0 = -2.303 RT \log_{10} K^0$$

where $R = 1.987$ cal/mole-degree
T = absolute temperature ($°C + 273$)
K^0 = equilibrium constant at standard temperature, pressure, and pH

Note that at the standard temperature of $25°C$, the term $2.303RT$ becomes $(2.303)(1.987)(298) = 1364$ cal/mole, so the equation can be written $\Delta G^0 = -1364 \log_{10} K^0$.

3. *To calculate the actual free-energy change,* use the expression:

$$\Delta G = \Delta G^0 + 2.303 RT \log_{10} Q$$

where R, T, and Q are defined as above and the concentrations of products and reactants used to calculate Q are expressed in moles per liter. Or, at $25°C$,

$$\Delta G = \Delta G^0 + 1364 \log_{10} Q$$

Equation 2-16 expresses a simple relationship between ΔG^0 and K^0. For a K^0 greater than 1, $\log_{10} K^0$ will be positive and, because of the minus sign, ΔG^0 will be negative. Now an equilibrium constant greater than 1 means that products predominate over reactants at equilibrium; in other words, the equilibrium point lies to the right for the reaction. Thus a negative ΔG^0 means that products outweigh reactants at equilibrium under standard conditions; i.e., the reaction can proceed spontaneously to the right in the standard state.

For a K^0 less than 1, on the other hand, $\log_{10} K^0$ will be negative and ΔG^0 will therefore be positive. By the converse of these arguments, a positive ΔG^0 means that reactants outweigh products at equilibrium under standard conditions and the reaction can go spontaneously to the left in the standard state. Thus, like the equilibrium constant, ΔG^0 provides information about the equilibrium point of a reaction: whether reactants or products predominate and to what extent. Bear in mind, however, that ΔG^0 is calculated for an arbitrary standard state and is therefore not a function of the actual concentrations present.

In contrast, ΔG tells how far from equilibrium a given reaction is at the concentrations actually existing—or, equivalently, the maximum amount of work that could be performed on the surroundings per mole of reactant consumed or product formed—if the reaction is allowed to proceed at the given concentrations. Hence ΔG is directly

Interpretation of ΔG and ΔG^0

dependent upon the concentrations present and will become smaller and eventually approach zero as the reaction nears its equilibrium point. These features of K^0, ΔG^0, and ΔG are summarized in Table 2-1.

  . An important feature of both ΔG and ΔG^0 is that, as pointed out earlier for ΔE, once the value of such a parameter is known for a given reaction under specified conditions, the value of that parameter for the same reaction but proceeding in the opposite direction can be obtained by simply changing the sign. For example, the free-energy change associated with the freezing of water at $-0.1°C$ is -0.5 cal/mole; this means, then, that the free-energy change for the melting of ice at the same temperature would be $+0.5$ cal/mole.

An Example   To illustrate the calculations involved in determining ΔG^0 and ΔG as summarized in Table 2-1, consider the cellular reaction that converts 3-phosphoglycerate to 2-phosphoglycerate (Equation 2-17) during the oxidation of glucose, as described later in Chapter 4:

$$\text{3-Phosphoglycerate} \rightleftharpoons \text{2-phosphoglycerate} \tag{2-17}$$

If the enzyme that catalyzes this interconversion is added to a solution of 3-phosphoglycerate at 25°C and pH 7 and the solution is incubated until equilibrium is reached, both 3-phosphoglycerate and 2-phosphoglycerate will be found to be present in an equilibrium ratio of 0.165:

$$K^0 = \frac{[\text{2-phosphoglycerate}]}{[\text{3-phosphoglycerate}]} = 0.165$$

Note that the equilibrium ratio of 0.165 is independent of the actual starting concentration of 3-phosphoglycerate and could equally well have been achieved by starting with 2-phosphoglycerate or any mixture of the two species.

    The standard free-energy change ΔG^0 can be calculated from K^0 as follows:

$$\Delta G^0 = -2.303RT \log_{10} K^0 = -(2.303)(1.987)(273 + 25) \log_{10}(0.165)$$
$$= (-1364)(-0.7825) = +1067 \text{ cal/mole} = +1.067 \text{ kcal/mole}$$

The positive value for ΔG^0 is therefore another way of expressing the fact that the reactant (3-phosphoglycerate) is the predominant species at equilibrium. A positive ΔG^0 also means that, under standard conditions, the reaction is nonspontaneous (thermodynamically impossible)

TABLE 2-1   The Meaning of ΔG^0 and ΔG

The Meaning of ΔG^0

ΔG^0 Negative ($K^0 > 1$)	ΔG^0 Positive ($K^0 < 1$)	$\Delta G^0 = 0$ ($K^0 = 1$)
Products predominate over reactants at equilibrium at standard temperature, pressure, and pH.	Reactants predominate over products at equilibrium at standard temperature, pressure, and pH.	Products and reactants are present equally at equilibrium at standard temperature, pressure, and pH.
Reaction goes spontaneously to right under standard conditions.	Reaction goes spontaneously to left under standard conditions.	Reaction is at equilibrium under standard conditions.

The Meaning of ΔG

ΔG Negative	ΔG Positive	$\Delta G = 0$
Reaction is thermodynamically feasible as written under conditions for which ΔG was calculated.	Reaction is not feasible as written under conditions for which ΔG was calculated.	Reaction is at equilibrium under conditions for which ΔG was calculated.
Work can be done by the reaction under conditions for which ΔG was calculated.	Energy must be supplied to drive reaction under conditions for which ΔG was calculated.	No work can be done nor is energy required by the reaction under conditions for which ΔG was calculated.

in the direction written; i.e., if you begin with both 2-phosphoglycerate and 3-phosphoglycerate present at concentrations of 1.0 M, net conversion of 3-phosphoglycerate to 2-phosphoglycerate cannot occur. Instead the reaction under standard conditions would proceed (in the presence of a suitable catalyst) to the left (since the ΔG^0 for that reaction is -1067 cal/mole), converting 2-phosphoglycerate into 3-phosphoglycerate until the equilibrium ratio of 0.165 was reached. Alternatively, if both species were added or removed continuously as necessary to maintain the concentrations of both at 1.0 M, then the reaction would proceed continuously to the left, with the liberation of 1.067 kcal of free energy per mole of 2-phosphoglycerate converted to 3-phosphoglycerate.

In a real cell, of course, it is unlikely that 3-phosphoglycerate and 2-phosphoglycerate would be present at either equilibrium or standard concentrations. In fact experimental values for the actual concentrations of these substances in human red blood cells are as follows:

3-phosphoglycerate: 61.2 μmoles/liter ($= 61.2 \times 10^{-6}$ M)
2-phosphoglycerate: 4.3 μmoles/liter ($= 4.3 \times 10^{-6}$ M)

Using these values we can calculate the actual ΔG for the reaction of Equation 2-17 as follows:[*]

$$\Delta G = \Delta G^0 + 2.303 RT \log_{10} [\text{2-phosphoglycerate}]/[\text{3-phosphoglycerate}]$$

$$= +1067 + 1364 \log_{10} \frac{4.3 \times 10^{-6}\ M}{61.2 \times 10^{-6}\ M}$$

$$= +1067 + (1364)(-1.153) = -410\ \text{cal/mole} = -0.41\ \text{kcal/mole}$$

The negative value for ΔG means that the conversion of 3-phosphoglycerate to 2-phosphoglycerate is thermodynamically possible under the conditions of concentration actually prevailing in human red blood cells and that the reaction will yield 0.41 kcal of free energy per mole of reactant converted to product. Thus although the conversion to 3-phosphoglycerate to 2-phosphoglycerate is thermodynamically impossible under standard conditions, the red blood cell maintains these two species at concentrations adequate to offset the positive ΔG^0 and thereby renders the reaction possible. This is, of course, essential if the red blood cell is to be successful in carrying out the glucose-degrading process (glycolysis; see Chapter 4) of which this reaction is part.

Multireaction Pathways

So far we have been concerned only with the thermodynamics of individual reactions. But individual reactions are really only of use to the cell insofar as they can be fitted together into a multistep *metabolic pathway* capable of carrying out some specific overall process, such as the oxidation of glucose or the synthesis of a particular amino acid. Our concern for thermodynamics must therefore extend beyond individual reactions to encompass the whole sequence of reactions that make up a complete functional pathway.

A useful fact in dealing with multistep pathways is that the free-energy changes of the individual steps are additive. For the three-step sequence or pathway shown in Equation 2-18, the overall free-energy change is simply the sum of the individual free-energy changes

---

[*]When calculating ΔG from Equation 2-11, it is important to express all concentrations of reactants and products in moles per liter, since the value used for the gas constant R (1.987 cal/mole-degree) presumes molar concentrations. It is true that in the particular example of Equation 2-17, concentration units cancel in the calculation of Q and the actual choice of units does not matter as long as the same units are used throughout. Often, however, Q is not a dimensionless quantity, and such cancellation does not occur. For the general reaction $A \rightarrow B + C$, for example, Q is calculated as $[B][C]/[A]$ and clearly has the units of concentration. In such a case the numerical value of Q will obviously depend upon the units in which the concentration is expressed. The same is true when calculating the equilibrium constant for such a reaction. To avoid errors, therefore, it is wise to express all concentrations in moles per liter before proceeding with thermodynamic calculations.

as illustrated in Equation 2-19 for ΔG^0 and in Equation 2-20 for ΔG:

$$A \rightleftharpoons B \rightleftharpoons C \rightleftharpoons D \tag{2-18}$$

$$\Delta G^0_{A \to D} = \Delta G^0_{A \to B} + \Delta G^0_{B \to C} + \Delta G^0_{C \to D} \tag{2-19}$$

$$\Delta G_{A \to D} = \Delta G_{A \to B} + \Delta G_{B \to C} + \Delta G_{C \to D} \tag{2-20}$$

Since the free-energy change of a system as a result of a multi-step reaction depends upon only the initial and final states of the system, it follows that the sums of the ΔG's along all alternative paths connecting the two states must be equal. If, for example, there were an alternate sequence of steps by which A could be converted to D via different intermediates, the sum of the ΔG's for the alternate steps would equal the sum of the ΔG's for the pathway shown in Equation 2-18. This principle enables ΔG's for reactions occurring in living cells to be determined experimentally in the laboratory even though the reaction pathways in the cell may be different or even unknown.

A common misconception that arises from this property of additivity, however, is that a multistep reaction sequence in a cell must necessarily be able to proceed if the ΔG for the overall reaction (obtained by summing ΔG values for the component reactions) is negative. This is quite an understandable misconception, since it would in fact be true if concentrations of all reactants, products, and intermediates in the pathway were free to rise and fall. Under such conditions, the negative overall ΔG value would ensure that the concentrations of all species would be adjusted to steady-state levels necessary to guarantee that the ΔG value for each step in the sequence would be negative. Frequently, however, cellular concentrations of reactants, products, and intermediates are not those dictated by the thermodynamic considerations of a specific reaction sequence but may in fact be held at quite different levels by other processes and pathways occurring simultaneously in the cell. And if the prevailing concentrations of any species are sufficient to render the ΔG for any single step in a metabolic pathway positive, then that particular reaction cannot occur, and the whole pathway will, as a consequence, be blocked at that point. In other words, for every metabolic pathway functioning within a cell it is necessary not only that the overall ΔG value be negative but also that the ΔG value for each individual component reaction be negative under the conditions of concentration actually prevailing within the cell. This is not, however, to say that the *standard* free-energy change must be negative for each step (or, for that matter, for any step). As we have seen in the foregoing example, it is possible by adjusting concentrations appropriately to ensure that ΔG is negative (and the reaction therefore feasible) even though ΔG^0 is positive.

Disequilibrium as a Condition of Life

The preceding discussion underscores the important point that the driving force behind thermodynamics is the tendency of reactions to move toward equilibrium. Indeed ΔG^0 and ΔG are really nothing more than means of quantitating how far and in what direction from equilibrium a reaction lies under the conditions dictated by standard or prevailing concentrations of products and reactants. Once at equilibrium, of course, the forward and backward rates become the same for a reaction, so that there is no net flow of matter in either direction. For all practical purposes, then, the reaction has stopped. But a living cell is characterized by reactions that are continuously under way. This means that the cell must be in a continual state of disequilibrium. Life might, in fact, be defined as the continual struggle to maintain a position far from equilibrium, since at equilibrium no net reactions are possible, no work can be done, and the thermodynamically improbable order of the living state cannot be maintained.

Thus life is possible only because life cells maintain themselves in a *steady state* of dynamic equilibrium. This situation could not persist, however, if the cell were isolated from its surroundings. All its reactions would gradually run to equilibrium and the cell would come to a state of minimum free energy, after which no further changes could occur and no work could be accomplished. A cell at equilibrium would be a dead cell. The living cell is able to maintain its disequilibrium state only by a continuous uptake of energy from its environment, whether in the form of light or preformed organic food molecules. This continuous uptake of energy and the flux of matter that accompanies it enable the maintenance of a disequilibrium in which the various reactants and products of cellular reactions are maintained at steady-state concentrations far enough from equilibrium to ensure that the thermodynamic drive toward equilibrium can be harnessed by the cell to perform useful work and thereby maintain and extend its activities. How this is accomplished will occupy our attention in coming chapters, as we look first (in Chapter 3) at principles of enzyme catalysis that determine the rates of cellular reactions and then move on (in Chapters 4 to 7) to the functional metabolic pathways that result from a series of such reactions acting in concert.

Practice Problems

2-1. Thermodynamic Spontaneousness: The thermodynamic spontaneousness of a reaction is of crucial importance to the cell and hence to the cell biologist. Especially important thermodynamic parameters are those that serve as measures of spontaneousness. Consider a system consisting of a mixture of ice and water, and assume that the heat of fusion (freezing) is absorbed by the surroundings in order to maintain the system at the specified temperature. Values for certain thermodynamic parameters at three different temperatures (but at constant pressure and volume) are given below for the reaction:

$$\text{Water} \rightleftharpoons \text{ice} \qquad (2\text{-}21)$$

Temperature (°C)	ΔS_{system} (cal/mole-°K)	$\Delta S_{surroundings}$ (cal/mole-°K)	ΔS_{total} (cal/mole-°K)	ΔE_{system} (cal/mole)
+0.1	−5.2613	+5.2595	−0.0018	−1437.2
0.0	−5.2581	+5.2581	0.0000	−1436.3
−0.1	−5.2549	+5.2567	+0.0018	−1435.4

(a) What sort of physical interpretation can you use to explain why ΔS_{system} is negative at all three temperatures? What can you say about ΔS_{system} as a measure of thermodynamic spontaneousness?

(b) What about ΔS_{total} as a measure of thermodynamic spontaneousness?

(c) What sort of physical interpretation can you give for the fact that ΔE_{system} is negative in all three cases? What can you say about ΔE_{system} as a measure of thermodynamic spontaneousness?

(d) Calculate ΔG_{system} for the reaction at each of the three temperatures. (You will have to use the value of $0°C = 273.16°K$ for the five-place accuracy demanded by this problem; slide rule accuracy will be inadequate.)

(e) What can you say about ΔG_{system} as a measure of thermodynamic spontaneousness?

2-2. Meaning of ΔG: The change in free energy ΔG for a reaction is a very useful parameter in thermodynamics. For each of the following statements, decide whether the statement is true or false. If it is false, modify it to make it true.

_____(a) ΔG is affected by the temperature at which the reaction is carried out, the rate at which the reaction proceeds, and the concentration of reactants and products.

_____(b) If a reaction occurs spontaneously (i.e., is thermodynamically possible), either its ΔG is negative or it has a positive ΔG but is followed or preceded in the metabolic sequence by a reaction with a negative ΔG.

_____(c) $\Delta G = \Delta G_{products} + \Delta G_{reactants}$

_____(d) If you know the ΔG for the reaction $A + B \rightarrow C + D$, you have enough information to predict ΔG for the reverse reaction at the same temperature and concentrations.

_____(e) Reactions for which ΔG^0 is positive are always endergonic and therefore cannot occur spontaneously.

2-3. Calculations: The conversion of fumarate to malate is a reaction in the TCA (Krebs) cycle that you will encounter in Chapter 5:

$$\text{Fumarate} + H_2O \rightleftharpoons \text{malate} \qquad (2\text{-}22)$$

The ΔG^0 for this reaction is -750 cal/mole at 25°C.

(a) What would be the ΔG value for this reaction proceeding under conditions where the concentrations of both fumarate and malate were maintained at 1 mM?

(b) If the fumarate concentration were maintained constant at 1 mM, what malate concentration would be required to bring the reaction to equilibrium? What would happen if the malate concentration were raised still higher?

2-4. More Calculations: An important reaction in the synthesis of the storage carbohydrates glycogen (in animals) and starch (in plants) involves the interconversion of the molecules glucose-6-phosphate and glucose-1-phosphate by movement of a phosphate group from carbon 6 of glucose to carbon 1:

$$\text{Glucose-6-phosphate} \rightleftharpoons \text{glucose-1-phosphate} \qquad (2\text{-}23)$$

The ΔG^0 value for this reaction at 25°C is $+1.8$ kcal/mole.

(a) Is this reaction exergonic or endergonic under standard conditions?

(b) If you start out with a solution containing 0.1 M glucose-6-phosphate and add an appropriate amount of the enzyme that catalyzes this reaction (phosphoglucomutase), will any glucose-1-phosphate be formed? If so, calculate the resulting equilibrium concentrations of glucose-6-phosphate and glucose-1-phosphate. If not, explain why not.

(c) Under what cellular conditions, if any, would this reaction produce glucose-1-phosphate continuously at a high rate?

2-5. Some Physiological Calculations: The following intermediates in the metabolism of glucose are found in human red blood cells (remember that 1 μmole $= 10^{-6}$ moles):

Fructose-1,6-diphosphate (FDP):	4.6 μmoles/liter
Dihydroxyacetone phosphate (DHAP):	4.9 μmoles/liter
Glyceraldehyde-3-phosphate (G3P):	2.6 μmoles/liter

In addition, red blood cells contain enzymes that catalyze the following reactions:

$$\text{G3P} + \text{DHAP} \rightleftharpoons \text{FDP} \qquad K^0 = 10^4 \qquad (2\text{-}24)$$

$$\text{G3P} \rightleftharpoons \text{DHAP} \qquad K^0 = 25 \qquad (2\text{-}25)$$

(a) Assuming standard conditions of temperature, pressure, and pH, calculate ΔG and ΔG^0 for each reaction. In which direction does each reaction run in the standard state? In the red blood cell?

(b) Calculate K^0, ΔG^0, and ΔG for the following reactions. Would you expect either of these to go spontaneously to the right in the red cell?

$$2\text{G3P} \rightleftharpoons \text{FDP} \qquad (2\text{-}26)$$

$$2\text{DHAP} \rightleftharpoons \text{FDP} \qquad (2\text{-}27)$$

(c) One of the reactions in part (b) will proceed spontaneously to the left in the red blood cell, but the other will not, even though both have a negative overall ΔG in that direction. Explain.

‡2-6. Understanding Helmholtz and Gibbs: If, for a system at constant

temperature, pressure, and volume, it is true that the change in internal energy ΔE is given by the expression

$$\Delta E = \Delta G + T\Delta S \qquad (2\text{-}28)$$

is it not also true that the total internal energy can be expressed as

$$E = G + TS \qquad (2\text{-}29)$$

‡2-7. Proof: Prove that ΔG^0's are additive for a stepwise sequence such as that of Equation 2-18.

‡2-8. Coupled Reactions: The direct phosphorylation of glucose, as represented by the following equation, is not thermodynamically spontaneous under cellular conditions:

Glucose + phosphate → glucose-6-phosphate + H_2O
$$\Delta G^0 = +3.3 \text{ kcal/mole at } 37°C \qquad (2\text{-}30)$$

(a) Calculate the equilibrium concentration of glucose-6-phosphate that would be reached if glucose and phosphate were each maintained in the cell at a concentration of $100 \ \mu M \ (= 100 \times 10^{-6} \ M)$. Is this a good way to form significant amounts of glucose-6-phosphate in the cell?

(b) One way to overcome the unfavorable conditions in part (a) and achieve synthesis of glucose-6-phosphate would be to greatly increase the concentration of glucose or phosphate or both to force the reaction to the right. Assuming that the phosphate concentration remains fixed at $100 \ \mu M$, how high would you have to raise the glucose concentration to produce an equilibrium concentration of $25 \ \mu M$ for glucose-6-phosphate? Is this a feasible way for a cell to make significant amounts of glucose-6-phosphate?

(c) In the cell, glucose-6-phosphate is actually synthesized by coupling Reaction 2-30 to the hydrolysis (splitting off) of phosphate from a high-energy compound called *adenosine triphosphate* (ATP), a molecule you will encounter in Chapter 4 because of its prominent role in cellular energy metabolism. The hydrolysis of ATP (Reaction 2-31) is coupled to the phosphorylation of glucose in such a way that part of the energy of that hydrolysis is used to drive the phosphorylation:

$$\text{Glucose + phosphate} \rightarrow \text{glucose-6-phosphate} + H_2O \qquad (2\text{-}30)$$

$$\underline{\text{ATP} + H_2O \rightarrow \text{ADP} + \text{phosphate}} \qquad (2\text{-}31)$$

$$\text{Glucose + ATP} \rightarrow \text{glucose-6-phosphate} + \text{ADP} \qquad (2\text{-}32)$$

The ΔG^0 values for Equations 2-30 and 2-31 are $+3.3$ kcal/mole and -7.3 kcal/mole, respectively. What is the ΔG^0 value for the overall reaction of Equation 2-32? Under these conditions, what concentration of glucose is required to produce an equilibrium concentration of $25 \ \mu M$ for glucose-6-phosphate, assuming an ATP/ADP ratio of 10 in the cell?

(d) One way to couple these reactions would be to have both the enzyme that catalyzes the first reaction (phosphatase) and the enzyme responsible for the second reaction (ATP hydrolase) present in the cell and to allow reaction 2 to drive reaction 1 by elevating the intracellular concentration of the common intermediate, inorganic phosphate. Can you think of any disadvantage to this mechanism?

(e) Actually the way the cell accomplishes this reaction is to use an entirely new enzyme, *hexokinase*, to effect the direct transfer of phosphate from ATP to glucose, with an enzyme-bound ATP-glucose complex as the common intermediate. What advantage can you see to this?

References

Blum, H. F., *Time's Arrow and Evolution*, 2nd ed. (Princeton, N.J.: Princeton University Press, 1955).

Carter, L. C., *Guide to Cellular Energetics* (San Francisco: Freeman, 1973).

Christensen, H. N., and R. A. Cellarius, *Introduction to Bioenergetics: Thermodynamics for the Biologist* (Philadelphia: Saunders, 1972).

Klotz, I. M., *Energy Changes in Biochemical Reactions* (New York: Academic Press, 1967).

Miller, G. T., Jr., *Energetics, Kinetics and Life— An Ecological Approach* (Belmont, Ca.: Wadsworth, 1971).

Morowitz, H. J., *Entropy for Biologists: An Introduction to Thermodynamics* (New York: Academic Press, 1970).

Wall, F. T., *Chemical Thermodynamics* (San Francisco: Freeman, 1965).

Enzyme Catalysis

3

The previous chapter introduced the change in free energy ΔG as a valuable thermodynamic signpost, indicating by its *sign* whether a reaction is thermodynamically possible in a given direction and by its *magnitude* how much energy can be obtained (or must be invested) as the reaction moves to equilibrium. At the same time, a cautionary note was raised concerning the meaning of the word "spontaneous," since a negative free-energy change simply indicates that the reaction is thermodynamically possible under the specified conditions and says nothing at all about whether it will actually take place. This depends in turn upon the availability of a suitable pathway or mechanism for the process and also often upon an input of energy to activate the process. Thus while we possess in ΔG the means for determining direction and extent of a reaction, we lack at present any indication of *rate*. And it is toward this end that the present discussion of enzyme catalysis is directed, for almost all the reactions and processes carried out by cells are mediated by protein catalysts called *enzymes*. It is only as we explore the nature of enzyme-catalyzed reactions that we begin to understand the principles involved in measuring and controlling rates of cellular reactions.

There are many reactions, both in cells and elsewhere in the physical universe, that occur either at very low rates or not at all, even though they are thermodynamically possible. One such example, already raised

The Activation Energy Barrier and the Metastable State

in Chapter 2 to make a point about directionality, is the combustion of paper (cellulose), which we can represent (nonstoichiometrically) as follows:

$$(C_6H_{10}O_5)_n + O_2 \rightarrow CO_2 + H_2O + \text{energy} \qquad (3\text{-}1)$$

Cellulose

We know that, once initiated, this reaction occurs readily with the liberation of much free energy as heat. Yet we also know that paper and oxygen molecules can (and routinely do) exist side by side without showing any detectable tendency to combine until we provide activating energy, usually in the form of heat from a match. We can represent this situation by drawing a free-energy diagram and including an *activation energy barrier* as shown in Figure 3-1. As indicated on the diagram, thermodynamics measures only the difference in free energy between the initial and the final states. For the actual reaction, however, there is an activation energy barrier, representing a finite additional energy content that at least some of the cellulose and oxygen molecules must possess before the reaction can be initiated. Thus the reactants coexist in a seemingly stable state until enough additional energy is supplied to get the reaction over this initial hump. Once initiated, the free energy released from the oxidation of one cellulose molecule provides the means whereby neighboring molecules can achieve the critical energy level, and the reaction becomes self-sustaining.

As it turns out, many thermodynamically feasible reactions share this property of requiring some activation energy before they will proceed in the direction predicted by thermodynamics. Thus many molecules that are thermodynamically unstable (i.e., that should react

Figure 3-1 Free-energy diagram for cellulose oxidation.

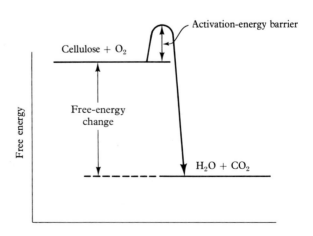

spontaneously) are rendered *metastable* because of a high energy of activation that serves as an effective barrier to equilibrium. In the case of living cells, this activation energy barrier and the resulting metastable state are crucial, because life by its very nature is a system maintained in a steady state a long way from equilibrium. Were it not for the metastable state, all reactions would run quickly and uncontrolledly to equilibrium and life as we know it would be impossible.

One way to overcome the energy barrier is to put enough energy into the system so that an adequate proportion of the potentially reactive molecules have sufficient energy to get over the activation energy barrier. Commonly, as illustrated by the combustion of paper, the activating energy is supplied as heat, which increases the kinetic energy of the molecules and facilitates their collision and interaction. Alternatively the input energy may be in the form of an electric spark. For example, the reaction of H_2 and O_2 to form water is potentially a violently exergonic reaction:

Overcoming the Barrier

$$2H_2 + O_2 \rightarrow 2H_2O + \text{energy} \qquad (3\text{-}2)$$

Yet, quite clearly, hydrogen and oxygen molecules can exist side by side in the metastable state. An electric spark, however, can provide the energy to overcome the barrier, and the two gases then react explosively with each other—which is why people no longer fill blimps with hydrogen gas!

The problem with using a match or spark to overcome an energy barrier is that this is not an isothermal (constant-temperature) solution to the activation problem, and living systems are characterized by and large by a requirement for a relatively constant temperature. Cells, therefore, would not be especially "happy" with the suggestion that we wish to overcome an activation barrier by touching a match to them. In addition, we can anticipate a later topic if we note that this approach would also suffer from a lack of specificity, since it would result in many activation energy barriers being indiscriminately overcome; yet a cardinal principle of cellular success lies in the ability to facilitate specific reactions under specific conditions while leaving other equilibrium barriers intact.

The alternative to overcoming an activation energy barrier by the input of additional energy is to lower the activation barrier, and this, of course, is the task of a catalyst. You have undoubtedly met the concept before: a *catalyst* is an agent that speeds up the rate of a reaction by lowering the energy of activation. A catalyst facilitates reactions in the absence of added energy by complexing transiently

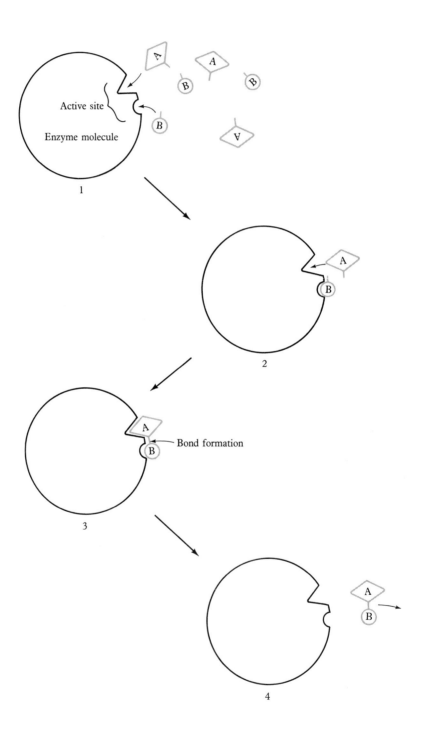

with the substrates and ordering them in an advantageous configuration with respect to each other. This enhances the likelihood of an interaction between them, as illustrated schematically in Figure 3-2. For a specific example of catalysis, consider the decomposition of hydrogen peroxide (H_2O_2) into water and oxygen:

$$2H_2O_2 \rightarrow 2H_2O + O_2 \qquad (3\text{-}3)$$

This is a thermodynamically favored reaction, but hydrogen peroxide exists in the metastable state because of the high activation energy barrier. One way to facilitate the reaction is by adding energy to overcome the barrier. (Indeed an electric spark can cause hydrogen peroxide to decompose explosively, accounting for its suitability in rocket propellants.) But if you instead add a small amount of ferric ion (Fe^{+++}) to an H_2O_2 solution under the appropriate conditions, you can demonstrate that the reaction proceeds 30,000 times faster than without the ferric ions. Clearly Fe^{+++} is a catalyst for this reaction, lowering the activation energy barrier and thereby ensuring that a significant proportion of the hydrogen peroxide molecules possess adequate energy to decompose at the ambient temperature without the input of added energy.

Your knowledge of chemistry should allow you to agree readily with the following basic properties of catalysts:

Enzymes as Biological Catalysts

1. A catalyst increases the rate of a reaction by lowering the activation

Figure 3-2   The role of an enzyme in catalyzing a chemical reaction. An enzyme facilitates the reaction it catalyzes by ordering the substrate molecules on its surface in a configuration that enhances the likelihood of their interaction. The enzyme shown here has an *active site* at which one molecule each of substrates *A* and *B* can bind. In step 1 the active site is open, but no substrate molecules have yet collided with it in the appropriate position for binding. In step 2, substrate *B* is in place and a molecule of substrate *A* is just arriving. With both substrates in place on the enzyme surface with their reactive groups in close proximity, their interaction is favored and covalent bond formation occurs (step 3). In step 4 the covalently bound *A-B* product leaves the active site, and the enzyme is free to accept further substrate molecules. Although substrate specificity appears in the diagram to be due to geometry alone, real enzymes depend also upon the specific chemical properties of the amino acids in the region of the active site, with ionic bonds, hydrogen bonds, and hydrophobic interactions involved in the actual substrate binding.

energy barrier, thereby allowing thermodynamically feasible reactions to proceed without an input of energy from the outside.

2. A catalyst acts by complexing with substrate molecules to facilitate their interaction and reaction.

3. A catalyst changes only the *rate* at which equilibrium is achieved; it does not affect the *position* of the equilibrium. In other words, a catalyst is not a thermodynamic genie that can push impossible reactions "uphill"; it can only facilitate the rate of "downhill" reactions.

These are three properties common to all catalysts, but in biological systems we can add a fourth, very important property of catalysts:

4. In cells, all catalysts are proteins called *enzymes*.

This means that virtually every reaction that occurs in cells is mediated by an enzyme, a protein catalyst which, as it turns out, has a high degree of specificity for that particular reaction.

A protein, as we shall see in Chapter 8, is a biological macromolecule made up of a genetically determined sequence of component *amino acids* linked together by peptide bonds. An important feature of protein chemistry is that the three-dimensional structure, the function, and the regulatory properties of a protein are all dictated by its amino acid sequence. This means that, by specifying a particular amino acid sequence, a gene specifies a particular protein (and, for an enzyme, a particular catalytic role) uniquely. Furthermore the dependence of specific functional activity upon proper structure is so intimate that the substitution of one incorrect amino acid in the sequence of an enzyme molecule frequently results in a serious impairment in the catalytic properties of that enzyme. Thus the ability of a cell to carry out a specific reaction sequence or metabolic pathway depends not only upon the presence of a specific enzyme to catalyze each reaction in the sequence but also upon each enzyme having the correct amino acid sequence.

The Specificity of Enzymes

An especially important point about enzymes concerns their high degree of *specificity*. Specificity is probably one of the most characteristic properties of a living system, and enzymes are dramatic examples of biological specificity.

Many inorganic catalysts are quite nonspecific in the sense that they will act on a variety of compounds that have some general chemical feature in common. Consider, for example, the hydrogenation of (addition of hydrogen to) an unsaturated $C=C$ bond:

$$\underset{\text{Ni or Pt}}{R-\overset{\overset{\displaystyle H}{|}}{C}=\overset{\overset{\displaystyle H}{|}}{C}-R' + H_2 \longrightarrow R-\overset{\overset{\displaystyle H}{|}}{\underset{\underset{\displaystyle H}{|}}{C}}-\overset{\overset{\displaystyle H}{|}}{\underset{\underset{\displaystyle H}{|}}{C}}-R'} \qquad (3\text{-}4)$$

This reaction can be carried out in the laboratory by using a platinum or nickel catalyst, as indicated. These inorganic catalysts are very nonspecific, however, in the sense that the same catalyst can be used to hydrogenate a wide variety of unsaturated compounds. In fact the use of nickel or platinum in this way finds commercial application in the hydrogenation of polyunsaturated vegetable oils (such as cottonseed oil) in the manufacture of solid cooking fat. Regardless of the exact structure of the molecule containing the C=C bond, it can be effectively hydrogenated in the presence of nickel or platinum.

By way of contrast, consider the biological example of hydrogenation shown in Equation 3-5 for the conversion of fumarate into succinate (a reaction we shall encounter as part of aerobic energy metabolism in Chapter 5):

$$\underset{\text{Fumarate}}{\overset{\text{HOOC}}{\underset{\text{H}}{\diagdown}}\underset{\overset{\displaystyle ||}{\underset{\displaystyle C}{}}}{\overset{\displaystyle C}{}}\underset{\text{COOH}}{\diagup}} \underset{\substack{\text{Succinate} \\ \text{dehydrogenase}}}{+ 2H^+ + 2e^- \rightleftharpoons} \underset{\text{Succinate}}{\overset{\text{HOOC}}{\underset{|}{}}\underset{|}{CH_2}\underset{|}{CH_2}\text{COOH}} \qquad (3\text{-}5)$$

This particular reaction is catalyzed in cells by an enzyme called succinate dehydrogenase (so called because it normally acts in the opposite direction). This dehydrogenase, like most enzymes, is highly specific—it will not add or subtract hydrogen from any compounds except the above pair. In fact the enzyme is so specific that it will not even recognize maleate, which is simply a stereoisomer of fumarate. Such a degree of specificity has as a corollary, of course, the requirement that a cell possess as many different enzymes as it has reactions to catalyze. For a typical cell this means that several thousand different enzymes are necessary to enable the cell to carry out its full metabolic program. That may, at first glance, appear to be extravagant in terms of enzymes to be synthesized, genetic information to be stored, and protein molecules to have on hand in the cell, but you should also be able to see the tremendous regulatory possibilities this suggests, a point to which we shall have occasion to return later.

$$\underset{\text{Maleate}}{\overset{\text{H}}{\underset{\text{H}}{\diagdown}}\underset{\overset{\displaystyle ||}{\underset{\displaystyle C}{}}}{\overset{\displaystyle C}{}}\underset{\text{COOH}}{\diagup}\overset{\text{COOH}}{\diagup}}$$

Enzyme Kinetics   Thus far the discussion of enzyme catalysis has been exclusively qualitative. We have dealt with the activation energy barrier and with catalysts as a means of lowering the barrier, and we have encountered enzymes as highly specific biological catalysts. Still lacking is a means of assessing rates of enzyme-catalyzed reactions and understanding the factors that influence reaction rates. Yet the rate at which its enzymes are capable of functioning is obviously of critical importance to a cell (and hence to a cell biologist). The rate at which a cell's enzymes are capable of functioning is, as one might expect, affected both by factors such as pH and temperature that influence protein structure and activity in general and also by concentrations of available substrates (reactants), accumulated products, inhibitors, and activators. It is as we turn our attention to these quantitative aspects of enzyme catalysis that we encounter *enzyme kinetics*. The word "kinetics" is of Greek origin (*kinetikos* = moving) and, as applied to chemical reactions, deals with reaction rates and the manner in which rates are influenced by a variety of factors, but especially by concentrations of substrates, products, and inhibitors. Our goal here is to consider the effects of substrate concentrations (and, for the intrepid worker of challenging problems, the effects of inhibitor concentrations) upon the kinetics of enzyme-catalyzed reactions. Initially an appreciation for enzyme kinetics will be developed by analogy, using as a model system a roomful of monkeys ("enzymes") shelling peanuts ("substrates"). From there we shall proceed to the actual enzyme-catalyzed reaction, as we seek to become conversant with the means whereby reaction rates can be assessed from known substrate concentrations and kinetic parameters. Our considerations will be restricted to *initial* reaction rates, measured over a period of time during which the substrate concentration has not yet decreased enough to affect the rate and the accumulation of product is still too small to cause any measurable inhibition.

Of Monkeys and Peanuts: A Model   We shall first attempt to understand enzyme kinetics by analogy, using monkeys and peanuts as a model system. As you proceed through this section, try to understand each step first in terms of the monkeys shelling peanuts, but then also in terms of an actual enzyme system. To facilitate this thought process, blocks of questions have been inserted into the text (the answers are in Appendix B); therefore this section is really a self-testing exercise on enzyme kinetics.

   As our model we shall use a troop of ten monkeys, all equally adept at finding and shelling peanuts. We shall assume that the monkeys are too full to eat any of the peanuts they shell but nonetheless have an irresistible compulsion to go on shelling. To make our model a little more rigorous, we should also introduce the restrictions that the peanuts are a new hybrid variety which can be readily stuck back together again and that the monkeys are in fact just as likely to put the peanuts back in the shells as they are to take them out. But these qualifications need not concern us here since we are interested only in initial conditions in which all the peanuts start out in the unshelled form. [*]

Consider next a room of fixed floor space with peanuts scattered equally about on the floor. The number of peanuts in the room will be varied as we move through the exercise, but in all cases there will be many more peanuts than monkeys. Since we know the number of peanuts and the total floor space, we can calculate the "concentration" of peanuts in the room. Initially the monkeys start out in an adjacent room. Each assay will start by opening the door and allowing the monkeys to enter the Peanut Gallery.

Questions

1. Explain the reason for each of the following restrictions:
   (a) The monkeys are all equally adept at shelling peanuts.
   (b) The monkeys eat no peanuts.
   (c) Shelling of peanuts requires no effort.
   (d) The monkeys are as likely to put peanuts back together again as they are to take them apart.

2. Why is the term "concentration" in quotation marks?

3. What assumption should be made with regard to the length of time it takes the monkeys to get through the door into the Peanut Gallery?

To develop a quantitative model, assume that at an initial peanut "concentration" of one peanut per square foot, the average monkey spends 9 seconds locating a peanut he wishes to shell and that once he has found one he requires 1 second to shell it. Each monkey thus requires a total of 10 seconds per peanut and can consequently shell peanuts at the rate of 0.1 peanut per second. Since there are ten

Monkeys, Peanuts, and Hyperbolas

---

[*] To make the analogy even better, we should also insist that having picked up a peanut, a monkey can either shell it or drop it and reach for another peanut instead (i.e., that the formation of an enzyme-substrate complex is reversible).

monkeys in the room, the rate (or velocity v) of peanut shelling for all monkeys is one peanut per second. All this can be tabulated as follows:

S = Concentration of Peanuts (peanuts/ft²)	Time Required per Peanut			Rate of Shelling	
	To Find (sec/peanut)	To Shell (sec/peanut)	Total (sec/peanut)	Per Monkey (peanut/sec)	Total (v) (peanut/sec)
1	9	1	10	0.1	1.0

For a second assay, herd all the monkeys back into the waiting room, sweep up the debris, and arrange peanuts about the Peanut Gallery at a "concentration" of three peanuts per square foot. Since peanuts are now three times as abundant as previously, the average monkey should find a peanut three times more quickly than before, such that the time spent finding a peanut is reduced to 3 seconds. But the average peanut, once found, will still take 1 second to shell, so that the total time per peanut will be 4 seconds and the velocity of shelling will be 0.25 peanut per second for one monkey, or 2.5 peanuts per second for the roomful of monkeys. Thus another line of entries has been generated for the data table:

S = Concentration of Peanuts (peanuts/ft²)	Time Required per Peanut			Rate of Shelling	
	To Find (sec/peanut)	To Shell (sec/peanut)	Total (sec/peanut)	Per Monkey (peanut/sec)	Total (v) (peanut/sec)
1	9	1	10	0.10	1.0
3	3	1	4	0.25	2.5

By continuing this process for higher and higher peanut "concentrations" and plotting the resultant velocities v as a function of the peanut concentration S, you can demonstrate that the relationship between v and S is that of a hyperbola, as shown in Figure 3-3.

Questions

4. Extend the table above by filling in the data corresponding to peanut "concentrations" of 9, 18, 90, and ∞ peanuts per square foot. Then graph the kinetics of this "reaction" by plotting v on the ordinate (suggested scale: 0 to 10 peanuts per second) versus S on the abscissa (suggested scale: 0 to 100 peanuts per square foot). What kind of curve does this generate? What happens to v as S approaches infinity?

5. The general mathematical expression that relates y to x for this sort of hyperbola is

$$y = \frac{ax}{b + x}$$

where a and b are constants. Considering the process whereby the table of v versus S was generated for the monkeys and peanuts in Question 4, can you show how the relationship between v and S suggested by this equation arises?

6. From the data table of Question 4 and the derivation of Question 5, explain why v approaches a finite upper limit as S approaches infinity. What is the physical meaning of the constant a in terms of monkeys and peanuts?

The analogy with monkeys and peanuts may aid in your understanding of the equations of enzyme kinetics, but it is obviously not the way the equations were originally derived. From studies of enzymes acting on substrates rather than monkeys acting on peanuts, two pioneering

From Monkeys and Peanuts to Michaelis and Menten

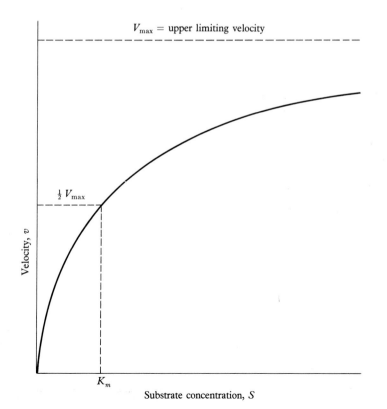

Figure 3-3 The dependence of the initial velocity v upon the substrate concentration S for an enzyme-catalyzed reaction.

enzymologists, L. Michaelis and M. L. Menten, postulated in 1913 a general theory of enzyme action and kinetics that has turned out to be basic to the quantitative analysis of all aspects of enzyme kinetics. To understand their approach, consider one of the simplest possible enzyme-catalyzed reactions, in which a substrate S is converted into a product P:

$$S \xrightarrow[\text{enzyme } [E]]{} P \qquad (3\text{-}6)$$

According to the Michaelis-Menten hypothesis the enzyme E in Equation 3-6 first reacts with the substrate S to form the transient enzyme-substrate complex ES, which then undergoes the actual catalytic reaction to form free enzyme and product P, as shown in the following sequence:

$$E_f + S \underset{k_2}{\overset{k_1}{\rightleftharpoons}} ES \underset{k_4}{\overset{k_3}{\rightleftharpoons}} P + E_f \qquad (3\text{-}7)$$

where E_f = free form of enzyme
S = substrate
ES = enzyme-substrate complex
P = product
k_1, k_2, k_3, k_4 = rate constants for indicated reactions

Beginning with the definition of *velocity* as the disappearance of substrate (or the appearance of product) per unit time (Equation 3-8),

$$v = \frac{-d[S]}{dt} = \frac{+d[P]}{dt} \qquad (3\text{-}8)$$

it is possible, in a series of straightforward algebraic manipulations, to arrive as Michaelis and Menten did at the following relationship between velocity and substrate concentration:

$$v = \frac{V_{\max}[S]}{K_m + [S]} \qquad (3\text{-}9)$$

where S is the *initial substrate concentration*, v is the *reaction velocity* at that substrate concentration, and $V_{\max}$ and K_m are two kinetic parameters we shall want to examine further. This is the *Michaelis-Menten equation*, which serves as the cornerstone upon which the imposing edifice of enzyme kinetics has been built.

The Meaning of $V_{\max}$ **and** K_m To appreciate the implications of this relationship, both for monkeys and for enzymes, and to examine the meaning of the parameters

V_{max} and K_m, it is useful to consider three special cases of substrate concentration: (1) very low substrate concentration ($S \ll K_m$), (2) very high substrate concentration ($S \gg K_m$), and (3) the special intermediate condition at which $S = K_m$.

Case 1: Very Low Substrate Concentration ($S \ll K_m$)

At very low substrate concentration, S becomes negligibly small compared to the constant K_m in the denominator of the Michaelis-Menten equation, so we can write:

$$v = \frac{V_{max}[S]}{K_m + [S]} \simeq \frac{V_{max}[S]}{K_m} \tag{3-10}$$

Thus at very low substrate concentration the reaction velocity is linearly proportional to the substrate concentration. In terms of the monkey and peanut model, this means that the rate of shelling at low peanut inputs is roughly proportional to the abundance of peanuts, because the overall rate at which peanuts are processed is limited primarily by the rate at which peanuts can be found, and this in turn increases as the peanut concentration increases. In terms of real enzymes, this relationship means that at low substrate concentrations there is a first-order relationship between velocity and substrate concentration (i.e., velocity is directly proportional to S).

Case 2: Very High Substrate Concentration ($S \gg K_m$)

At very high substrate concentration, K_m becomes negligibly small compared to S in the denominator of Equation 3-9, so we can write:

$$v = \frac{V_{max}[S]}{K_m + [S]} \simeq \frac{V_{max}[S]}{[S]} = V_{max} \tag{3-11}$$

For the monkeys, this means that in a world saturated with peanuts, the rate of peanut processing is limited ultimately not by the rate at which the peanuts can be found but by the finite rate at which they can be shelled. For enzymes, this relationship means that at *saturating* (nonlimiting) substrate concentrations, there is a *zero-order* dependence of velocity on substrate concentration (i.e., velocity is independent of variation in S and is therefore constant). Moreover Equation 3-11 provides us with the following definition of V_{max}, one of the two kinetic parameters in the Michaelis-Menten equation:

V_{max} is the *maximum velocity*, or the upper limiting value that the initial reaction velocity v approaches as the substrate concentration S approaches infinity.

In other words V_{max} is the velocity at saturating substrate concentrations. For our monkeys, V_{max} is therefore determined by (1) the length of time required for one monkey to shell one peanut in a world full of peanuts and (2) how many such monkeys are shelling peanuts. For enzymes, V_{max} is similarly a function of (1) the time required for the actual catalytic reaction plus subsequent release of product from the enzyme surface at saturating substrate concentrations and (2) how many such enzyme molecules are present.

Case 3: $S = K_m$

So far, then, we have seen the reason for first-order reaction kinetics at low substrate concentrations and for zero-order kinetics at high concentrations. We have also formulated a definition for V_{max} but have yet to explore the meaning of the second kinetic parameter, K_m. Note, however, that whatever its meaning K_m appears to have something to do with determining how low one must assume the substrate concentration to be to ensure first-order kinetics—or, alternatively, how high a concentration is required to ensure zero-order kinetics. Thus K_m seems to be some sort of benchmark on the concentration scale that determines how high is high and how low is low. To explore its meaning more exactly, consider the special case where S is chosen to be equal to K_m. Under these conditions the Michaelis-Menten equation can be written as follows:

$$v = \frac{V_{max}[S]}{K_m + [S]} = \frac{V_{max}[S]}{2[S]} = \frac{V_{max}}{2} \qquad (3\text{-}12)$$

This now defines K_m as that specific substrate concentration at which the reaction proceeds at one-half of its maximum (upper limiting) velocity. This specific concentration is constant for a given enzyme catalyzing a specific reaction under fixed conditions and is called the Michaelis constant (hence the designation K_m) in honor of the enzymologist who first elucidated its meaning. Thus we can now define the second kinetic parameter explicitly:

K_m is the *Michaelis constant*, or the specific substrate concentration required to ensure that the enzyme-catalyzed reaction which that substrate undergoes is proceeding at one-half of its maximum velocity.

The Michaelis constant is sometimes difficult to understand at first, usually because of a reluctance to remember that K_m is a *concentration*. It is *not* "one-half of the maximum velocity"; it is the substrate concentration necessary to drive the reaction at one-half of the maximum velocity. Some of the difficulty might be alleviated if you try thinking of it as the "Michaelis concentration" to reinforce the idea that it is a benchmark along the concentration scale.

Questions

7. From the graph of Question 4, explain in your own words what is meant by "first-order kinetics" and why the dependence of v on S is first-order only at low substrate concentrations. Similarly, explain the meaning of "zero-order kinetics" and why high substrate concentrations are required.

8. From the derivation of Question 5, explain the relationship between V_{max} and (1) the time required for one monkey to shell one peanut (or one enzyme molecule to catalyze one reaction and release the products) and (2) the number of monkeys (or enzyme molecules) available.

9. From the data table and graph of Question 4, determine the actual values of V_{max} and K_m for the "reaction" of monkeys shelling peanuts. Note that, most likely, you began with the known value of v at infinite substrate concentration. How likely is that for the experimental determination of V_{max} and K_m for real enzymes?

10. If you did not know the value of v at $S = \infty$, why would it be difficult to determine V_{max} and K_m quantitatively from a plot of v versus S such as the one you constructed in Question 4?

The Double-Reciprocal Plot

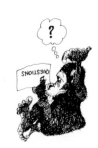

The classic hyperbolic plot of v versus S generated by an enzyme-catalyzed reaction that follows Michaelis-Menten kinetics is a faithful representation of the dependence of velocity on substrate concentration, but it is not especially useful for the quantitative determination of the key kinetic parameters, K_m and V_{max}. This is because it is a non-linear plot that cannot be readily extrapolated to infinite substrate concentration as is required to determine the critical parameter, V_{max}. To circumvent this problem, two clever enzymologists, H. Lineweaver and D. Burk, discovered that the hyperbolic relationship of the Michaelis-Menten equation could be converted into a linear function by taking the reciprocal of both sides of Equation 3-9 and simplifying the resulting expression to the following form:

$$\frac{1}{v} = \frac{1}{V_{max}} + \frac{K_m}{V_{max}}\left[\frac{1}{S}\right] \tag{3-13}$$

This is the *Lineweaver-Burk equation.* Its utility is illustrated in Figure 3-4: if we plot $1/v$ versus $1/S$, the resulting *double-reciprocal plot* is linear, with a y intercept of $1/V_{max}$, a slope of K_m/V_{max}, and an x intercept of $-1/K_m$.

Figure 3-4   The double-reciprocal plot for an enzyme-catalyzed reaction.

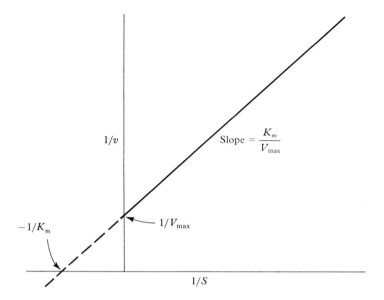

Note that once the double-reciprocal plot has been constructed, V_{max} can be determined directly from the y intercept and K_m from the x intercept, while the slope can be used to confirm or check on both values. In actual practice, then, an enzymologist interested in characterizing a specific enzyme-catalyzed reaction by its V_{max} and K_m values would proceed as follows:

1. Set up a series of reaction mixtures, each containing a fixed amount of enzyme, but with substrate concentration increasing across the series.

2. Determine the initial velocity v for each reaction mixture by measuring velocity before significant depletion of substrate or significant accumulation of product occurs.

3. Calculate the reciprocals of each substrate concentration and its resulting reaction velocity and plot the reciprocals of velocity, $1/v$, as a function of the reciprocals of substrate concentrations, $1/S$. (A calculator or computer is frequently used to determine the best linear fit for the available data points.)

4. From the double-reciprocal plot, determine V_{max} as the reciprocal of the y intercept and K_m as the negative reciprocal of the x intercept (or from the known slope once V_{max} has been determined).

Questions

11. Derive the double-reciprocal relationship of Equation 3-13 from the Michaelis-Menten expression of Equation 3-9. Prove that the y intercept of the double-reciprocal plot is $1/V_{max}$ and that the x intercept is $-1/K_m$.

12. Calculate $1/S$ and $1/v$ for each peanut "concentration" on the data table of Question 4 and plot $1/v$ versus $1/S$. Determine K_m from the x intercept and V_{max} from the y intercept, and verify that the ratio K_m/V_{max} is equal to the slope of the line.

13. Why is it logical that a plot of $1/v$ should approach $1/V_{max}$ as $1/S$ approaches zero?

Enzyme Inhibition

In the foregoing discussion our monkeys have been allowed to go about their business with nothing to hinder or inhibit them. Unfortunately monkeys are not always so lucky, and neither are enzymes. Many enzyme-catalyzed reactions are subject to inhibition, meaning that there exist substances (either occurring naturally in the cell or added in the test tube by diabolical enzymologists) that reduce the rate of the reaction at a given substrate concentration. Obviously it is important for the cell biologist or enzymologist to be able to detect and quantitate such inhibitory effects, since they bear critically on the rates at which reactions can proceed under the actual conditions prevailing in cells. A detailed analysis of enzyme inhibition is beyond the scope of the present discussion. Suffice it here to note that the study of enzyme inhibitors represents another important application of the double-reciprocal plot, since the various classes of inhibitors have characteristic effects on the slope and position of the graph. Thus the double-reciprocal plotting technique is useful not only for determining K_m and V_{max} values from experimental data but also as a diagnostic tool in discriminating between the various kinds of inhibitors and in quantitating inhibitory effects.

Allosteric Inhibition

Before leaving the topic of enzyme catalysis, a few words should be included about a specific kind of inhibition that plays an important role in the regulation of enzyme activities and hence of metabolic pathways in cells. Our view of enzyme inhibition would be distorted indeed if we were to conclude that the presence of an inhibitor and the resulting reduction in enzyme activity are invariably detrimental to the best interests of the cell. In fact quite the contrary is true: inhibition of enzyme activity is one of the most efficient ways cells have of regulating their metabolic activities. Far from simply running at maximum rates, cellular pathways must be continuously regulated and adjusted to keep them tuned to the needs of the cell, and an important aspect of that regulation and adjustment lies in the ability to inhibit enzyme activities with specificity and precision.

As an example, consider the pathway whereby a cell converts some precursor A into some final product P via the intermediates B, C, and D, as mediated by enzymes E_1, E_2, E_3, and E_4:

$$A \xrightarrow{E_1} B \xrightarrow{E_2} C \xrightarrow{E_3} D \xrightarrow{E_4} P \qquad (3\text{-}14)$$

Product P could, for example, be an amino acid needed for cellular protein synthesis, and A could be some common cellular component that serves as the starting point for the specific pathway leading to P.

If allowed to proceed at an unrestricted rate, this pathway clearly has the capacity to convert large amounts of A to P, with possible deleterious consequences for the cell due to an excessive accumulation of P or depletion of A. Clearly the pathway functions in the best interests of the cell not at its maximum possible rate but at a rate carefully tuned to the cellular need for P. Somehow, then, the enzymes of this pathway must be sensitive to the cellular level of the product P. This regulation of a metabolic pathway by its end product is possible because the end product is an inhibitor of one or more enzymes (usually the first one) in the pathway:

feedback inhibition
of E_1 by P

$$A \xrightarrow{E_1} B \xrightarrow{E_2} C \xrightarrow{E_3} D \xrightarrow{E_4} P \qquad (3\text{-}15)$$

This phenomenon, called *feedback inhibition*, is one of the most common mechanisms used by cells for adjusting pathway rates to cellular needs.

Feedback inhibition of a cellular pathway by its end product has been aptly compared to the thermostatic regulation of a furnace. In both cases, when the level of the end product (a compound like P in one case, temperature in the other) rises above some critical level, it causes a shutdown of the machinery (the metabolic pathway or the furnace) generating that product. It is clear that, in our example, P corresponds to the heat and the metabolic pathway to the furnace, but what about the thermostat? What serves as the sensing device to render the activity of the pathway sensitive to regulation by its end product? Assuming, as is almost always the case, that the key regulatory enzyme is the first one in the pathway, the question can be phrased more explicitly: how can the enzyme E_1, which recognizes A as substrate and converts it to B, be feedback-regulated by P, which is neither its substrate nor its product? The answer is that such key regulatory enzymes always have *two* sites, one to recognize and bind the substrate and the other to recognize and respond to the feedback inhibitor or other regulatory substance. The site at which substrate binds and the actual catalytic reaction occurs is called the *active site* of the enzyme, since it is responsible for the actual enzyme activity. The regulatory site is always physically distinct from the active site and is called the *effector* or *modulator site*. As shown in Figure 3-5, the distinctive feature of such enzymes is that just as the thermostat functions as a switch to turn the furnace off and on, so the effector site acts as a switch to control catalytic activity at the active site. In our example, the active site of enzyme E_1 is functional ("turned on") when the effector site is free due to low cellular levels of product P, but it is inactivated ("turned off") when the effector site is occupied by a molecule of P.

Generally we can say that such regulatory enzymes have two alternative configurations, one active and one inactive, and that binding of the regulatory substance at the effector site serves as the means of switching from one form to the other. Such enzymes are called *allosteric enzymes* (from the Greek *allo* meaning "other" or "different" and *steric* referring to shape or form). Although introduced in the context of inhibition, allosteric enzymes may be subject to either negative regulation (inhibition, usually by the end product of a pathway) or positive regulation (activation, usually by the substrate). Indeed some regulatory enzymes possess one or more specific positive effector sites and one or more specific negative effector sites simultaneously, rendering the enzyme sensitive to control by a number of cellular substances. Allosteric enzymes are almost always located strategically at the first step (or sometimes at a branch point) of a metabolic pathway; by controlling the activity of the first step in a pathway, the rest of the pathway is effectively controlled as well. We shall have opportunities to examine allosteric regulation of metabolic pathways in Chapter 5.

To summarize, we return full circle to the theme raised in the introduction to this chapter—namely, that thermodynamics allows us to assess the *feasibility* of a reaction but says nothing about the likelihood

Rate Down, Route to Go

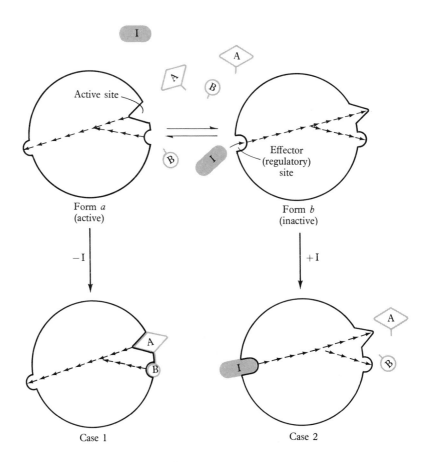

of the reaction actually occurring at a reasonable rate in the cell. We should, however, now be in a position to appreciate the valuable contribution that enzyme kinetics makes in providing us with the tools for determining the *rate* at which enzyme-catalyzed reactions will occur under actual cellular conditions, including substrate concentration and the presence and levels of allosteric effectors. Thus we turn to thermodynamics for *directionality* but to enzyme kinetics for *rate*.

Still remaining, however, is the question of *route*. By way of analogy, I cannot determine when (or even if) you will arrive in New York if you simply specify that the gasoline in your tank is capable of being combusted exergonically in the cylinders of your car and that you intend to drive at a speed of 55 miles per hour. Lacking, of course, is a specified route; I would have to know where you are starting from and what intermediate cities you intend to pass through

Figure 3-5   Mode of action of an allosteric (regulatory) enzyme. In addition to its active site, a regulatory enzyme possesses one or more *effector sites,* which serve as recognition and binding sites on the enzyme surface for *effector molecules* known either to inhibit or to activate the enzyme-catalyzed reaction. A regulatory enzyme can exist in two mutually exclusive but interconvertible forms, one active in catalysis and the other not. An allosteric inhibitor stabilizes the enzyme in the inactive form, whereas an allosteric activator stabilizes the active form of the enzyme. The regulatory enzyme shown here catalyzes the reaction of substrates *A* and *B* and is inhibited in this by the allosteric effector, *I.* Form *a* is the active configuration of the enzyme, with the active site open to incoming substrate molecules. Form *b* is inactive, without effective binding sites for *A* and *B.* In the absence of the allosteric inhibitor *I* (case 1), enzyme form *a* is favored, the substrate binding sites are open, and the enzyme is active. In the presence of *I* (case 2), binding of *I* to the effector site maintains the enzyme in the inactive form, without effective binding sites for *A* and *B.* Since both the binding of *I* and the interconversion of enzyme forms *a* and *b* are reversible processes, removal of *I* will restore enzyme activity, as *I* dissociates from the effector site and form *b* reverts to form *a.*

on your way. Which is why, armed with our newly acquired abilities in thermodynamics and enzyme kinetics, we proceed next to the consideration of metabolic pathways (routes) in the chapters of Part 2.

3-1. Need for Enzymes: You should now be in a position to appreciate the difference between the thermodynamic feasibility of a reaction and the likelihood that it will actually proceed. Whether a spontaneous reaction will in fact proceed depends upon the availability of a mechanism to get from the initial state to the final state.

Practice Problems

(a) Many reactions that are thermodynamically possible do not occur because of an activation energy barrier. In the physical world this barrier is often overcome by an initial input of heat (kindling a fire, for example). Explain why this is an effective mechanism. That is, what does it accomplish at the molecular level?

(b) Why is thermal activation not commonly used in the biological world as a means of overcoming the activation barrier?

(c) An alternative solution is to lower the barrier. What does it mean at the molecular level to say that a catalyst lowers the activation barrier?

(d) Organic chemists commonly use inorganic catalysts in their reactions, whereas the cell uses proteins called enzymes for the same purpose. What advantages can you see to the use of enzymes? Can you think of any disadvantages?

3-2. Enzyme Kinetics: Just as thermodynamics provides us with a measure (ΔG) of whether a given reaction *can* occur, so enzyme kinetics allows us to say something about whether it *will* occur and at what rate. Consider, for example, the reaction whereby the simple sugar glucose is phosphorylated to initiate the glycolytic pathway (to be encountered in Chapter 4). The reaction is catalyzed by the enzyme *hexokinase* and uses as its phosphate donor a compound called adenosine triphosphate, ATP:

$$\text{Glucose} + \text{ATP} \xrightarrow[\text{hexokinase}]{\text{Mg}^{++}} \text{glucose-6-phosphate} + \text{ADP} \quad (3\text{-}16)$$

(Mg^{++} is needed for enzyme activity, and ADP is the dephosphorylated form of ATP.) The ΔG^0 value for this reaction is -4 kcal/mole (see Problem 2-8), so the reaction is highly exergonic in the direction written. Whether it in fact proceeds in that direction, however, depends upon the availability of the appropriate catalyst (the hexokinase enzyme), and the rate in turn is dictated by the concentrations of substrates (ATP and glucose) to which the enzyme has access. To determine $V_{\max}$ and K_m of hexokinase for glucose, a sample of hexokinase was assayed at a number of glucose concentrations (in micromoles per liter) in the presence of Mg^{++} and 5 mM ATP. (At a concentration of 5 mM, ATP is saturating and will not limit the rate of the reaction.) At each glucose concentration, the initial rate of reaction was determined by measuring the amount of glucose-6-phosphate (in arbitrarily defined units) formed per minute. The following data were obtained:

Glucose Concentration (μmoles/liter)	Rate of Glucose-6-phosphate Formation (units/min)
10	0.010
20	0.017
40	0.027
50	0.030
100	0.040

(a) Plot v (rate of glucose-6-phosphate formation) versus S (glucose concentration). Why is it that when the glucose concentration is doubled (i.e., in going from 20 to 40 or from 50 to 100 μM), the increase in rate is always less than twofold?

(b) Calculate $1/v$ and $1/S$ for each entry on the data table and plot $1/v$ versus $1/S$.

(c) Determine K_m from the x-axis intercept of the double-reciprocal plot.

(d) Determine $V_{\max}$ from the y-axis intercept of the double-reciprocal plot.

(e) Referring to the graph for part (a), what is the physical meaning of $V_{\max}$? Of K_m?

3-3. **More Enzyme Kinetics:** To study the kinetics of the enzyme phosphoglyceromutase, which converts 3-phosphoglycerate into 2-phosphoglycerate in the glycolytic pathway (see Chapter 4), a cell biology student extracted the enzyme from a normal culture of the bacterium *Escherichia coli*. She then incubated a standard amount of enzyme with various concentrations of 3-

phosphoglycerate and determined the initial velocity v (in micromoles per minute) at various substrate concentrations S (in millimoles per liter). She also prepared and studied the enzyme from a mutant strain of *E. coli* that appeared to be impaired in its ability to carry out glycolysis. The double-reciprocal plots for the data obtained with the enzymes from the normal and mutant cultures are shown in Figure 3-6.

(a) Calculate K_m and V_{max} for both enzymes.

(b) If you know the intracellular concentration of 3-phosphoglycerate in the *E. coli* cells to be $5 \times 10^{-6}\ M$, how might you use these data to explain the observed impairment in the ability of the mutant strain to carry out glycolysis?

3-4. Double-Reciprocal Plots: An alternative form of the Lineweaver-Burk double-reciprocal equation (Equation 3-13) can be obtained by multiplying both sides of that equation by the term $(V_{max})(v)$.

(a) What is the resulting equation?

(b) What would you plot on the x and y axes to obtain a straight line?

(c) How would you determine K_m and V_{max} from your plot?

3-5. Kinetics Again: Phosphofructokinase is the enzyme that catalyzes the following reaction from the glycolytic pathway:

$$\text{Fructose-6-phosphate} + \text{ATP} \xrightarrow{\text{phosphofructokinase}}$$

$$\text{fructose-1,6-diphosphate} + \text{ADP} \qquad (3\text{-}17)$$

Assume that phosphofructokinase is incubated in each of two test tubes in the presence of excess (i.e., a nonlimiting concentration of) ATP. The starting concentration of fructose-6-phosphate is 0.1 mM in tube 1 and 0.2 mM in tube 2. The reaction in tube 2 is found to have an initial velocity 1.5 times that for tube 1. What is the K_m of the enzyme for fructose-6-phosphate?

‡**3-6. Allosteric Regulation:**   As an actual example of an allosterically regulated pathway, consider the sequence of cellular reactions required to synthesize the amino acid isoleucine from the starting compound threonine (which happens, in this case, to be an amino acid also):

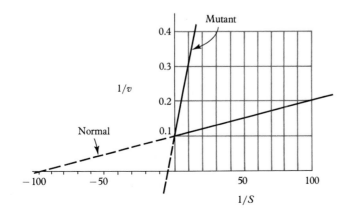

Figure 3-6   Double-reciprocal plots for the mutant and normal enzymes of Problem 3-3.

$$
\begin{array}{c}
\underset{\displaystyle \text{Threonine}}{
\begin{array}{l}
\text{CH}_3 \\
| \\
\text{H—C—OH} \\
| \\
\text{H—C—NH}_2 \\
| \\
\text{COOH}
\end{array}}
\xrightarrow{\;E_1\;}
\alpha\text{-ketobutyrate}
\end{array}
$$

α-ketobutyrate $\xrightarrow{\;E_2\;}$ α-acetohydroxybutyrate $\xrightarrow{\;E_3\;}$ α,β-dihydroxy-β-ethylvalerate $\xrightarrow{\;E_4\;}$ α-keto-β-ethylvalerate $\xrightarrow{\;E_5\;}$ Isoleucine   (3-18)

Isoleucine:

$$
\begin{array}{l}
\text{CH}_3 \\
| \\
\text{CH}_2 \\
| \\
\text{H—C—CH}_3 \\
| \\
\text{H—C—NH}_2 \\
| \\
\text{COOH}
\end{array}
$$

Isoleucine is an allosteric inhibitor of the first (but not the remaining) enzymes in this sequence.

(a) Inhibition of the first step in a sequence by an allosteric effector is a common regulatory strategy employed by cells. Explain why this is an effective and economical way of regulating this pathway.

(b) Feedback-inhibited pathways are said to be self-regulating. What does this mean, and why is it a desirable feature of a pathway like that leading to isoleucine?

(c) Allosteric regulation depends upon the ability of the regulated enzyme to exist in two alternative conformational states, one in which the active site has a high affinity for the substrate and another in which it has a low affinity. Binding of the allosteric effector at its recognition site is thought to stabilize the enzyme in one form or the other, depending upon whether the effector functions as an inhibitor or an activator. Explain the regulation of threonine deaminase (E_1) by isoleucine in these terms.

(d) What would be the effect of (i.e., how would you recognize) a mutation in the gene coding for threonine deaminase that resulted in the inability to bind threonine at the active site? What about a mutation that abolished recognition of isoleucine at the effector site?

‡3-7. Derivation of the Michaelis-Menten Equation: According to the Michaelis-Menten hypothesis, an enzyme-catalyzed reaction proceeds via a transient enzyme-substrate complex, as represented by Equation 3-7. Starting with the definition of velocity as the disappearance of substrate or the appearance of product per unit time (Equation 3-8), derive the Michaelis-Menten equation,

$$
v = \frac{V_{\max}[S]}{K_m + [S]}
$$

The following points should help you in your derivation:

1. Begin by expressing the rate equations $d[S]/dt$, $d[P]/dt$, and $d[ES]/dt$ in terms of concentrations and rate constants. Confine your derivation to the initial stages of the reaction, where $[P]$ is essentially zero.

2. Assume a steady state, at which the enzyme-substrate complex is being broken down at the same rate at which it is being formed, such that the net rate of change, $d[ES]/dt$, is zero.

3. Note that the total amount of enzyme present, E_t, is the sum of the free form E_f plus the amount of complexed enzyme ES:

$$E_t = E_f + ES \tag{3-19}$$

4. Note, when you get that far, that V_{max} and K_m can be defined as follows:

$$V_{max} = k_3[E_t] \tag{3-20}$$

$$K_m = \frac{k_2 + k_3}{k_1} \tag{3-21}$$

References

Bernard, S. A., *The Structure and Function of Enzymes* (New York: Benjamin, 1968).

Dickerson, R. E., and I. Geis, *The Structure and Action of Proteins* (Menlo Park, Ca.: Benjamin, 1969).

Frieden, E., "The Enzyme-Substrate Complex," *Sci. Amer.* (Aug. 1959), *201*:119.

Hammes, G. G., and C. W. Wu, "Regulation of Enzyme Activity," *Science* (1971), *172*:1205.

Koshland, D. E., Jr., "Protein Shape and Biological Control," *Sci. Amer.* (Oct. 1973), *229*:52.

Pardee, A. B., "Control of Metabolic Reactions by Feedback Inhibition," *The Harvey Lectures, 1969–70* (New York: Academic Press, 1971).

Phillips, D. C., "The Three-Dimensional Structure of an Enzyme Molecule," *Sci. Amer.* (Nov. 1966), *215*:78.

Westley, J., *Enzymic Catalysis* (New York: Harper & Row, 1969).

PART 2

Sources of Energy

Anaerobic Energy Metabolism

4

In the preceding two chapters we considered in some detail the powerful tools available to the cellular biologist in predicting the feasibility, direction, and rate of a given reaction. Specifically, thermodynamics allows us to predict in which direction a specific reaction will proceed under a given set of conditions and how much free energy can be derived from or must be put into the system during the reaction. Similarly enzyme kinetics enables us to determine the rate at which the reaction will actually proceed. Thus the foregoing chapters furnish the tools necessary to answer three crucial questions for any reaction or process: (1) in what direction, (2) to what extent, and (3) at what rate.

We come now to a consideration of a fourth important question as we attempt to understand by what *route* a process occurs. So far we have considered only individual reactions and individual enzymes working in isolation. But to accomplish any major task, a cell needs many such reactions occurring in an organized manner. This, in turn, requires many enzymes working in concert, since a given enzyme can catalyze only a single chemical manipulation. When we consider the sum total of all chemical reactions occurring within a cell, we speak of the *metabolism* of the cell. The overall metabolism consists of a large number of *metabolic pathways*, each designed to accomplish a specific task and therefore to answer the question "by what route" for a specific process. In biochemical terms, then, life at the cellular level might well be defined as a network of integrated and carefully

Introduction to Metabolic Pathways

regulated metabolic pathways, each contributing to the sum of all the activities a cell must carry out.

Metabolic pathways are of two types. Pathways concerned with the synthesis of cellular components are termed *anabolic* pathways (from the Greek *ana* = up); those involved in the degradation of cellular constituents are called *catabolic* pathways (from the Greek *kata* = down). Anabolic pathways usually involve an increase in atomic order (decrease in entropy), are often reductive in nature, and are almost always energy-requiring (endergonic). Catabolic pathways, on the other hand, are usually energy-liberating (exergonic) because they involve the oxidative release of chemical bond energies and a decrease in atomic order (increase in entropy). To a first approximation, a cell carries out anabolic reactions to accomplish the growth and repair characteristic of all living systems, and it carries out catabolic reactions to furnish the energy needed to drive the anabolic reactions. Crucial to the thermodynamic success of a cell, then, is the efficient coupling of energy-yielding to energy-requiring processes. This coupling is possible because of a common intermediate—a common chemical entity that serves as the "currency" of the cellular energy economy.

ATP: Currency of the Biological Realm

Clearly, if a cell is to obtain energy in one form (be it from the sun or a chemical bond) and use it in another form (to move a muscle, form a bond, or transport a molecule), there must be some way of *coupling* the energy-yielding and the energy-requiring processes—some way of temporarily storing the energy derived from an exergonic reaction for use in driving an otherwise endergonic reaction. There must, in short, be some common intermediate—some common energy currency. In the biological world, this common intermediate is the high-energy phosphate bond in general and a high-energy phosphorylated compound called ATP in specific. The term "ATP" stands for *adenosine triphosphate*, a molecule of paramount importance as the universal energy currency of all cells. Since ATP is involved in almost every cellular energy transaction, it is obviously essential to understand its structure and function.

As shown in Figure 4-1, ATP is a complex molecule containing an aromatic base called *adenine,* a five-carbon sugar called *ribose,* and a chain of three phosphate groups linked to each other by acid anhydride bonds and to the ribose by an ester bond.* The compound formed by linking adenine and ribose is called *adenosine.* Adenosine may occur in the cell in the unphosphorylated form or with one, two, or three phosphates attached to carbon 5 of the ribose, forming, respectively, adenosine monophosphate (AMP), diphosphate (ADP), and triphosphate (ATP).

The ATP molecule is well suited for the important role it plays

---

* An ester is a compound formed between an alcohol (R—OH) and an acid by the removal of water. For a carboxylic acid (R—COOH), ester formation can be represented as follows:

$$R—OH + HO—\overset{\overset{\displaystyle O}{\|}}{C}—R' \quad \xrightarrow{\;H_2O\;} \quad R—O—\overset{\overset{\displaystyle O}{\|}}{C}—R'$$

Alcohol   Acid   Ester

An *acid anhydride* is formed by the removal of water from two acids. For two carboxylic acid molecules, acid anhydride formation is as follows:

$$R—\overset{\overset{\displaystyle O}{\|}}{C}—OH + HO—\overset{\overset{\displaystyle O}{\|}}{C}—R' \quad \xrightarrow{\;H_2O\;} \quad R—\overset{\overset{\displaystyle O}{\|}}{C}—O—\overset{\overset{\displaystyle O}{\|}}{C}—R'$$

Acid   Acid   Acid anhydride

Note that the two R groups of an acid anhydride may, but need not, be the same.

The chemistry of the phosphate group is best understood by keeping in mind that a phosphate anion is the ionized form of phosphoric acid, H_3PO_4. The most common phosphate-containing organic compounds in cells are either phosphate esters or anhydrides. Phosphate esters are formed by linking phosphoric acid to an alcohol, whereas acid anhydrides involve either two phosphoric acid molecules, as in *pyrophosphate*, or phosphoric acid and a carboxylic acid, forming a *mixed anhydride*. These classes of compounds are represented as follows:

$$R—OH + HO—\overset{\overset{\displaystyle O}{\|}}{\underset{\underset{\displaystyle OH}{|}}{P}}—OH \xrightarrow{\;H_2O\;} R—O—\overset{\overset{\displaystyle O}{\|}}{\underset{\underset{\displaystyle OH}{|}}{P}}—OH$$

Alcohol   Phosphoric   Phosphate
           acid   ester

$$HO—\overset{\overset{\displaystyle O}{\|}}{\underset{\underset{\displaystyle OH}{|}}{P}}—OH + HO—\overset{\overset{\displaystyle O}{\|}}{\underset{\underset{\displaystyle OH}{|}}{P}}—OH \xrightarrow{\;H_2O\;} HO—\overset{\overset{\displaystyle O}{\|}}{\underset{\underset{\displaystyle OH}{|}}{P}}—O—\overset{\overset{\displaystyle O}{\|}}{\underset{\underset{\displaystyle OH}{|}}{P}}—OH$$

Phosphoric   Phosphoric   Pyrophosphate
   acid   acid

$$R—\overset{\overset{\displaystyle O}{\|}}{C}—OH + HO—\overset{\overset{\displaystyle O}{\|}}{\underset{\underset{\displaystyle OH}{|}}{P}}—OH \xrightarrow{\;H_2O\;} R—\overset{\overset{\displaystyle O}{\|}}{C}—O—\overset{\overset{\displaystyle O}{\|}}{\underset{\underset{\displaystyle OH}{|}}{P}}—OH$$

Carboxylic   Phosphoric   Mixed
   acid   acid   anhydride

Note, then, that the bond linking the first (innermost) phosphate group of ATP to the ribose of adenosine is a phosphate ester bond, whereas the links between successive phosphates are acid anhydride bonds.

Figure 4-1 The structure of adenosine triphosphate (ATP).

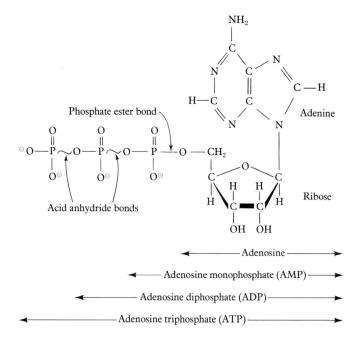

as an intermediate in cellular energy metabolism because of the unstable, energy-rich nature of the acid anhydride bond that links the third (outermost) phosphate to the second. (The bond between the second and the first phosphates has the same properties, but we are at present interested primarily in the terminal anhydride bond.) The anhydride bonds that link the phosphate groups of ATP together are high in energy because of *charge repulsion* between the adjacent phosphate groups and also because of *resonance stabilization* of both products of hydrolysis (ADP and inorganic phosphate). Charge repulsion is easy to understand: each of the three phosphate groups of ATP bears a negative charge at the near-neutral pH of most cells, and these negative charges repel one another, making the molecule less stable than it would otherwise be. Even more important, however, is the phenomenon of resonance stabilization, which warrants a brief explanation.

To understand resonance stabilization, we must first realize that for molecules or ions like carboxylic acids or phosphate groups that contain atoms with unshared electrons attached directly to a double bond, the π electrons are not localized; they are spread out over all the atoms bearing unshared electrons. For example, we can easily

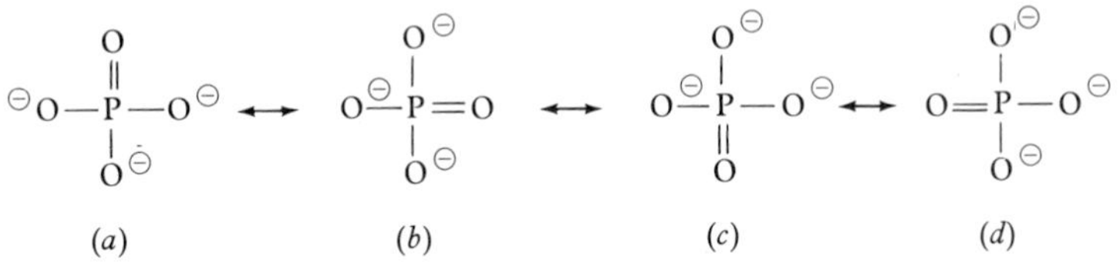

$$
\begin{array}{cccc}
\overset{\text{O}}{\underset{\overset{|}{\text{O}}^{\ominus}}{\overset{\|}{\underset{|}{{}^{\ominus}\text{O}-\text{P}-\text{O}^{\ominus}}}}} &
\overset{\text{O}^{\ominus}}{\underset{\overset{|}{\text{O}}^{\ominus}}{\overset{|}{{}^{\ominus}\text{O}-\text{P}=\text{O}}}} &
\overset{\text{O}^{\ominus}}{\underset{\overset{\|}{\text{O}}}{\overset{|}{{}^{\ominus}\text{O}-\text{P}-\text{O}^{\ominus}}}} &
\overset{\text{O}^{\ominus}}{\underset{\overset{|}{\text{O}}^{\ominus}}{\overset{|}{\text{O}=\text{P}-\text{O}^{\ominus}}}} \\
(a) & (b) & (c) & (d)
\end{array}
$$

Figure 4-2 Contributing resonance structures for the phosphate anion. The true structure is a resonance hybrid, with the extra electrons delocalized over all four P—O bonds.

write a structure for the phosphate ion (PO_4^{---}) in which each atom has an octet of electrons, as shown in structure (a) of Figure 4-2. Three oxygen atoms are connected to the phosphorus atom by σ bonds, and the fourth is connected by both σ and π bonds. There is no reason why any particular oxygen atoms should differ from the other three, however, so structures (b), (c), and (d) of Figure 4-2 would seem equally likely. In fact there is nothing special about any one particular oxygen, and compounds or ions of this type cannot be represented satisfactorily by a single valence-bond structure. They are represented instead by a series of *contributing structures* connected by double-headed arrows as in Figure 4-2. The true structure is an average of the contributing structures, a *resonance hybrid* with the extra electrons delocalized over all four P—O bonds. When the electrons of such a compound are maximally delocalized in this way, the molecule is in its most stable (lowest energy) configuration and is said to be *resonance-stabilized*. If, on the other hand, one or more of the oxygen atoms of a phosphate anion becomes involved in a linkage to an organic compound, that oxygen atom is no longer available for electron delocalization and the molecule becomes "locked" in a higher-energy configuration, with extra electrons delocalized over fewer than the maximum number of oxygen atoms. This means that a phosphorylated compound (such as a phosphate ester) is always higher in energy than the corresponding nonphosphorylated form and will liberate energy upon hydrolysis (addition of water across the bond) because of the greater electron delocalization possible in the phosphate ion, which is one of the two products of hydrolysis. And when both the products of hydrolysis can undergo resonance stabilization, the hydrolysis will be highly exergonic indeed. This is the case for an acid anhydride bond such as that which joins the terminal phosphate of ATP to the rest of the molecule. Since both the ADP and the phosphate resulting from ATP hydrolysis undergo resonance stabilization, the hydrolysis (Equation 4-1) is a highly exergonic reaction, with a standard free-energy change (ΔG^0) of -7.3 kcal/mole. In fact, under cellular conditions the actual free-energy change (ΔG) is even greater, often being in the range of -10 to -14 kcal/mole because of the high ATP/ADP ratio that prevails in most cells.[*]

$$ATP + H_2O \rightleftharpoons ADP + P_i \qquad (4\text{-}1)$$

That it is the products of hydrolysis which are resonance-stabilized underscores the need to emphasize what otherwise tends to be a common misconception. Although the anhydride bond itself is often referred to as "high energy" or "energy rich," what is actually meant is not the bond itself but the *hydrolysis* of that bond. To say that ATP is a high-energy compound should therefore always be interpreted as simply a shorthand way of saying that it possesses one or more bonds, the hydrolysis of which is strongly exergonic.

The *ATP/ADP system* represents a reversibly chargeable means of conserving, transferring, and releasing energy within the cell. The ATP is the "charged" form; whenever a cell is carrying out an exergonic reaction such as those described in this and the succeeding two chapters, it couples the reaction to the ATP/ADP system and uses the available free energy to drive Reaction 4-1 to the left, generating the "high-energy" bond of ATP. Conversely, when the cell needs energy to drive an otherwise endergonic reaction such as those discussed in Part 3, it uses the energy of that anhydride bond to provide the driving force by allowing Reaction 4-1 to move to the right. Thus we are dealing with an energy-storing system that is charged during the oxidative metabolism of foodstuffs (or, as we shall see in Chapter 7, during the photosynthetic trapping of solar energy) and discharged during the performance of cellular work. The whole system is therefore a continuous, dynamic cycle. The many chemical steps in this charging and discharging of the ATP/ADP system in the cell are

---

[*]Note that "P_i" is used in this and many subsequent equations as a shorthand representation for inorganic phosphate, the ionized form of phosphoric acid as it exists in cells. Because it has three ionizable protons, phosphoric acid can exist in a variety of ionization stages:

| Phosphoric acid | Dihydrogen phosphate | Monohydrogen phosphate | Phosphate anion |

As you might expect, the relative proportion of each form present in solution is dependent upon the pH, with the dihydrogen and monohydrogen forms predominating at the near-neutral pH of most cells. Textbooks, including this one, tend to represent phosphate groups in either the fully protonated or fully ionized forms, but you should keep in mind that what is really intended is the equilibrium mixture of the various forms dictated by the prevailing pH.

catalyzed by enzyme systems operating as parts of coherent metabolic pathways. This is the basic principle of cellular energy metabolism, and central to it in all cells everywhere is the substance ATP.

We have in hand almost all the pieces necessary to take up the theme for this part of the book—the means whereby the energy of oxidizable foodstuffs, or of sunlight, is conserved during cellular energy metabolism as ATP molecules (which are then available for the various energy-requiring processes to be discussed in Part 3). We return, therefore, to the phototrophs and the chemotrophs initially encountered in Chapter 1 and to the overall task of understanding how solar energy is trapped into chemical bond energies by the phototrophs and how this chemical bond energy is released to meet the cellular energy needs of the chemotroph.

Chemotrophic Energy Metabolism

Let us take this second problem first and pose it in a general form: how do chemotrophic cells that are dependent upon the chemical bond energies of oxidizable food molecules make use of that energy? Or, more personally, what are the metabolic processes by which the cells of your own body make use of the food you eat to meet your energy needs?

To say that food molecules such as sugar, starch, fat, or protein have energy stored in their chemical bonds is another way of saying that all these molecules are oxidizable organic compounds and that their oxidation is a highly exergonic process. We have, therefore, to inquire into the meaning of oxidation.

The Meaning of Oxidation

As soon as we call these foodstuffs oxidizable organic compounds, we raise the question of what it means to say something is oxidizable. The answer provided by chemistry is that an atom or molecule is oxidizable if it is possible to remove one or more electrons from it. The ferrous ion (Fe^{++}) is an oxidizable species since it can lose an electron to become a ferric ion (Fe^{+++}):

$$Fe^{++} \xrightarrow{\text{oxidation}} Fe^{+++} + e \qquad (4\text{-}2)$$

In organic and biological chemistry, oxidation is defined in exactly the same way: it is the removal of electrons. The only difference is that the oxidation of organic molecules frequently results in the removal of both electrons and hydrogen ions (protons). Consider, for example, the oxidation of an alcohol to an aldehyde:

$$R-CH_2-OH \xrightarrow{\text{oxidation}} R-\overset{\displaystyle H}{\underset{\displaystyle |}{C}}=O + 2e^- + 2H^+ \qquad (4\text{-}3)$$

This is clearly an oxidation, since electrons are removed. And since an electron plus a proton (H^+) is the equivalent of a hydrogen atom, what is occurring is in effect the removal of the equivalent of hydrogen atoms. Thus for cellular reactions involving organic molecules we can define oxidation in a secondary way as the removal of hydrogen atoms. Or, put another way, biological oxidation reactions are frequently *dehydrogenation* reactions:

$$R-CH_2-OH \xrightarrow[\text{dehydrogenation}]{\text{oxidation or}} R-\overset{\displaystyle H}{\underset{\displaystyle |}{C}}=O + 2[H] \qquad (4\text{-}4)$$

Note that the hydrogen atoms are enclosed in brackets in Equation 4-4 to indicate that free hydrogen atoms are not actually released as such; we shall discuss later the actual forms in which H^+ and e^- are transferred within cells.

Just as oxidation is defined as the removal of electrons, frequently manifested as a dehydrogenation, so reduction may be defined in general as the addition of electrons, often accompanied in biological reactions by the addition of protons. Thus the overall effect is a *hydrogenation* process:

$$R-\overset{\displaystyle H}{\underset{\displaystyle |}{C}}=O + 2[H] \xrightarrow[\text{hydrogenation}]{\text{reduction or}} R-CH_2-OH \qquad (4\text{-}5)$$

Each of the preceding reactions is really only a half-reaction, representing either an oxidation or a reduction event. In actuality, of course, oxidation and reduction reactions are always coupled; any time an oxidation occurs, a concomitant reduction must also take place, since the electrons removed from one molecule must be added to another molecule. Equation 4-6 illustrates this point in general form: AH_2 is clearly being oxidized (losing electrons or, equivalently, hydrogen atoms to form A) and B is obviously being reduced (gaining electrons or hydrogen atoms to form BH_2):

$$AH_2 + B \xrightarrow{\text{oxidation/reduction}} A + BH_2 \qquad (4\text{-}6)$$

Biological Oxidation   The significance of oxidation for our present discussion is that this is the means whereby chemotrophs derive the energy needed for cellular activities. The biological world displays a rich diversity of oxidative processes by which energy can be obtained; there are, in other words, a great many variations on the common oxidative theme depicted by Equation 4-6. Basically, however, organisms and cells can be classified according to (1) the kind of oxidizable molecules they depend upon for energy (i.e., the nature of the AH_2 of Equation 4-6) and (2) the nature of the electron acceptor used in oxidation reactions (i.e., the B of Equation 4-6).

Thinking first in terms of oxidizable molecules, it should be obvious that many different kinds of fuel molecules can be used by cells. Some microorganisms can use reduced inorganic compounds (such as reduced forms of iron, nitrogen, or sulfur) as energy sources and therefore play key roles in the inorganic economy of the biosphere. But most organisms depend for energy upon organic food molecules, with carbohydrates, fats, and proteins as the three major categories. To simplify our discussion and provide a unifying metabolic theme, we shall use as our prototype the six-carbon sugar glucose ($C_6H_{12}O_6$), since this is the single most important substrate for biological oxidation in many (though by no means all) chemotrophic cells.

In terms of the electron acceptors used in biological energy metabolism, we recognize an important dichotomy between those organisms that use oxygen as the ultimate electron acceptor and those that do not. In this respect cells fall into three categories: *strict aerobes* have an absolute requirement for oxygen, *strict anaerobes* cannot tolerate the presence of oxygen, and *facultative* cells can adapt their metabolism to either aerobic ($+O_2$) or anaerobic ($-O_2$) conditions.

The nature of cellular energy metabolism is very much dependent upon whether or not oxygen is used as the terminal electron acceptor. Maximum energy yields from oxidizable organic substrates are in general realized only under aerobic conditions. Consider, for example, the overall expression for the complete oxidation of glucose as shown in Equation 4-7. With oxygen as the electron acceptor, the reaction is highly exergonic, with a ΔG^0 value of -686 kcal/mole. As we shall see later, this is a much greater free-energy change than can be tapped by alternative metabolic possibilities carried out under anaerobic conditions.

Respiration and Fermentation

$$C_6H_{12}O_6 + 6O_2 \xrightarrow[\text{metabolism}]{\text{aerobic}} 6CO_2 + 6H_2O + \text{energy} \qquad (4\text{-}7)$$

Aerobic energy metabolism, such as the process summarized in Equation 4-7, is called *respiration*. Anaerobic metabolism, in which energy is extracted from organic substrates without the involvement of oxygen, is called *fermentation*. We shall encounter pathways for the fermentation of glucose later in this chapter.

In this and the following chapter, we shall be concentrating on the fermentative and respiratory metabolism of the single compound glucose, but keep in mind that cells can use a wide variety of other substrates as well. Some of these alternative sources of energy are considered in Chapter 6, since they turn out to share many features of the pathway for glucose utilization. For others, particularly the variety of intriguing fermentation processes characteristic of anaerobic microorganisms, you will need to venture beyond the confines of this book.

Some Sugar Chemistry Since glucose has been chosen as the prototype for a discussion of chemotrophic energy metabolism, we need some background in carbohydrate chemistry before we can appreciate our starting material appropriately. All carbohydrates are biological compounds that contain the elements C, H, and O in the approximate ratios of $1:2:1$. The most common carbohydrates are either sugar molecules or polymers of sugars, so the next question obviously becomes, What is a sugar?

A sugar can be defined as an aldehyde or ketone that also has two or more hydroxyl groups. The simple sugars can have a variable number of carbon atoms (from three to seven, usually). If we use n to designate the number of carbon atoms in the molecule, then the general formulas for the two types of sugars are as follows:

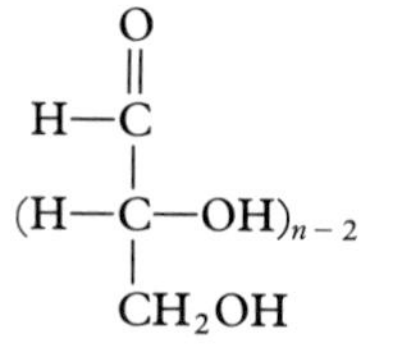

$$
\begin{array}{cc}
\begin{array}{c}
\text{O} \\
\parallel \\
\text{H--C} \\
| \\
(\text{H--C--OH})_{n-2} \\
| \\
\text{CH}_2\text{OH}
\end{array}
&
\begin{array}{c}
\text{CH}_2\text{--OH} \\
| \\
\text{C=O} \\
| \\
(\text{H--C--OH})_{n-3} \\
| \\
\text{CH}_2\text{OH}
\end{array}
\\
\text{Aldosugar} & \text{Ketosugar} \\
\text{(polyhydroxy aldehyde)} & \text{(polyhydroxy ketone)}
\end{array}
$$

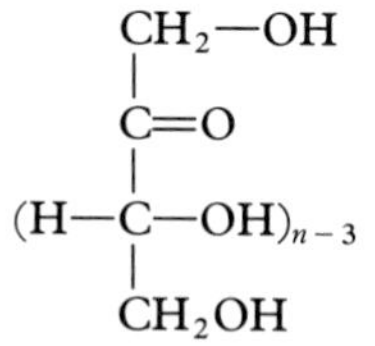

Within these categories sugars are named generically according to the number of carbon atoms: triose = 3 carbons; tetrose = 4 carbons; pentose = 5 carbons; hexose = 6 carbons; and heptose = 7 carbons. We have already met an example of a pentose in the ribose of ATP (Figure 4-1).

The single most widespread sugar in the biological world is the aldosugar glucose, which is represented by the formula $C_6H_{12}O_6$ and has the following structure:

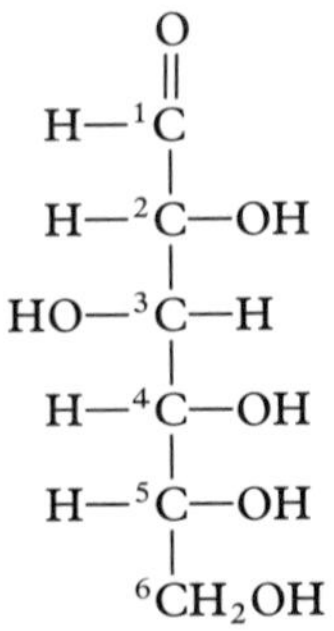

$$
\begin{array}{c}
\overset{\displaystyle O}{\underset{\displaystyle \|}{}} \\
H-{}^{1}C \\
| \\
H-{}^{2}C-OH \\
| \\
HO-{}^{3}C-H \\
| \\
H-{}^{4}C-OH \\
| \\
H-{}^{5}C-OH \\
| \\
{}^{6}CH_2OH
\end{array}
$$

Glucose
(open-chain representation)

In keeping with the general rule for numbering carbon atoms in organic molecules, the carbons of glucose are numbered beginning with the most oxidized end of the molecule. Note that there are four asymmetric carbon atoms in the glucose molecule (carbons 2, 3, 4, and 5). There are therefore $2^4 = 16$ different possible stereoisomers of the molecule, but we shall concern ourselves only with glucose, the isomer shown above. The representation shown above is the Fischer projection of glucose, in which the H— and —OH groups are intended to be projecting slightly out of the plane of the paper. This structure indicates glucose to be an open-chain compound. Actually, however, glucose exists in the cell in a dynamic equilibrium between the open-chain configuration and a ring form, with the ring form the predominant (thermodynamically favored) structure. The ring form results from the addition of the hydroxyl group on carbon 5 across the C=O bond on carbon 1 to form an *internal hemiacetal*.[*]

In terms of the open-chain model depicted above, the equilibrium between the open-chain and hemiacetal ring forms of glucose can be represented as follows:

---

[*] A hemiacetal is formed when an alcohol is added across the C=O double bond of an aldehyde. The general reaction is as follows:

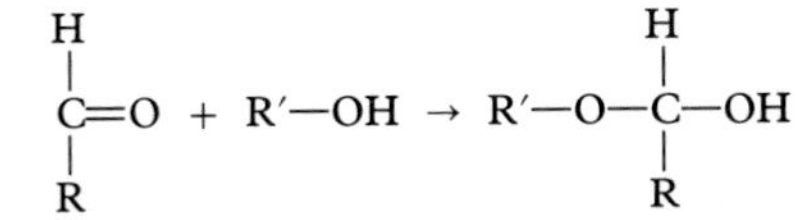

$$
\begin{array}{ccc}
H & & H \\
| & & | \\
C{=}O \;+\; R'{-}OH \;\rightarrow\; & R'{-}O{-}C{-}OH \\
| & & | \\
R & & R
\end{array}
$$

Aldehyde   Alcohol   Hemiacetal

In its open-chain configuration, glucose contains both an aldehyde group (on carbon 1) and several hydroxyl groups, one of which (on carbon 5) is positioned favorably to add across the C=O bond on carbon 1 to form an intramolecular or internal hemiacetal as shown in Equation 4-8.

Open-chain form   Hemiacetal ring form
for glucose   for glucose

$$(4\text{-}8)$$

This is basically an unsatisfactory representation, however, since it gives no indication of the spatial relationship of different parts of the molecule (and, in fact, makes it appear extremely unlikely that a hemiacetal involving carbons 1 and 5 would form spontaneously). A better representation is the following:

Open-chain form   Hemiacetal ring form
for glucose   for glucose

$$(4\text{-}9)$$

Since the equilibrium for glucose in solution lies far to the right, the ring form predominates in cells, and most textbook authors choose the ring form of Equation 4-9 as their means of representing glucose.

Glucose is an example of a simple sugar consisting of a single unit. Such a molecule is called a *monosaccharide* (from the Greek *mono* = single and *saccharide* = sugar). Frequently several or many such monosaccharide units can be strung together to form *oligosaccharides* or *polysaccharides* (from the Greek *oligo* = few; *poly* = many).

Common examples of oligosaccharides are the disaccharides, consisting of two monosaccharides linked together. If both are glucose units, then the resulting disaccharide is called *maltose*. If glucose is linked to fructose, the result is *sucrose* (common table sugar); if glucose and galactose are the two monosaccharides to be linked, the result is *lactose* (milk sugar).

It is also possible to link many monosaccharide units together to form long-chain polysaccharides. Long chains of glucose molecules commonly occur in cells: in plants the resulting molecule is *starch*; in animals the product is *glycogen*.

Our choice of glucose as the prototype substrate for our discussion of chemotrophic energy metabolism is based on the prominence of this sugar in the biosphere and its importance as an energy source for many, though not all, chemotrophic cells. Cells obtain glucose from one or more of the following sources: (1) direct uptake or ingestion of glucose or of oligosaccharides like sucrose or maltose that can be broken down to yield glucose; (2) breakdown of ingested or stored polysaccharides (principally starch and glycogen); (3) conversion of other foodstuffs into glucose; and (4) photosynthetic sugar synthesis.

With that introduction to the chemistry of glucose, we are ready now to turn to the mechanisms whereby cells metabolize this sugar and extract a portion of its free energy for their own use. As already noted, glucose can be used as a source of energy under either anaerobic or aerobic conditions. Figure 4-3 depicts the overall process of glucose oxidation and indicates its division into three phases or stages, one common to both anaerobic and aerobic metabolism and the other two unique to aerobic conditions. These three phases can be characterized as follows:

The Biological Oxidation of Glucose

1. *Glycolysis* is common to both the aerobic and the anaerobic metabolism of glucose. It involves the splitting of the 6-carbon sugar into two 3-carbon units and the subsequent oxidation of both 3-carbon units to pyruvate, with conservation of energy as ATP. Under anaerobic conditions the pyruvate is fermented to lactate or ethanol, from which the cell can extract no further energy. In the presence of oxygen, however, the pyruvate can be oxidized further.

2. The *tricarboxylic acid cycle* (TCA cycle) effects the complete oxidation of all three carbon atoms of pyruvate to CO_2, provided that oxygen is available to allow the concomitant reoxidation of the coenzymes that serve as electron carriers in the oxidative events of the cycle.

3. *Electron transport* couples to *oxidative phosphorylation* to accomplish the stepwise transport of electrons from reduced coenzyme carriers to molecular oxygen, with concomitant generation of ATP.

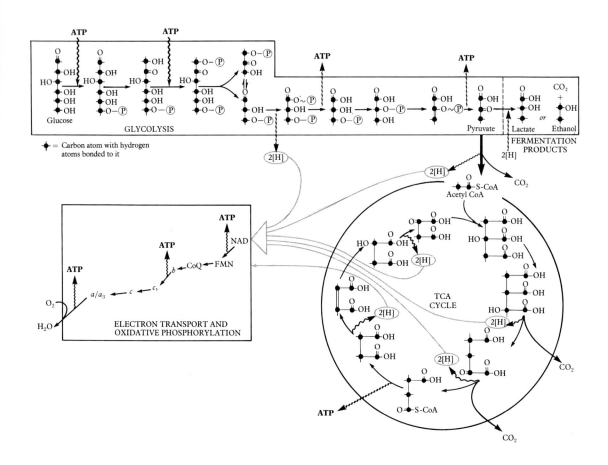

Figure 4-3 The three phases of the oxidative metabolism of glucose. Glucose metabolism begins with *glycolysis* (upper box), a ten-step sequence that degrades the six-carbon starting compound to two molecules of the three-carbon compound, pyruvate, with a net yield of two ATPs per glucose. Under anaerobic conditions the pyruvate is *fermented* further, either to lactate or to ethanol and carbon dioxide. Under aerobic conditions the pyruvate is instead oxidized completely to carbon dioxide by the *tricarboxylic acid cycle* (circle, lower right). The hydrogen ions and electrons removed in the process of oxidation are passed via coenzymes to the *electron transport chain* (box, lower left). As the electrons flow to carriers of successively lower energies, the flow of electrons is coupled to the generation of ATP in the process of *oxidative phosphorylation*. Eventually the electrons are passed to molecular oxygen, which is thereby reduced to water. The glycolytic pathway is presented in greater detail in Figure 4-4. Details of aerobic metabolism are to be found in Figures 5-2 (TCA cycle) and 5-4 (electron transport).

This is the stage at which oxygen actually enters the overall scheme of glucose metabolism, but both stages 2 and 3 are actually dependent upon aerobic conditions, as we shall see in Chapter 5.

The remainder of this chapter is devoted to the anaerobic metabolism, or *fermentation*, of glucose. This means that our discussion will be restricted to the first of the three processes discussed above, that of *glycolysis*. But since the glycolytic pathway functions as part of aerobic energy metabolism as well, it is of great general significance. Indeed the ten-step glycolytic sequence is the single most ubiquitous process in all energy metabolism. Four general properties of glycolysis can be noted at the outset:

Glycolysis and Fermentation: The Anaerobic Metabolism of Glucose

1.   It is a nearly *universal* process in that it occurs in almost all cells; almost every living cell, be it microorganism, plant, or animal, carries out these same essential reactions.

2.   It is *anaerobic* in that the pyruvate produced by the pathway can be converted to lactate (or to ethanol and CO_2), allowing the overall process to occur in the absence of oxygen.

3.   It occurs in the *cytosol* rather than being compartmentalized into a specific eukaryotic organelle.

4.   It is a *primitive* process in the sense that it probably arose early in biological history, before the appearance of oxygen in the atmosphere and before the advent of eukaryotic organelles.

The glycolytic pathway is shown in the context of overall chemotrophic energy metabolism in Figure 4-3 and is presented in detail in Figure 4-4. The very term suggests the essence of the process, since the Greek roots are *glyco* for "sugar" and *lysis* for "splitting." In that context it is appropriate to begin our consideration of the pathway with the fourth step (labeled as Reaction Gly-4 both in Figure 4-4 and in the text below), since it is at this point in the sequence that the actual splitting (of a 6-carbon compound into two 3-carbon compounds) occurs. Reaction Gly-4 is of key importance in the glycolytic scheme for two reasons: (1) it is the actual sugar-splitting step for which the process is named; and (2) it gives rise to glyceraldehyde-3-phosphate, the sole molecule in the pathway actually capable of being oxidized. If we view the purpose of this step to be the breakdown of a 6-carbon molecule into smaller, more manageable pieces, then we might invoke three conditions such a breakdown ought to satisfy: (1) the products of the split should be identical or interconvertible for the sake of efficiency (so that a pathway designed to handle one of the pieces can also be used for the other piece); (2) this in turn suggests that the simplest split of a 6-carbon compound would be into two

Figure 4-4 The glycolytic pathway. Glycolysis converts the starting six-carbon sugar, glucose, into two molecules of the three-carbon compound, pyruvate, with a yield of two ATPs per glucose. The glycolytic pathway is common to both anaerobic and aerobic glucose metabolism, with pyruvate as the key branching point. Under anaerobic conditions pyruvate is reduced fermentatively either to lactate or to ethanol and carbon dioxide, and further metabolism is not possible. Aerobically, pyruvate is instead oxidized and decarboxylated to acetate, which enters the tricarboxylic acid cycle (Figure 5-2) as acetyl coenzyme A.

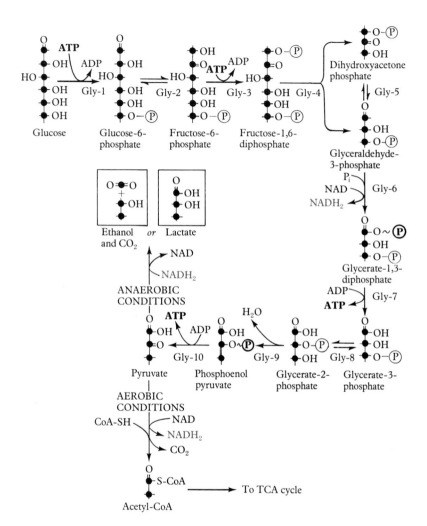

3-carbon pieces; and (3) since, for reasons we shall encounter shortly, the 3-carbon fragments should be phosphorylated, the 6-carbon starting compound should be doubly phosphorylated in a manner such that each of the resulting 3-carbon compounds retains one of the phosphate groups. Now look carefully at the structure of fructose-1,6-diphosphate and at the manner in which it is split into smaller pieces. Notice how nicely these three conditions are fulfilled. Fructose-1,6-diphosphate is an essentially symmetric molecule, with a phosphate group projecting from each end. Upon cleavage it gives rise to two monophosphorylated 3-carbon compounds that, as it turns out, can be readily interconverted and subsequently oxidized with release of energy.

Once we understand that the actual splitting reaction at step 4 of the glycolytic sequence requires a doubly phosphorylated sugar molecule, then the first three steps of the pathway seem especially logical: they are the preliminary maneuvers required to produce such a doubly phosphorylated molecule by attaching a phosphate group to each end (that is, to carbons 1 and 6) of the starting compound, glucose.

In terms of the chemistry involved, the phosphorylation of carbon atom 6 can be accomplished readily, since that carbon atom already possesses a free hydroxyl group that can be readily linked to a phosphate group to form a phosphate ester. Thermodynamically, however, the direct addition of inorganic phosphate to glucose (Equation 4-10) would be very unfavorable, primarily because of the reduced possibilities for resonance within the phosphate group when one of its oxygen atoms is linked covalently to a sugar molecule. As a result the cell cannot phosphorylate glucose by this means. Instead the addition of phosphate at carbon 6 is achieved by transfer of a phosphate group from ATP, thereby effectively coupling glucose phosphorylation (Equation 4-10; $\Delta G^0 = +3.3$ kcal/mole) to ATP hydrolysis (Equation 4-11; $\Delta G^0 = -7.3$ kcal/mole). The difference in bond energies between the acid anhydride bond of ATP that is broken and the phosphate ester bond of glucose-6-phosphate that is formed gives the overall transfer reaction (Equation Gly-1) a ΔG^0 of -4.0 kcal/mole, which renders it not only favorable but essentially irreversible in the direction written.

$$\text{Glucose} + \text{P}_i \longrightarrow \text{glucose-6-phosphate} \qquad (4\text{-}10)$$

$$\text{ATP} \longrightarrow \text{ADP} + \text{P}_i \qquad (4\text{-}11)$$

$$\text{Glucose} + \text{ATP} \xrightarrow[\text{hexokinase}]{} \text{glucose-6-phosphate} + \text{ADP} \qquad (\text{Gly-1})$$

This coupled reaction is actually the first step in the glycolytic pathway and has therefore been identified as Reaction Gly-1. (This pathway-specific nomenclature is used so that reactions belonging to a particular sequence can be identified readily within the context of a chapter. Enzyme names are indicated for each step to underscore the fact that every reaction we encounter is catalyzed by a specific enzyme.)[*] As indicated by Equation Gly-1, then, the process of

---

[*] It is important to keep in mind that every cellular reaction is in fact catalyzed by a specific enzyme. It is equally important, however, that you avoid getting so caught up in the intricacies of enzyme nomenclature that you miss the simplicity and beauty of the reactions they catalyze. The enzyme names are artificial conventions and may be arbitrary and confusing, but the chemistry of the reaction, if properly understood, is always logical and predictable.

glycolysis begins with an ATP-driven phosphorylation of glucose on carbon 6, catalyzed by the enzyme *hexokinase*.

With one phosphate group already in place, we can now turn our attention to carbon 1 of the glucose molecule, since a doubly phosphorylated sugar is needed. Here, however, we encounter a problem, for the carbonyl group (or, in the ring form, the hemiacetal group) of carbon 1 is not so readily phosphorylated as an alcohol group. The problem is resolved by the reversible conversion of glucose, an aldosugar, into fructose, the corresponding ketosugar (Equation Gly-2). As Figure 4-4 illustrates, this conversion involves the movement of the carbonyl group from carbon 1 to carbon 2, with the appearance or carbon 1 of a free hydroxyl group: [*]

$$\text{Glucose-6-phosphate} \underset{\text{isomerase}}{\rightleftharpoons} \text{fructose-6-phosphate} \quad \text{(Gly-2)}$$

With a hydroxyl group now available on carbon 1, the way is clear for a second phosphorylation. This proceeds in a manner entirely analogous to that already discussed for the initial phosphorylation reaction, by transfer of a high-energy phosphate group from ATP:

$$\text{Fructose-6-phosphate} + \text{ATP} \xrightarrow{\text{phosphofructokinase}}$$

$$\text{fructose-1,6-diphosphate} + \text{ADP} \quad \text{(Gly-3)}$$

The enzyme responsible for this reaction, phosphofructokinase, is the key regulatory enzyme in the glycolytic sequence. In addition to an active site specific for the substrates, the enzyme also has allosteric binding sites for both ATP and ADP. This renders the enzyme sensitive to the prevailing ATP/ADP balance of the cell and allows the rate of glycolysis to be adjusted to the energy needs of the cell. We shall return to this topic in Chapter 5 when the regulation of cellular energy metabolism is taken up.

We can summarize the first segment of glycolysis by noting that,

---

[*]As the name of the enzyme suggests, Reaction Gly-2 is an *isomerization* reaction involving an internal rearrangement of functional groups. The chemically inclined might like to note that the mechanism involves an enol intermediate as follows:

$$
\begin{array}{ccc}
\overset{\text{O}}{\overset{\|}{\text{H}-\text{C}}} & \text{H}-\text{C}-\text{OH} & \overset{\text{H}}{\overset{|}{\text{H}-\text{C}-\text{OH}}} \\
| & \| & | \\
\text{H}-\text{C}-\text{OH} \rightarrow & \text{C}-\text{OH} \rightarrow & \text{C}=\text{O} \\
| & | & | \\
\text{R} & \text{R} & \text{R} \\
\text{Aldosugar} & \text{Enol} & \text{Ketosugar}
\end{array}
$$

in three steps (two phosphorylations and an intervening rearrangement), glucose has been converted into the fructose-1,6-diphosphate
required for the actual sugar-splitting reaction that lies just ahead. The
net result of these first three steps is the conversion of the glucose
molecule to a higher-energy, activated intermediate. This conversion
has been accomplished at the expense of two ATP molecules per
glucose molecule. It may seem strange that a pathway functioning to
produce ATP begins by consuming ATP, but this is in fact necessary
to activate the glucose molecule and so to prepare it for its imminent
dismantling. As we shall see, the overall yield more than compensates
for this investment, such that the initial ATP input is reminiscent
of the "pump priming" familiar to economists.

As a result of the energy input, fructose-1,6-diphosphate is more **The Actual Cleavage**
reactive than glucose and can, as already anticipated, be broken down **Step of Glycolysis**
into two interconvertible three-carbon compounds:

Fructose-1,6-diphosphate $\underset{\text{aldolase}}{\overset{\longrightarrow}{\rightleftharpoons}}$

$\qquad$ glyceraldehyde-3-phosphate + dihydroxyacetone phosphate $\qquad$ (Gly-4)

It is useful to note carefully the fate of individual carbon atoms in
this reaction, since the numbering of individual carbon atoms changes
as a result of the cleavage. This change in number is clearly of no
consequence to the carbon atoms, but it is a source of enough confusion to make it worthwhile noting explicitly the fate of each carbon
atom in the starting hexose molecule:

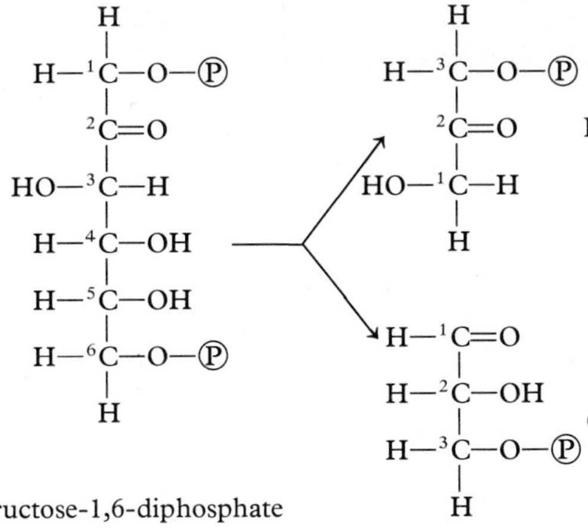

Fructose-1,6-diphosphate

Dihydroxyacetone phosphate

Glyceraldehyde-3-phosphate

Note that the symbol Ⓟ is being used for organically linked phosphate.

The fructose-1,6-diphosphate molecule (like the glucose from which it is derived) is numbered from the most oxidized end, such that the carbonyl group is at carbon 2. The two resulting triose phosphate molecules are also numbered from their own most oxidized ends. The result, however, is that each carbon atom in fructose-1,6-diphosphate undergoes a change in number upon cleavage. In particular, the two terminal carbons of fructose-1,6-diphosphate (carbons 1 and 6) give rise to the two number 3 carbon atoms of the trioses, and the two internal carbons of fructose diphosphate (carbons 3 and 4) give rise to the two number 1 carbon atoms of the trioses.

The two products of the hexose cleavage of step 4 are both monophosphorylated and are readily interconvertible by an isomerization reaction entirely analogous to the interconversion of glucose-6-phosphate and fructose-6-phosphate:

$$
\begin{array}{ccc}
& \text{H} & \\
& | & \\
\text{H—C—OH} & & \text{H—C}\!=\!\text{O} \\
| & & | \\
\text{C}\!=\!\text{O} & \underset{\text{triose isomerase}}{\rightleftarrows} & \text{H—C—OH} \qquad \text{(Gly-5)} \\
| & & | \\
\text{H—C—O—Ⓟ} & & \text{H—C—O—Ⓟ} \\
| & & | \\
\text{H} & & \text{H} \\
\text{Dihydroxyacetone} & & \text{Glyceraldehyde-} \\
\text{phosphate} & & \text{3-phosphate}
\end{array}
$$

This interconvertibility is of real economic significance to the cell. It means that the pathway which exists for metabolism of one of these two compounds (glyceraldehyde-3-phosphate, as it turns out) is, upon addition of this step, also utilized for the other compound, without the involvement of additional enzymatic machinery. As we follow the fate of glyceraldehyde-3-phosphate in detail, keep in mind that the dihydroxyacetone phosphate is subjected to the same metabolism; in the final summing up, all reactions from this point on must be multiplied by 2 if we are to account for the metabolism of the entire glucose molecule.

In the first five steps of glycolysis, then, the original glucose molecule has been doubly phosphorylated and split into two interconvertible molecules, each of which can therefore be subjected to the same further steps in metabolism. Note also that the net ATP yield thus far is −2, since two ATPs have been consumed per glucose. We are, however, about to see how this ATP debt is turned into a profit, as we encounter the two energy-yielding events of glycolysis.

The stage is now set for an energy-yielding event that is at the very heart of glycolysis: the oxidation of glyceraldehyde-3-phosphate to glycerate-3-phosphate. This is a highly exergonic reaction; if carried out as a direct oxidation with an inorganic catalyst, it would yield much heat. The cell, however, couples the oxidative event both to the generation of ATP and to the reduction of the coenzyme *nicotinamide adenine dinucleotide*, abbreviated NAD. The reaction sequence occurs in two steps, each catalyzed by a separate enzyme, with a high-energy diphosphorylated compound as an intermediate:

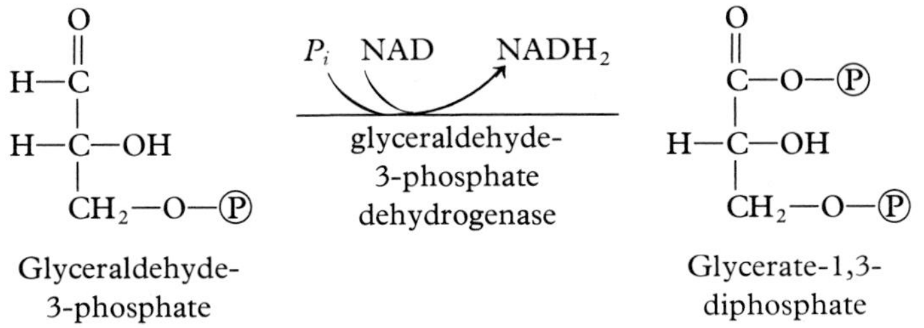

The First Energy-yielding Event of Glycolysis

$$
\begin{array}{c}
\text{O} \\
\parallel \\
\text{H—C} \\
| \\
\text{H—C—OH} \\
| \\
\text{CH}_2\text{—O—}\textcircled{P}
\end{array}
\qquad
\underset{\substack{\text{glyceraldehyde-}\\\text{3-phosphate}\\\text{dehydrogenase}}}{\xrightarrow{P_i \quad \text{NAD} \quad \text{NADH}_2}}
\qquad
\begin{array}{c}
\text{O} \\
\parallel \\
\text{C—O—}\textcircled{P} \\
| \\
\text{H—C—OH} \\
| \\
\text{CH}_2\text{—O—}\textcircled{P}
\end{array}
\qquad \text{(Gly-6)}
$$

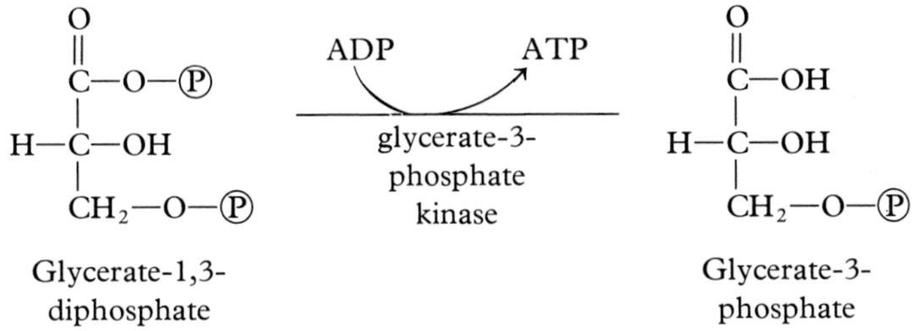

Glyceraldehyde-3-phosphate   Glycerate-1,3-diphosphate

$$
\begin{array}{c}
\text{O} \\
\parallel \\
\text{C—O—}\textcircled{P} \\
| \\
\text{H—C—OH} \\
| \\
\text{CH}_2\text{—O—}\textcircled{P}
\end{array}
\qquad
\underset{\substack{\text{glycerate-3-}\\\text{phosphate}\\\text{kinase}}}{\xrightarrow{\text{ADP} \quad \text{ATP}}}
\qquad
\begin{array}{c}
\text{O} \\
\parallel \\
\text{C—OH} \\
| \\
\text{H—C—OH} \\
| \\
\text{CH}_2\text{—O—}\textcircled{P}
\end{array}
\qquad \text{(Gly-7)}
$$

Glycerate-1,3-diphosphate   Glycerate-3-phosphate

The first step in this sequence is the actual oxidative event, with the aldehyde group on carbon 1 of glyceraldehyde-3-phosphate oxidized to the carboxylic acid group of glycerate. The two electrons removed from the aldehyde in this oxidation are transferred to the coenzyme NAD, which is converted to $NADH_2$, a reduced and therefore energy-rich form.

Since this is our first encounter with the coenzyme NAD, a brief digression is in order to introduce the molecule. This is the most common, but by no means the only, coenzyme used in biological oxidation-reduction reactions. Its structure is shown in Figure 4-5. It consists, as the name implies, of two nucleotides, each containing a heterocyclic ring compound linked to a phosphorylated ribose. One of the two component nucleotides is, in fact, adenosine monophosphate (AMP; see Figure 4-1); the other is nicotinamide monophosphate (NMP). The two are linked together by a pyrophosphate bridge. The

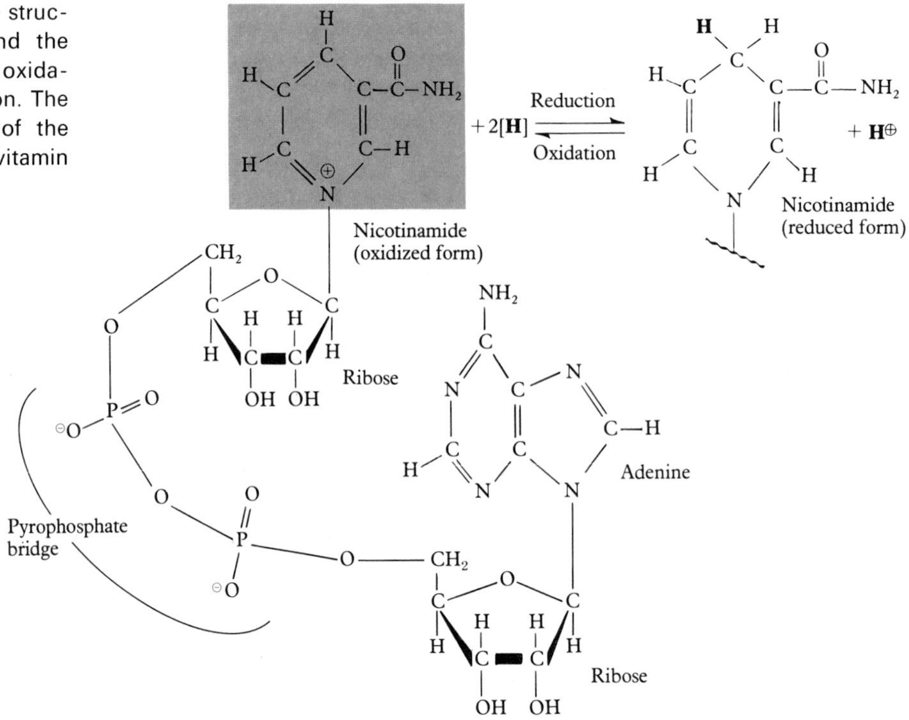

Figure 4-5 The structure of NAD and the chemistry of its oxidation and reduction. The shaded portion of the coenzyme is the vitamin nicotinamide.

Nicotinamide (oxidized form)

$+ 2[\mathbf{H}]$ Reduction / Oxidation $+ \mathbf{H}^{\oplus}$

Nicotinamide (reduced form)

Ribose

Pyrophosphate bridge

Adenine

Ribose

coenzyme NAD serves as an electron acceptor in biological oxidation reactions by adding two electrons and one proton to the ring of the nicotinamide portion of the molecule, thereby generating the reduced form, $NADH_2$. (Purists will recognize from Figure 4-5 that what we write here as $NAD/NADH_2$ ought really to be written as $NAD^+/NADH + H^+$, but the convention $NAD/NADH_2$ will suffice for our purposes.) Also worth a passing mention is that nicotinamide is a derivative of nicotinic acid, one of the essential B vitamins that must be present in the diet of humans and other vertebrates. It is precisely its involvement in energy metabolism as part of the vital coenzyme NAD that makes this substance essential in the diet of any organism that cannot manufacture its own supply.

The oxidation of glyceraldehyde-3-phosphate (Reaction Gly-6) is sufficiently exergonic to allow not only the reduction of NAD (itself an energy-storing process, as we shall see in the next chapter) but also the production of ATP. This is possible because the actual oxidative step is coupled with the uptake of inorganic phosphate (abbreviated P_i) from the medium to form not the glycerate-3-phosphate that would otherwise be expected as the product of oxidation, but instead a diphosphorylated intermediate. This intermediate has not only the expected phosphate group on carbon 3 but also a phosphate group on carbon 1. It is the newly added phosphate on carbon 1 that is of interest here, because it is a high-energy phosphate bond. The phosphate group on carbon 3 is an *ester* linkage, which is relatively low in energy ($\Delta G^0 = -3.3$ kcal/mole). That on carbon 1, however, is an acid anhydride linkage, involving as it does a phosphoric acid group (derived from inorganic phosphate) and a carboxylic acid group (generated by the oxidation of the aldehyde). As discussed earlier, such an acid anhydride linkage is a high-energy bond, because both the products of hydrolysis undergo resonance stabilization. It is, in fact, sufficiently energy-rich ($\Delta G^0 = -11.8$ kcal/mole) that the phosphate group can be transferred exergonically to ADP to generate ATP, as shown in Reaction Gly-7. We arrive, therefore, at the expected glycerate-3-phosphate, but the mechanism involving the diphosphorylated intermediate is essential if some of the free energy of the oxidation is to be trapped as ATP.

The sequence of Reactions Gly-6 and Gly-7 serves as a beautiful prototype of coupled reactions in bioenergetics, and you should go over it until you understand the simple but elegant way in which an otherwise heat-liberating reaction has been harnessed to conserve energy as ATP. Note also that each of the two molecules of glyceraldehyde-3-phosphate derived from glucose via the cleavage reaction of Gly-4 is oxidized in this manner, such that the ATP yield from the oxidation is two ATPs per starting glucose. The two ATPs initially invested in the phosphorylation reactions of Gly-1 and Gly-3 are therefore recouped and the net ATP yield through this stage is zero. Note, then, that we have used up seven of the ten reactions of glycolysis to convert one molecule of glucose to two molecules of glycerate-3-phosphate—but still have nothing to show for it in terms of net ATP generation.

The Second Energy-yielding Event of Glycolysis

If net ATP production is to be realized, it has to involve the phosphate group on the third carbon of glycerate-3-phosphate. But that phosphate group is bonded to the glycerate by a phosphate ester linkage of relatively low energy. The last three steps of glycolysis, however, involve what is in effect an intramolecular rearrangement of energy, energizing this ester bond by converting it to a high-energy phosphoenol bond. Crucial to that process is the movement of the phosphate group from its original position on carbon 3 to a new location on carbon 2:[*]

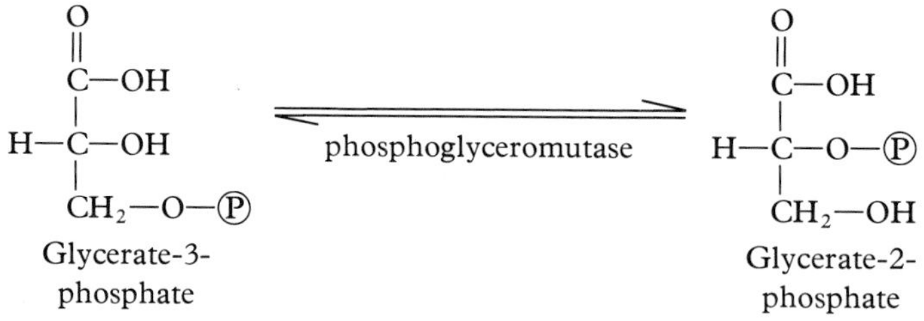

$$
\begin{array}{ccc}
\underset{\substack{| \\ \text{CH}_2-\text{O}-\text{\textcircled{P}}}}{\overset{\displaystyle \overset{\text{O}}{\underset{\|}{}}}{\underset{\text{C}-\text{OH}}{\text{C}-\text{OH}}}} & \underset{\text{phosphoglyceromutase}}{\xrightleftharpoons{\hspace{2cm}}} & \underset{\substack{| \\ \text{CH}_2-\text{OH}}}{\overset{\displaystyle \overset{\text{O}}{\underset{\|}{}}}{\underset{\text{H}-\text{C}-\text{O}-\text{\textcircled{P}}}{\text{C}-\text{OH}}}}
\end{array}
$$

(Gly-8)

Glycerate-3-phosphate   Glycerate-2-phosphate

In the next step a double bond is created by removal of water from glycerate-2-phosphate, generating the compound phosphoenol pyruvate:

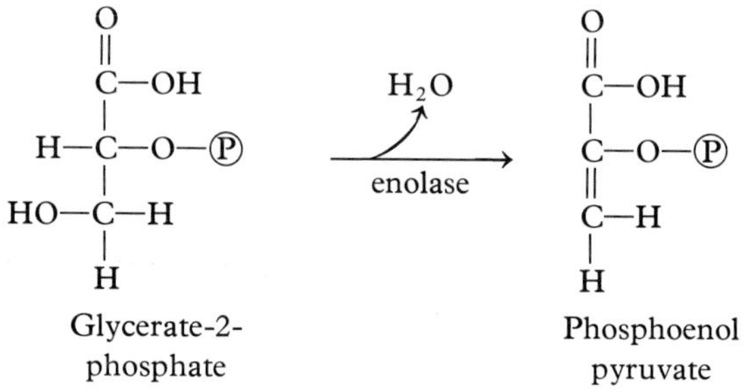

$$
\text{Glycerate-2-phosphate} \quad \xrightarrow[\text{enolase}]{\text{H}_2\text{O}} \quad \text{Phosphoenol pyruvate}
$$

(Gly-9)

Glycerate-2-phosphate   Phosphoenol pyruvate

With the generation of the double bond, the phosphate group on carbon 2 takes on what might be defined as the distinguishing characteristic of a high-energy phosphate bond—a phosphate group adjacent to a double bond. In fact, the ΔG^0 for the hydrolysis of this phosphate group is -14.8 kcal/mole, making this one of the highest-energy phosphate bonds known in biological systems.

---

[*] What may otherwise seem a mystifying bit of phosphate sleight of hand in the translocation of Reaction Gly-8 becomes completely logical chemistry when you realize that the conversion proceeds via a cyclic phosphodiester bridge:

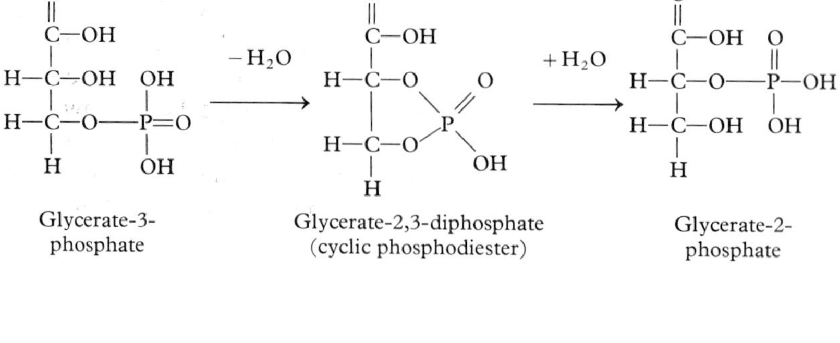

$$
\text{Glycerate-3-phosphate} \xrightarrow{-\text{H}_2\text{O}} \text{Glycerate-2,3-diphosphate (cyclic phosphodiester)} \xrightarrow{+\text{H}_2\text{O}} \text{Glycerate-2-phosphate}
$$

Glycerate-3-phosphate   Glycerate-2,3-diphosphate (cyclic phosphodiester)   Glycerate-2-phosphate

To understand what makes this such a high-energy (i.e., unstable) bond, it is necessary to look at the parent compound, pyruvate, which can exist in either the keto or the enol form:

$$
\begin{array}{ccc}
\begin{array}{l}
O \\
\parallel \\
C\text{—}OH \\
\mid \\
C{=}O \\
\mid \\
CH_3
\end{array}
&
\rightleftharpoons
&
\begin{array}{l}
O \\
\parallel \\
C\text{—}OH \\
\mid \\
C\text{—}OH \\
\parallel \\
CH_2
\end{array}
\\
\text{Keto form} & & \text{Enol form} \\
\text{of pyruvate} & & \text{of pyruvate}
\end{array}
\qquad (4\text{-}12)
$$

The equilibrium lies very far to the left, meaning that the keto form is by far the more stable; the enol form is a highly unstable, thermodynamically improbable configuration. That being so, any compound in which pyruvate is "frozen" in the enol form will be highly unstable because of the strong tendency of the enol form to return to the keto form. Thus when water is removed from glycerate-2-phosphate and the product is pyruvate "locked" into the enol form by the presence of a phosphate group on carbon 2, that product will be especially unstable: in addition to the usual free energy released when the phosphate becomes free to resonate, there is also free energy released when the pyruvate becomes free to return to the keto form.

Since the phosphate bond of phosphoenol pyruvate is very high in energy (free energy of hydrolysis, really), the next step is entirely reasonable, involving as it does the transfer of that high-energy phosphate to ADP, thereby generating another molecule of ATP:

$$
\text{Phosphoenol pyruvate} + \text{ADP} \xrightarrow{\text{pyruvate kinase}} \text{pyruvate} + \text{ATP} \quad (\text{Gly-10})
$$

A point of frequent confusion in the three steps from glycerate-3-phosphate to pyruvate comes with the realization that what began as a fairly low-energy ester bond has somehow become a very high-energy phosphoenol bond that can be used to generate ATP. The source of this energy is readily understandable, however, when these three steps are viewed as an intramolecular oxidation-reduction process in which carbon 2 is oxidized from the alcohol level of glycerate-3-phosphate to the keto level of pyruvate while carbon 3 is reduced from the alcohol level to the hydrocarbon ($-CH_3$) level. The energy of this internal oxidation is conserved, concentrated in the high-energy phosphate bond of phosphoenol pyruvate, and used to generate ATP. Again, then, ATP synthesis is coupled to an oxidative event, although in this case the oxidation occurs intramolecularly and the actual driving force for the generation of ATP is not an oxidative event directly but rather the instability of the enol form of pyruvate.

Recall that each molecule of glucose with which the pathway starts leads to two molecules of phosphoenol pyruvate and hence to two molecules of ATP generated at this second phosphorylation event. Whereas the first phosphorylation event (oxidation of glyceraldehyde-3-phosphate) was required to recoup the initial ATP investment, the ATP yield of this second event represents the net yield of the glycolytic pathway: two ATPs per glucose molecule. As an overall equation, therefore, glycolysis (from glucose to pyruvate) can be summarized as follows:

$$C_6H_{12}O_6 + 2NAD + 2ADP + 2P_i \xrightarrow[\text{Gly-10}]{\text{Gly-1 through}}$$
Glucose

$$2C_3H_4O_3 + 2NADH_2 + 2ATP \qquad (4\text{-}13)$$
Pyruvate

Pyruvate as a Branching Point

Pyruvate is a key intermediate in cellular energy metabolism, primarily because of the central position it occupies at the crossroads of several metabolic pathways. It is, as should now be clear, the immediate product of the glycolytic pathway. What then becomes of the pyruvate depends both on the conditions under which metabolism is proceeding and on the specific organism involved.

The key environmental feature is the availability of oxygen. In aerobic cells (either strict or facultative), pyruvate serves as a point of departure for further metabolism, in which molecular oxygen is used as an electron acceptor and the pyruvate is oxidized completely to carbon dioxide. Aerobically, then, glycolysis is but the first phase of respiratory metabolism, which results eventually in the complete oxidation of glucose, with a much greater ATP yield than is possible by glycolysis alone.

An important feature of glycolysis, however, is that it can also be carried out under completely anaerobic conditions (as in the strict anaerobes or in facultative cells operating in the absence of oxygen). Indeed it is thought to have evolved under environmental conditions characterized by lack of oxygen. Under anaerobic conditions, pyruvate is not the end product of glycolysis, since it is necessary to reoxidize the $NADH_2$ generated in the oxidation of glyceraldehyde-3-phosphate.

The most common mechanism for reoxidation of $NADH_2$ under anaerobic conditions is by transfer of the electrons to pyruvate, reducing the pyruvate to lactate, and oxidizing the $NADH_2$ to NAD:

$$
\begin{array}{c}
\underset{\text{Pyruvate}}{\overset{\displaystyle\overset{O}{\underset{\|}{}}}{\underset{\displaystyle CH_3}{\underset{|}{\overset{\displaystyle C-OH}{\underset{|}{\overset{\displaystyle |}{C=O}}}}}}} + NADH_2 \xrightarrow[\text{dehydrogenase}]{\text{lactate}}
\underset{\text{Lactate}}{\overset{\displaystyle\overset{O}{\underset{\|}{}}}{\underset{\displaystyle CH_3}{\underset{|}{\overset{\displaystyle C-OH}{\underset{|}{\overset{\displaystyle |}{H-C-OH}}}}}}} + NAD \qquad (4\text{-}14)
\end{array}
$$

Since the two molecules of NAD that are reduced (per glucose) in the oxidation of glyceraldehyde-3-phosphate are reoxidized in the above manner, the overall equation for the metabolism of glucose to lactate under anaerobic conditions becomes simply

$$Glucose + 2ADP + 2P_i \rightarrow 2\,lactate + 2ATP \qquad (4\text{-}15)$$

This process (anaerobic glycolysis terminating in lactate) is called *lactate fermentation* and is the major energy-yielding pathway in many bacterial and animal cells operating under anaerobic or relatively anaerobic conditions. The classic example of a system that depends upon lactate fermentation is skeletal muscle, which upon strenuous exertion temporarily depletes its supply of both circulating and stored oxygen and must then depend upon anaerobic glycolysis to meet its energy needs.

Lactate is the most common but not the only possible end product of anaerobic glycolysis. In some cells, of which yeast is the classic example, $NADH_2$ is also reoxidized by transfer of electrons (hydrogen atoms) to pyruvate, but in this case the end result is somewhat different because the oxidative event is preceded by a decarboxylation such that the end products are ethanol and CO_2 instead of lactate:

$$
\underset{\text{Pyruvate}}{C-OH,\ C=O,\ CH_3} \xrightarrow[\text{decarboxylase}]{CO_2\ \ \text{Pyruvate}} \underset{\text{Acetaldehyde}}{H,\ C=O,\ CH_3} \xrightarrow[\text{dehydrogenase}]{NADH_2\ \ NAD\ \ \text{Alcohol}} \underset{\text{Ethanol}}{CH_2-OH,\ CH_3} \quad (4\text{-}16)
$$

Since the end product of glycolysis in this case is ethanol (alcohol), the process is called *alcoholic fermentation*. This process is of tremendous economic significance, both for the baker, who depends upon yeast for the evolution of the CO_2 needed to make his dough rise, and for the brewer, whose preoccupation is with the other end product, the ethanol.

From a thermodynamic point of view, however, the two fermentation processes are very similar. In both cases the $NADH_2$ is reoxidized and therefore does not appear in the overall equation, and the energy yield is still two ATPs. As a general equation for anaerobic glycolysis we can therefore write:

$$\text{Glucose} + 2ADP + 2P_i \rightarrow \begin{Bmatrix} 2 \text{ lactate } or \\ 2 \text{ ethanol} + 2CO_2 \end{Bmatrix} + 2ATP \qquad (4\text{-}17)$$

For the anaerobic cell, no additional ATP is available from the glucose and the end products of fermentation are passed by the cell to its surroundings. Further metabolism and greater energy yields are possible only under aerobic conditions. But before turning to that topic in Chapter 5, we can conclude the current discussion with a brief look at the energy yields of anaerobic glycolysis.

Energy Yields and Efficiency of Glycolysis

The fermentation of glucose to lactate is characterized by a standard free-energy change of -47 kcal/mole of glucose. Since the complete oxidation of glucose to CO_2 and H_2O under aerobic conditions (Equation 4-7) has a standard free-energy change of -686 kcal/mole, it is clear that anaerobic fermentation taps only about 7 percent ($47/686 \times 100$ percent) of the free energy potentially available from glucose. Hence the price that the anaerobe must pay for the privilege of extracting energy from glucose in the absence of oxygen is the wasting of more than 90 percent of the available free energy, which remains entrapped in the end product, lactate. As a corollary the anaerobic cell must therefore consume much larger quantities of glucose (or another fuel) per unit time to accomplish the same amount of cellular work as an aerobic cell. The advantages of aerobic metabolism to the cell are therefore profound, and the metabolism to be discussed in the following chapter is accordingly very important.

As a final question, one might ask how well the anaerobe does with the 47 kcal/mole of glucose available to it from glycolysis. The energy is, of course, partially conserved in the generation of two ATPs per glucose. Based on standard free-energy changes (which are of limited significance to the living cell), the two ATPs, with a ΔG^0 of -7.3 kcal/mole, represent an efficiency of energy conservation of about 31 percent ($14.6/47 \times 100$ percent). This compares favorably with the efficiencies of artificial machines, which seldom exceed 25 percent. It is, if anything, a low estimate of the efficiency of anaerobic glycolysis, because the ΔG value for ATP hydrolysis under cellular conditions is usually substantially higher than the ΔG^0 value.

This, then, is glycolysis. As complex as it may seem upon first encounter, it represents the simplest possible way in which the glucose molecule can be degraded in dilute solution at temperatures compatible with life, with a large portion of its energy yield conserved as ATP. Together with other fermentation processes that utilize a variety of alternative substrates, the anaerobic pathways described here are admirably adapted to meet the energy needs of cells functioning in the absence of oxygen. But all we have seen so far pales in comparison with the potential for energy release and conservation in the presence of oxygen, for aerobic respiration is the capstone of bioenergetics and the mainspring of cellular energy metabolism for most forms of life. And for that, we take the $NADH_2$ and pyruvate of Equation 4-13 and proceed to Chapter 5.

Practice Problems

4-1. True or False: Answer each of the following as true (T) or false (F).

(a) The principal food molecules used for chemotrophic energy metabolism are carbohydrates, fats, and proteins.

(b) To initiate glycolysis, glucose is phosphorylated and oxidized to a molecule that can be cleaved into two interconvertible three-carbon sugars.

(c) The acid anhydride bond between the second and third phosphates of ATP is much more energy-rich (i.e., has a more negative-free energy of hydrolysis) than that between the first and second phosphates because of the greater possibilities for resonance stabilization upon hydrolysis.

(d) Of the eleven steps required to convert glucose into lactate, only two are actually oxidation reactions that couple to ATP synthesis.

(e) The lactate/pyruvate ratio is likely to be higher in muscle following strenuous exertion than in resting muscle.

4-2. Glycolysis: The glycolytic pathway can terminate in any of three alternative products: (1) pyruvate, (2) lactate, or (3) ethanol ($+ CO_2$). For each of these end products, describe the conditions under which it would be expected as the main product and write a balanced overall equation starting with glucose (include all ATPs, NADs, and so on as appropriate). What is significant about the net appearance of NAD and $NADH_2$ in one of your equations but not in the other two?

4-3. ATP Synthesis: The oxidation of glyceraldehyde-3-phosphate to glycerate-3-phosphate occurs in glycolysis via a diphosphorylated intermediate and requires a two-step reaction sequence (Equations Gly-6 and Gly-7). Recently a Biocore student at Standard State University discovered a mutant microorganism capable of converting glyceraldehyde-3-phosphate directly into glycerate-3-phosphate:

$$(4\text{-}18)$$

Glyceraldehyde-3-phosphate   Glycerate-3-phosphate

Because this mutation eliminated the intermediate and shortened the glycolytic pathway by one step, the student proclaimed the mutant to be a significant evolutionary advance in anaerobic energy metabolism. Do you agree with that opinion? Why or why not?

4-4. Galactose Metabolism: The glycolytic pathway is usually written starting with glucose, since this is the single most important sugar for most organisms. But in fact cells can utilize a variety of sugars. The diet of young mammals, for example, consists almost entirely of milk, which contains as its principal carbohydrate the disaccharide lactose. When lactose is hydrolyzed in the intestine, it yields one molecule each of the hexoses glucose and galactose, so the cells of these animals must metabolize just as much galactose as glucose. Galactose is metabolized by phosphorylation and conversion to glucose. The reaction sequence is a bit complicated, though, because the conversion to glucose (an *epimerization* reaction on carbon 4) occurs while the sugar is attached to the carrier uridine diphosphate (UDP, a close relative of ADP). The reactions are as follows:

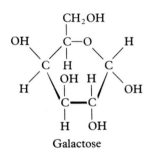

Galactose

$$\text{Galactose} + \text{ATP} \rightarrow \text{galactose-1-phosphate} + \text{ADP} \qquad (4\text{-}19)$$

$$\text{Galactose-1-phosphate} + \text{UDP-glucose} \rightarrow$$
$$\text{glucose-1-phosphate} + \text{UDP-galactose} \qquad (4\text{-}20)$$

$$\text{UDP-galactose} \rightarrow \text{UDP-glucose} \qquad (4\text{-}21)$$

$$\text{Glucose-1-phosphate} \rightarrow \text{glucose-6-phosphate} \qquad (4\text{-}22)$$

(a) Write an equation for the overall conversion of galactose to glucose-6-phosphate.

(b) How do you think the ΔG^0 for the overall reaction of part (a) compares with that for the hexokinase reaction (Equation Gly-1)?

(c) If you know that the epimerase reaction (Equation 4-21) has an absolute requirement for the coenzyme NAD and involves 4-ketoglucose as an enzyme-bound intermediate, can you suggest a logical reaction sequence to explain the conversion of galactose to glucose?

(d) The congenital disease *galactosemia* is due to a genetic absence of the enzyme that catalyzes Reaction 4-20. The symptoms of galactosemia, which include mental disorders and cataracts of the eye, are thought to result from high levels of galactose in the blood. Why does this seem a reasonable postulate?

4-5. Glycerol Metabolism: Glycerol is another compound that is metabolized by being converted into an intermediate in the glycolytic pathway.

(a) If you know that the intermediate to which glycerol is converted is dihydroxyacetone phosphate and that two enzyme-catalyzed reactions are necessary for the conversion, can you suggest a reaction sequence that converts glycerol to dihydroxyacetone phosphate using enzymes very similar to those in the glycolytic pathway with which you are already familiar?

(b) Write a balanced overall equation for the conversion of one molecule of glycerol into one molecule of pyruvate.

‡4-6. Pathway of Carbon in Glycolysis: Compounds labeled with radioactive carbon atoms at specific positions in the molecule have been very useful in the

elucidation of metabolic pathways. (For example: in glucose-1-^{14}C, only carbon atom 1 is radioactively labeled with an atom of ^{14}C.) We shall use the technique "backward" here, in that we supposedly already know the pathway and are now chasing labeled carbon atoms around simply to convince ourselves that we really do know it. For each of the starting compounds below, indicate where in the final lactate (or pyruvate) molecule the labeled carbon atom will appear. (Lactate and pyruvate, like all organic molecules, are numbered beginning from the most oxidized end, so the carboxyl carbon is number 1 and the methyl carbon is number 3.)

(a) Glucose-1-^{14}C.

(b) Galactose-6-^{14}C (see Problem 4-4).

(c) Glycerol-2-^{14}C (see Problem 4-5).

(d) Fructose-5-^{14}C.

(e) Glucose-3,4-^{14}C.

(f) Where in the starting glucose molecule would the radioactive carbon atom have to be located in order to ensure that all the radioactivity is liberated as $^{14}CO_2$ during alcoholic fermentation by a yeast culture?

‡ 4-7. Arsenate Poisoning: Arsenate ($HAsO_4^{--}$) is a potent poison in almost all living systems. Among other effects arsenate is known to uncouple the phosphorylation event from the oxidation of glyceraldehyde-3-phosphate. This is because the enzyme involved (glyceraldehyde-3-phosphate dehydrogenase) can utilize arsenate instead of inorganic phosphate, forming glycerate-1-arseno-3-phosphate. This is a highly unstable compound, which immediately undergoes nonenzymatic decomposition into glycerate-3-phosphate and free arsenate. What is the consequence of this for the organism?

References

Bagley, S., and D. E. Nicolson, *An Introduction to Metabolic Pathways* (Oxford: Blackwell, 1970).

Barker, H. A., *Bacterial Fermentation* (New York: Wiley, 1956).

Barker, R., *Organic Chemistry of Biological Compounds* (Englewood Cliffs, N.J.: Prentice-Hall, 1971).

Breslow, R., *Organic Reaction Mechanisms* (New York: Benjamin, 1965).

Kaleker, H. M., *Biological Phosphorylations* (Englewood Cliffs, N.J.: Prentice-Hall, 1969).

White, E. H., *Chemical Background for the Biological Sciences* (Englewood Cliffs, N.J.: Prentice-Hall, 1970).

Aerobic Energy Metabolism

5

We saw in Chapter 4 that the fate of the pyruvate produced by the glycolytic pathway varies with cell type and environmental conditions. If oxygen supplies are limited or the cell is incapable of aerobic metabolism, then energy metabolism is restricted to anaerobic fermentation, with reduction of the pyruvate to lactate or ethanol. This provides for the regeneration of NAD from $NADH_2$ and allows further glycolysis, but the net energy yield is low (two ATPs per glucose) and most of the energy of the glucose molecule remains untapped in the lactate or ethanol that is returned to the medium as a waste product. Anaerobic glycolysis is thus a primitive process and is wasteful in the sense that it gives the cell access to less than 10 percent of the free energy stored in glucose. In addition it causes the buildup in the environment of toxic end products and usually produces nothing that can be used as starting material for photosynthesis (and therefore does not complete the chemotrophic portion of the biosphere cycle shown in Figure 1-1). It is not surprising, then, that processes or conditions capable of further utilization of the energy of pyruvate have come to play so prominent a role in chemotrophic energy metabolism. The process is aerobic *respiratory metabolism*, the condition required is the availability of molecular oxygen (O_2), and the resulting complete combustion of glucose (or other oxidizable substrates) to CO_2 and H_2O is the most important source of energy for most aerobic chemotrophs.

The oxygen required for aerobic metabolism acts as a terminal electron acceptor, thereby providing for the reoxidation of coenzyme molecules. It is these coenzymes that carry out the actual stepwise oxidation of organic intermediates derived from pyruvate. Respiratory metabolism can be considered to consist of two separate but tightly integrated processes: (1) the actual carbon metabolism necessary to oxidize pyruvate completely to CO_2 with transfer of electrons (as hydrogen atoms) to coenzyme molecules and (2) the subsequent re-oxidation of the reduced coenzyme molecules by transfer of the electrons to oxygen, with concomitant generation of ATP.

To put these two processes into perspective, recall from Chapter 4 that, under aerobic conditions, glycolysis represents the first phase of overall glucose metabolism. The two component parts of respiratory metabolism complete the picture—the carbon metabolism involved in further oxidation of pyruvate is the *tricarboxylic acid cycle* already described as phase 2, and phase 3 is the *electron transport* scheme necessary to reoxidize coenzyme molecules at the expense of molecular oxygen, accompanied by phosphorylation of ADP. Figure 4-2 sum-marized all three phases of chemotrophic metabolism. Now we are ready to look at the aerobic components in more detail.

The Tricarboxylic Acid Cycle

In the presence of oxygen, pyruvate is not reduced to lactate or ethanol but is instead oxidized completely to CO_2. This oxidation is accomplished stepwise in a cyclic process that is at the very heart of aerobic metabolism. The cyclic series of reactions is initiated with the formation of an organic acid (citrate) having three carboxylic acid groups, and the process is accordingly called the tricarboxylic acid (TCA) cycle. Alternatively it is also called the *Krebs cycle* in honor of Sir Hans Krebs who, more than any other single scientist, was responsible for the elucidation of this pathway and went on to receive a Nobel prize for his efforts. The postulation of the TCA cycle by Krebs in 1937 was, in fact, a brilliant piece of experimenta-tion and reasoning that ranks among the classic investigations of modern biology.

Preparatory to the actual TCA cycle, there is an initial reaction in which pyruvate is both oxidized and decarboxylated (CO_2 removed), forming acetate, a two-carbon compound. The coenzyme NAD serves as the electron acceptor, and the excess energy of oxidation is used not to generate an ATP (though this would be energetically feasible) but rather to "energize" or "activate" the acetate molecule for further metabolism by linking it via a high-energy sulfur bond to a molecule called *coenzyme A*:

$$
\begin{array}{l}
\overset{\displaystyle O}{\underset{\displaystyle \|}{}} \\
C-OH \\
| \\
C=O \\
| \\
CH_3
\end{array}
\quad + \text{ coenzyme A—SH}
$$

Pyruvate

NAD   NADH$_2$ CO$_2$

pyruvate dehydrogenase

(5-1)

$$
\text{coenzyme A—S—}\overset{\displaystyle O}{\underset{\displaystyle |}{\overset{\displaystyle \|}{C}}} \\
\underset{\displaystyle CH_3}{}
$$

Acetyl coenzyme A

As indicated by the shaded letter, carbon 1 of pyruvate is liberated as CO_2 such that carbons 2 and 3 of pyruvate become, respectively, carbons 1 and 2 of acetate. The actual oxidation occurs on carbon 2 of pyruvate, which is oxidized from a keto to a carboxylic acid group (possible only because of the concomitant elimination of carbon 1 as CO_2). The resulting acetate molecule is immediately picked up by coenzyme A.

As shown in Figure 5-1, coenzyme A is a complicated molecule containing the vitamin *pantothenic acid*. Indeed it is the involvement of this compound in an essential coenzyme as well as the inability of humans and other vertebrates to synthesize it that make pantothenic acid a required dietary constituent. We can ignore the detailed molecular structure of coenzyme A and concentrate solely on the free *sulfhydryl* (—SH) *group* with which the molecule terminates, for it is this sulfhydryl group that can form a high-energy *thioester bond* with organic acids. (As Figure 5-1 illustrates, a thioester is simply an ester in which the linking oxygen atom is replaced by a sulfur atom.) Just as NAD is a coenzyme adapted for transfer of electrons, so coenzyme A is adapted as a carrier of organic acids like acetate.

The critical feature of the thioester linkage involved in this carrier function is that it is a high-energy bond. The thioester bond of acetyl coenzyme A is in fact even slightly higher in energy ($\Delta G^0 = -7.5$ kcal/mole) than the terminal acid anhydride linkage of ATP ($\Delta G^0 = -7.3$ kcal/mole). Thus the acetyl group picked up by coenzyme A upon oxidation of pyruvate is an activated, high-energy form, the utility of which will become apparent shortly.

The oxidative decarboxylation of pyruvate (Equation 5-1) can be regarded as the aerobic gateway to the TCA cycle and occurs twice per original glucose molecule. The net result of metabolism of one

Figure 5-1 The structure of coenzyme A and the chemistry of thioester formation. The shaded portion of the coenzyme is the vitamin pantothenic acid.

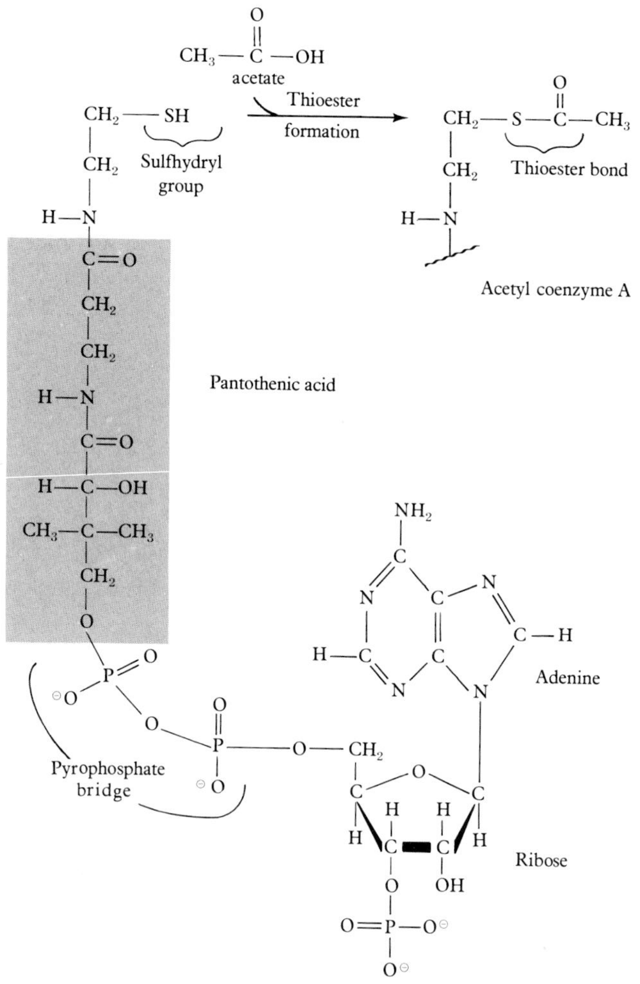

glucose molecule through this gateway step can therefore be summarized by doubling Equation 5-1 and adding it to Equation 4-13 for the glycolytic oxidation of glucose to two molecules of pyruvate:

Glycolysis (Equation 4-13):   glucose + 2NAD + 2ADP + 2P$_i$ →
2 pyruvate + 2NADH$_2$ + 2ATP

Gateway step (2 × Equation 5-1):   2 pyruvate + 2CoA—SH + 2NAD →
2 acetyl CoA + 2NADH$_2$ + 2CO$_2$

Summary: glucose + 4NAD + 2ADP + 2P$_i$ + 2CoA—SH →
2 acetyl CoA + 2CO$_2$ + 4NADH$_2$ + 2ATP   (5-2)

All is now in readiness for the TCA cycle itself, the details of which are shown in Figure 5-2. The cycle begins with the entry of acetate in its activated form, acetyl CoA. As the thioester, the acetate

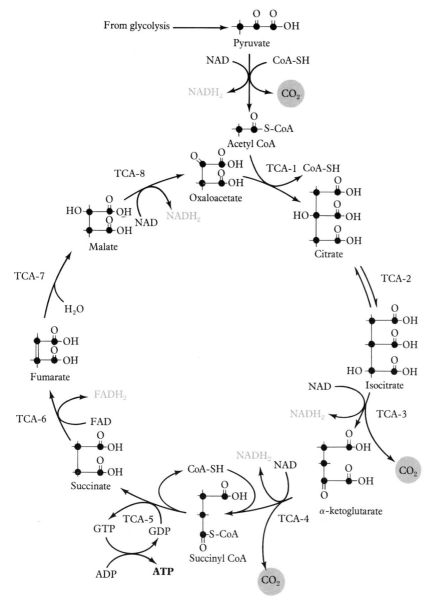

Figure 5-2 The tricarboxylic acid cycle. Pyruvate from the glycolytic pathway of Figure 4-4 is oxidatively decarboxylated to acetate, which enters the cycle in its activated form, acetyl CoA. Per turn of the cycle, two carbon atoms enter (as acetate), two fully oxidized carbon atoms leave (as CO_2), one ATP is generated, and four coenzyme molecules are reduced. The reduced coenzymes ($3NADH_2$ and $FADH_2$) are reoxidized via the electron transport chain with concomitant generation of ATP, as shown in Figure 5-4.

is sufficiently energized to add across the carboxyl group of oxaloacetate, a molecule with two carboxylic acid groups and a keto group:[*]

$$
\underset{\text{Acetyl coenzyme A}}{CH_3-\overset{\overset{\displaystyle O}{\|}}{C}-S-CoA} + \underset{\text{Oxaloacetate}}{\overset{\overset{\displaystyle O}{\|}}{\underset{\displaystyle CH_2-COOH}{C-COOH}}} + H_2O \xrightarrow[\substack{\text{citrate} \\ \text{synthase}}]{\quad CoA-SH \quad} \underset{\text{Citrate}}{HO-\underset{\displaystyle CH_2-COOH}{\overset{\displaystyle CH_2-COOH}{C}}-COOH}
$$

(TCA-1)

The product of this condensation is citrate, one of the tricarboxylic acids for which the TCA cycle is named (in fact, yet another name for this pathway is the *citrate* or *citric acid cycle*). Boldface (**C**) and shaded letters indicate the position of the incoming acetyl carbons in the resulting citrate molecule; these shadings will also appear in subsequent equations to facilitate your understanding of the chemistry of the TCA cycle.

The TCA cycle itself is essentially a disassembly-line-in-the-round dedicated to successive oxidations such that, per turn of the cycle, the entrance of two carbons as acetate is balanced by the release of two fully oxidized carbons as CO_2. Of the eight reactions that make up the cycle, four are oxidation reactions, one is the initial condensation necessary to introduce the incoming acetate group, one is used to generate ATP directly, and two are reactions designed to prepare the molecule for the next oxidative event.

The conversion of citrate to isocitrate (Equation TCA-2), which occurs by movement of a hydroxyl group, is such a preparatory reaction, since the transfer of the hydroxyl group of citrate to the next carbon atom converts a nonoxidizable tertiary alcohol (citrate) into an oxidizable secondary alcohol (isocitrate):[†]

---

[*] For the chemically inquisitive, the reaction whereby activated acetate adds across the double bond to form citrate is an aldol-type reaction, involving nucleophilic addition across a C=O bond. The reaction is initiated by the removal of a hydrogen from the methyl carbon of the acetyl group and continues with the addition of the resulting anion ($^-CH_2-CO-S-CoA$) across the carbonyl bond of oxaloacetate.

[†] A *secondary alcohol* has its hydroxyl group on a carbon atom that has a single hydrogen atom attached to it. A *tertiary alcohol* bears its hydroxyl group on a carbon atom that has no hydrogen atoms attached to it. If the secondary alcohol is an α-hydroxy acid (that is, a carboxylic acid with the hydroxyl group on the adjacent carbon atom), it is a potentially oxidizable compound, as illustrated in the subsequent oxidation of isocitrate in Reaction TCA-3. A tertiary alcohol like citrate, on the other hand, cannot be oxidized. The actual mechanism for the conversion of citrate to isocitrate involves successive dehydration and rehydration reactions in which the elements of water are removed from citrate to generate an unsaturated intermediate, to which the —H and —OH of water are added back again but in the opposite positions; see Problem 5-9.

$$
\begin{array}{ccc}
\begin{matrix} CH_2-COOH \\ | \\ HO-C-COOH \\ | \\ CH_2-COOH \end{matrix}
& \underset{\text{aconitase}}{\rightleftharpoons}
& \begin{matrix} CH_2-COOH \\ | \\ H-C-COOH \\ | \\ HO-CH-COOH \end{matrix} \quad \text{(TCA-2)} \\
\text{Citrate} & & \text{Isocitrate}
\end{array}
$$

The hydroxyl group is now the target of the first of the four oxidative events of the TCA cycle. Isocitrate is oxidized to the corresponding six-carbon keto compound, with NAD serving as the electron acceptor. The six-carbon product is unstable, however, and undergoes spontaneous decarboxylation of the "middle" carboxyl group (enclosed in the box) to form the five-carbon compound α-ketoglutarate (the α means that the keto group is on the carbon atom adjacent to the carboxylic acid group):

$$
\begin{array}{ccc}
\begin{matrix} CH_2-COOH \\ | \\ H-C-\boxed{COOH} \\ | \\ HO-CH-COOH \end{matrix}
& \xrightarrow[\text{isocitrate dehydrogenase}]{\text{NAD} \quad \text{NADH}_2 \quad CO_2}
& \begin{matrix} CH_2-COOH \\ | \\ CH_2 \\ | \\ C-COOH \\ \| \\ O \end{matrix} \quad \text{(TCA-3)} \\
\text{Isocitrate} & & \alpha\text{-Ketoglutarate}
\end{array}
$$

A comparison of the structure of α-ketoglutarate with that of pyruvate reveals a striking similarity—both are α-keto acids with the general formula

$$
\begin{matrix} O \\ \| \\ R-C-COOH \end{matrix}
$$

It is not surprising, therefore, that the mechanism for the oxidation of α-ketoglutarate is the same as for pyruvate, complete with decarboxylation and uptake of the resulting succinate by coenzyme A as a thioester (the carboxyl group giving rise to the CO_2 is again enclosed in a box):

$$
\begin{array}{ccc}
\begin{matrix} CH_2-COOH \\ | \\ CH_2 \\ | \\ C-\boxed{COOH} \\ \| \\ O \end{matrix} + CoA-SH
& \xrightarrow[\substack{\alpha\text{-ketoglutarate} \\ \text{dehydrogenase}}]{\text{NAD} \quad \text{NADH}_2 \quad CO_2}
& \begin{matrix} CH_2-COOH \\ | \\ CH_2 \\ | \\ C-S-CoA \\ \| \\ O \end{matrix} \\
\alpha\text{-Ketoglutarate} & & \text{Succinyl CoA} \\
& & \text{(TCA-4)}
\end{array}
$$

At this point we are already halfway around the cycle and can pause briefly to take stock. Note first of all that the net carbon balance

of the cycle is already satisfied: two carbon atoms entered as acetyl
CoA and two carbon atoms (though of course not the same two) have
now been lost as CO_2. We have also encountered two of the four
oxidative steps of the TCA cycle and have two molecules of $NADH_2$
ready for eventual reoxidation via the electron transport chain. In
addition we have in succinyl CoA a compound that, like acetyl CoA,
has a high-energy thioester bond.

Unlike acetyl CoA, however, succinyl CoA is not destined to be
condensed onto another molecule but is instead used to generate an
ATP. Indeed the rationale behind the formation of succinyl CoA rather
than free succinate in Reaction TCA-4 is specifically to conserve the
energy of α-ketoglutarate oxidation in a form suitable for ATP production. The actual high-energy intermediate generated upon hydrolysis of
succinyl CoA in Reaction TCA-5 is not ATP, however, but a closely
related compound called *guanosine triphosphate* or GTP. As we shall
see in Chapter 8, GTP has a structure very much like ATP and is
formed by phosphorylation of the corresponding diphosphate, GDP.
Since the cell has an enzyme (nucleoside diphosphokinase) capable of
interconverting ATP and GTP (Equation 5-3; not an integral part of
the TCA sequence), the net result of succinyl CoA hydrolysis can be
viewed as the generation of one molecule of ATP, as shown in Equation
TCA-5a:

$$\text{Succinyl CoA} + \text{GDP} + P_i + H_2O \xrightarrow[\text{succinyl thiokinase}]{} \text{succinate} + \text{CoA—SH} + \text{GTP} \qquad \text{(TCA-5)}$$

$$\text{GTP} + \text{ADP} \xrightarrow[\substack{\text{nucleoside} \\ \text{diphosphokinase}}]{} \text{GDP} + \text{ATP} \qquad \text{(5-3)}$$

$$\text{Succinyl CoA} + \text{ADP} + P_i + H_2O \rightarrow \text{succinate} + \text{CoA—SH} + \text{ATP} \qquad \text{(TCA-5a)}$$

The free succinate resulting from this reaction is now oxidized by
removal of a hydrogen atom from each of the two internal carbon
atoms, generating a double bond:

$$\begin{array}{ccc}
\text{CH}_2\text{—COOH} & \xrightarrow[\text{Succinate dehydrogenase}]{\text{FAD} \quad \text{FADH}_2} & \text{CH—COOH} \\
| & & \| \\
\text{CH}_2\text{—COOH} & & \text{CH—COOH}
\end{array} \qquad \text{(TCA-6)}$$

Succinate   Fumarate

Of the four oxidative steps in the TCA cycle, this one is unique in
that both electrons come from carbon atoms. The reaction is not
sufficiently energetic to allow transfer of electrons to NAD, so the

acceptor for this dehydrogenation is a related but lower-energy coenzyme, *flavin adenine dinucleotide* (FAD), whose structure is shown in Figure 5-3. Coenzyme FAD shares several features with NAD: both are dinucleotides involving AMP, both have a B vitamin as part of their structure (nicotinamide in NAD; riboflavin in FAD), and both are reversibly oxidizable.

The fumarate produced in Reaction TCA-6 is now hydrated to produce malate. Note that since fumarate is a symmetric molecule, the hydroxyl group has an equal chance of adding to either of the internal carbons.* Put another way, the shaded carbons being used to keep track of the most recent acetate group to enter the cycle are randomized at this step between the "upper" and "lower" two carbon atoms of malate and are therefore omitted from this point on.

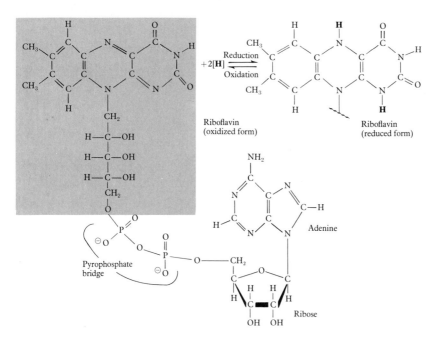

Figure 5-3 The structure of flavin adenine dinucleotide (FAD) and the chemistry of its oxidation and reduction. The shaded portion of the coenzyme is the vitamin riboflavin.

*The discerning chemist will note that the citrate of Equation TCA-2 is also a symmetric molecule and its hydroxyl group ought by the same argument to be moved either "up" or "down" the molecule, randomizing carbon atoms at an early stage in the cycle. The enzyme which catalyzes that reaction is, however, capable of distinguishing between the two ends of the molecule and, by requiring three-point attachment between enzyme and substrate, moves the hydroxyl group in one direction only. By contrast the enzyme responsible for Reaction TCA-6 lacks this property and cannot tell the two inner carbon atoms apart.

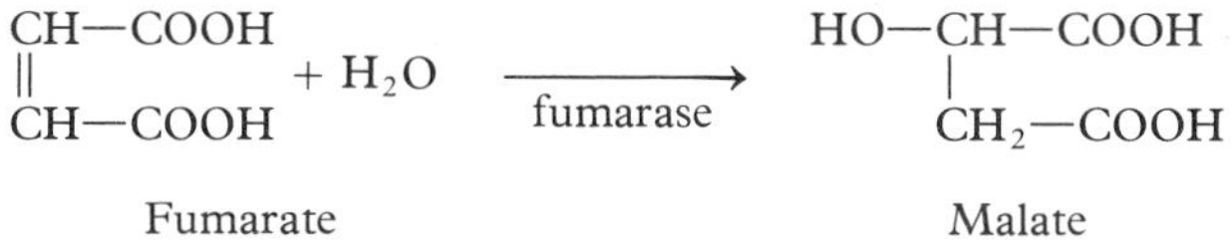

$$
\begin{array}{l}
\text{CH—COOH} \\
\| \qquad\qquad + H_2O \\
\text{CH—COOH}
\end{array}
\xrightarrow[\text{fumarase}]{}
\begin{array}{l}
\text{HO—CH—COOH} \\
| \\
\text{CH}_2\text{—COOH}
\end{array}
\qquad \text{(TCA-7)}
$$

Fumarate   Malate

The hydroxyl group of malate is the target of the final oxidative step in the cycle. Again NAD serves as the electron acceptor and the product is the corresponding keto compound, oxaloacetate:

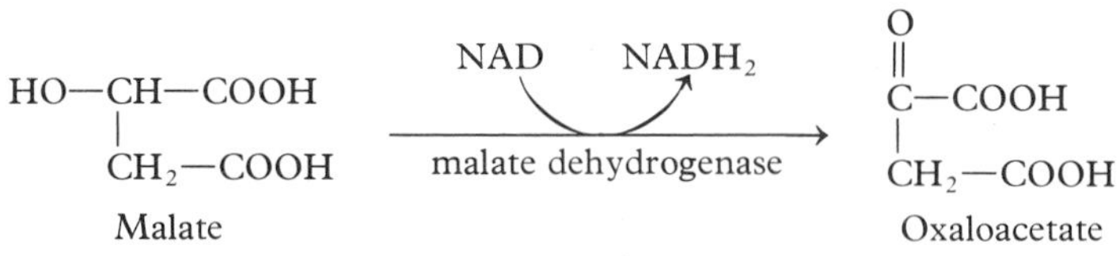

$$
\begin{array}{l}
\text{HO—CH—COOH} \\
| \\
\text{CH}_2\text{—COOH}
\end{array}
\xrightarrow[\text{malate dehydrogenase}]{\overset{\text{NAD} \quad \text{NADH}_2}{\curvearrowright}}
\begin{array}{l}
\overset{\displaystyle O}{\overset{\displaystyle \|}{\text{C}}}\text{—COOH} \\
| \\
\text{CH}_2\text{—COOH}
\end{array}
\qquad \text{(TCA-8)}
$$

Malate   Oxaloacetate

The Cycle in Summary With the regeneration of oxaloacetate, the compound destined to have the next acetyl CoA added across its double bond, one full round of the cycle is complete.

The TCA cycle may be summarized by noting the following characteristics:

1. Acetate (as acetyl CoA) enters the cycle by condensation onto a four-carbon acceptor molecule.
2. Decarboxylation occurs at two steps in the cycle so that the input of a two-carbon acetate unit is balanced by the loss of two carbons as CO_2 (though the two carbons released in any given cycle do not actually come from the acetate added during that cycle).
3. Oxidation occurs at four steps, with NAD serving as the electron acceptor in three cases and FAD being used in one case.
4. ATP is generated (via GTP) in one step.
5. The cycle is completed upon regeneration of the original acceptor.

By summing the eight component reactions of the cycle (Equations TCA-1 through TCA-8), we arrive at the following overall equation for the TCA cycle:

$$
\begin{aligned}
\text{Acetyl CoA} + 3H_2O + 3\text{NAD} + \text{FAD} + \text{ADP} + P_i \rightarrow \\
2CO_2 + \text{CoA—SH} + 3\text{NADH}_2 + \text{FADH}_2 + \text{ATP}
\end{aligned}
\qquad (5\text{-}4)
$$

Since the cycle must turn twice to metabolize both the acetyl CoA molecules derived from one glucose, the overall equation on a per-glucose basis is just twice the above:

$$
\begin{aligned}
2 \text{ acetyl CoA} + 6H_2O + 6\text{NAD} + 2\text{FAD} + 2\text{ADP} + 2P_i \rightarrow \\
4CO_2 + 2\text{CoA—SH} + 6\text{NADH}_2 + 2\text{FADH}_2 + 2\text{ATP}
\end{aligned}
\qquad (5\text{-}5)
$$

And if we now add to this equation the overall equation for all reactions through the gateway step (Equation 5-2), the overall equation for the entire sequence from glucose through the TCA cycle becomes

$$\text{Glucose} + 6H_2O + 10NAD + 2FAD + 4ADP + 4P_i \rightarrow$$
$$6CO_2 + 10NADH_2 + 2FADH_2 + 4ATP \qquad (5\text{-}6)$$

Looking at this summary equation, one is struck by the fact that we have not yet seen much evidence for the substantially greater ATP yield that is supposed to be characteristic of respiratory metabolism. As it stands the equation indicates only a modest enhancement in ATP yield through the TCA cycle (an extra two ATPs per glucose), which hardly seems to justify the metabolic jungle we have just been through. Where, one might ask, is all the energy?

The answer, of course, is that it is stored in the reduced coenzyme molecules, which are high-energy compounds in their own right— the free energy liberated upon reoxidation of these reduced coenzymes by molecular oxygen is substantial indeed. For the release of that energy we must look to the last phase of chemotrophic energy metabolism, for it is as the electrons are transferred stepwise from the coenzyme carriers to molecular oxygen that the coupled generation of ATP occurs. It is this final train of oxidations that accounts for the vast majority of the ATPs produced during the complete oxidation of glucose.

Before leaving the TCA cycle, though, its central position in energy metabolism should be underscored. In the foregoing discussion we have considered only the simple sugar glucose as a substrate for chemotrophic energy metabolism. This is in one sense entirely appropriate, since glucose is the single most important source of energy for the chemotrophic world. However, it is also important to stress that other sugars and other kinds of food molecules (including fats and proteins) can be metabolized by cells to yield ATP. And it is a unifying feature of cellular energy metabolism that the glycolytic pathway and especially the TCA cycle serve as the common framework around which nearly all the catabolic pathways of cells are constructed. Most common sugars, for example, are readily converted into intermediates along the glycolytic pathway (the problems for this chapter provide several examples). Oligo- and polysaccharides are degraded to yield either glucose directly or related sugars that can be converted efficiently into metabolizable intermediates.

As we shall see in Chapter 6, fats and proteins are also catabolized by pathways that feed into the mainstream glycolytic and TCA pathways. Fats, for example, are degraded by a process called β-oxidation to acetyl CoA units, which then enter the TCA cycle directly. Protein metabolism is more complex, since proteins consist of 20 different kinds of amino acids, each of which has a characteristic catabolic fate. In general, however, they are converted eventually to a variety of intermediates in the TCA cycle (some directly to acetyl CoA) for subsequent metabolism. Thus the glycolytic pathway and the TCA cycle, far from being obscure routes for the catabolism of a single sugar, represent the mainstream of cellular energy metabolism into which many side paths flow.

Finally it should be noted that in eukaryotic cells all the enzymes of the TCA cycle (as well as those of the gateway step from pyruvate to acetyl CoA and of the electron transport chain) are compartmentalized in the *mitochondrion*. This means that the flow of pyruvate from the glycolytic pathway to the TCA cycle represents not only the availability of molecular oxygen but also the physical transport of the pyruvate from the cytoplasm to the mitochondrion.

Electron Transport and Oxidative Phosphorylation

In the first two stages of chemotrophic metabolism (glycolysis and the TCA cycle), all six carbon atoms of glucose are oxidized completely to CO_2. To accomplish that oxidation, 12 pairs of electrons and hydrogen ions are removed successively and transferred to the coenzymes NAD and FAD. Concomitant with the continued reduction of coenzymes during the functioning of the glycolytic pathway and the TCA cycle, there must clearly be provision for the reoxidation of these coenzymes, since a recycling is obviously necessary if they are to be continuously available for the oxidation of further glucose molecules. And it is the third stage of energy metabolism that provides for this continued reoxidation of reduced coenzyme molecules. This process is accomplished by the stepwise transfer of electrons (hydrogen atoms, really) from the reduced coenzymes to the ultimate acceptor, molecular oxygen. The overall equations for this electron transport process can be written as follows:

$$NADH_2 + \tfrac{1}{2}O_2 \rightarrow NAD + H_2O \qquad \Delta G^0 = -52.6 \text{ kcal/mole} \qquad (5\text{-}7)$$

$$FADH_2 + \tfrac{1}{2}O_2 \rightarrow FAD + H_2O \qquad \Delta G^0 = -43.4 \text{ kcal/mole} \qquad (5\text{-}8)$$

Several important features of these overall reactions are worth noting. First of all they provide a means for regenerating the oxidized forms of the coenzymes, which can thereby be continuously recycled into the glycolytic pathway and the TCA cycle to function in the oxidation of further food molecules. The entire process of chemotrophic energy metabolism must, in fact, be regarded as a continuous, dynamic, and integrated process. It may be convenient to regard component phases separately in our discussion, but it would be a mistake to think of them as occurring in isolation or in ordered sequence, since all three phases—glycolysis, TCA cycle, and electron transport—must occur in a continuous, highly integrated manner to accomplish respiratory metabolism.

Note also that it is only here, in the third stage of energy metabolism, that molecular oxygen actually appears as the final electron acceptor. In both the glycolytic pathway and the TCA cycle, it is always a coenzyme that serves as the immediate electron acceptor in the oxidation of an organic compound. The ultimate acceptor for all these electrons, however, is oxygen, since all coenzyme molecules are eventually reoxidized at the expense of molecular oxygen. It is also in this ultimate transfer of electrons from coenzymes to oxygen that the water—which we recognize, along with CO_2, as one of the end products of aerobic energy metabolism—is generated. The CO_2 is generated in the TCA cycle, but the water arises only upon reduction of oxygen at the end of the electron transport system.

The most important aspect of Equations 5-7 and 5-8, however, is the amount of free energy available from the oxidation of $FADH_2$ and $NADH_2$, for it is in the stepwise transfer of electrons from these reduced coenzyme molecules to molecular oxygen that most of the ATP yield of respiratory metabolism is realized. As indicated by the ΔG^0 values, the oxidation of a coenzyme is a highly exergonic reaction. In fact most of the free energy originally present in the oxidizable bonds of the glucose molecule is still conserved in the reduced coenzyme molecules generated during glycolysis and the TCA cycle and is tapped by the cell only during this third and final phase of respiratory metabolism.

The standard free-energy changes that accompany the direct oxidation of $NADH_2$ and $FADH_2$ by oxygen are much larger than those usually seen in biological reactions, and they are sufficient to drive the synthesis of at least several ATPs. Hence it should not come as a surprise to find that electrons are not passed directly from reduced coenzymes to oxygen in the living cell. Instead the transfer is accomplished stepwise by using a series of reversibly oxidizable electron acceptors. In this way the total free-energy difference between $NADH_2$ and oxygen is parceled out along a series of intermediates and is released in a series of smaller increments that couple wherever possible to the synthesis of ATP.

The electron transport chain is therefore just a shorthand biochemical way of describing a series of reactions in which a pair of electrons is passed through a series of intermediates, moving from the most willing donor molecules (reduced coenzymes) to the most willing acceptor, molecular oxygen. There are several different ways in which this chain of intermediates can be represented. A common representation is that shown in Figure 5-4, in which the intermediates are arranged in order from the electron donors ($NADH_2$ and $FADH_2$) shown on the left to the ultimate acceptor (O_2) indicated on the right. As Figure 5-4 indicates, the electron transport chain consists of the following intermediates:

1. The coenzymes $NADH_2$ and $FADH_2$, which enter the chain in the reduced form from the glycolytic pathway and the TCA cycle

2. A riboflavin-containing electron carrier called *flavin mononucleotide* (FMN), complexed with a protein to form a *flavoprotein*

3. A compound called *coenzyme Q* with the structure shown in Figure 5-5

4. A series of proteins called *cytochromes*, designated as *b*, c_1, *c*, and a complex of *a* and a_3 in order of decreasing energy[*]

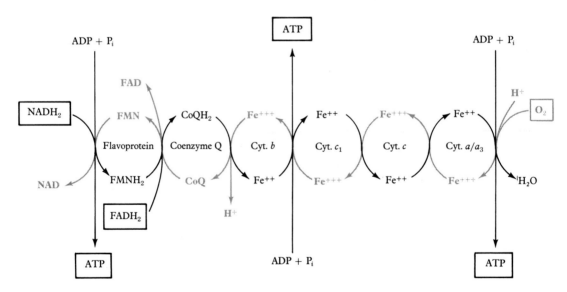

Figure 5-4 The electron transport chain and the coupled generation of ATP by oxidative phosphorylation. The chain consists of a series of reversibly oxidizable carrier molecules that transfer electrons in a stepwise manner from reduced coenzymes ($NADH_2$ and $FADH_2$ on the left) to the ultimate electron acceptor, molecular oxygen (on the right). Electron transfer is accompanied by a flow of protons (H^+) from $NADH_2$ through coenzyme Q but not in later steps involving cytochromes only. Generation of ATP is coupled to electron transport such that three ATPs are formed per molecule of $NADH_2$ reoxidized, but only two ATPs per molecule of $FADH_2$ reoxidized. For the energetics of electron transport see Figure 5-7.

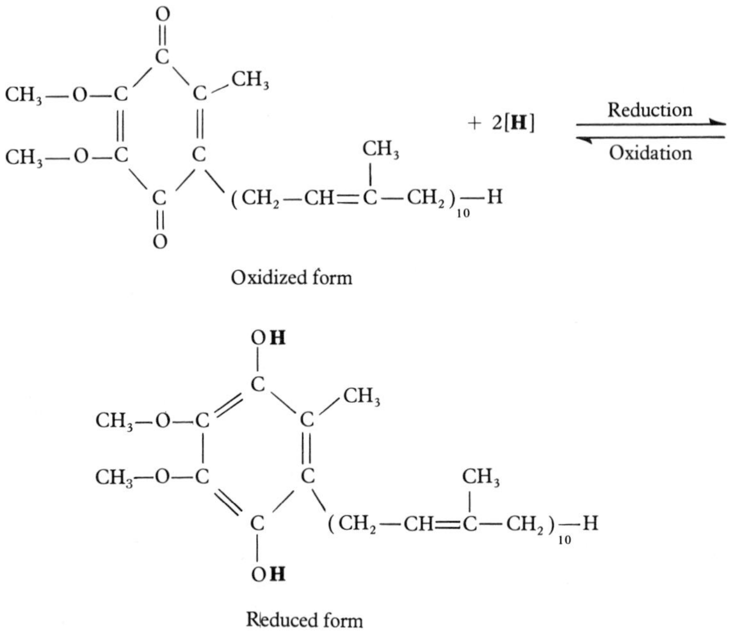

Figure 5-5 The structure of coenzyme Q and the chemistry of its oxidation and reduction.

The cytochromes that make up most of the transport chain are proteins with a nonprotein *heme group* attached to them. The structure of heme is shown in Figure 5-6; like that of the hemoglobin in your red blood cells, the heme group of cytochromes contains an iron atom that can exist in either the ferrous (Fe^{++}) or ferric (Fe^{+++}) form. It is the iron atom of the heme group that allows the cytochrome proteins in the transport chain to serve as electron carriers. Thus while NAD, FAD, FMN, and coenzyme Q transfer electrons by oxidation and reduction of organic compounds, the cytochromes simply use an iron atom. And since the ferrous/ferric conversion represents a one-electron event whereas all previous oxidation reactions are two-electron events, it is necessary to have two iron atoms (and hence two cytochrome molecules) reduced in order to oxidize a single molecule of $NADH_2$ or $FADH_2$.

Of all the electron-transferring intermediates that participate in respiratory metabolism, only the cytochrome a/a_3 complex (also called *cytochrome oxidase*) is capable of direct transfer of electrons to molecular

---

*Obviously the cytochromes were recognized and named before their relative positions in the transport chain were assigned. The nomenclature is, in fact, based on the absorption spectra of the individual cytochromes.

oxygen. Almost every electron extracted from any oxidizable organic molecule anywhere in the cell must eventually pass through cytochrome a/a_3, for this is the only link between respiratory metabolism and the oxygen that makes it all possible. It is no wonder, then, that cyanide ion (CN^-) and carbon monoxide (CO), which compete with oxygen for the binding site on cytochrome a/a_3, are such deadly poisons for all aerobic forms of life.

When the electron transport scheme is represented as in Figure 5-4, the order of the intermediates in the transport process can be indicated but not the actual energy changes between successive intermediates. For that reason it is especially useful to plot the intermediates on an energy scale as in Figure 5-7. In this way the decline in free energy as electron pairs flow down the respiratory chain can be readily visualized and actual changes in free energy between successive elements of the chain can be calculated.

Standard Reduction Potentials

As can be seen from Figure 5-7, each component in the chain has a characteristic *standard reduction potential* E_0 assigned to it. This is a measure of the tendency of a reducing agent to lose electrons. The more negative the E_0 value for a given component, the greater the tendency of the reduced form of the component to lose electrons. Conversely the more positive the E_0 value, the greater the tendency of the oxidized form of the component to gain electrons. (The point of reference for the standard reduction potential scale, by the way, is the

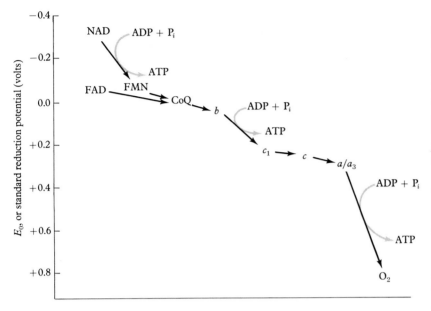

Figure 5-7 The energetics of electron transport. Each of the intermediates in the electron transport chain of Figure 5-4 is depicted here at a position appropriate to its energy level, as measured by its standard reduction potential. The large difference in reduction potential between NAD/NADH$_2$ (−0.32 volt) and O$_2$/H$_2$O (+0.82 volt) is divided into a series of stepwise electron transfers, three of which are sufficiently energetic to drive the phosphorylation reaction that generates ATP.

H$_2$/2H$^+$ couple, to which is assigned the value $E_0 = 0.0$ volt.) Thus the NAD/NADH$_2$ couple, with an E_0 value of -0.32 volt, has a strong tendency to lose electrons from the reduced form, whereas the O$_2$/H$_2$O pair, with an E_0 of $+0.82$ volt, has a strong tendency to gain electrons and become reduced. Coenzyme NADH$_2$ is therefore an excellent electron donor and O$_2$ is a willing electron acceptor, and it is this situation that provides the driving force for electron transport.

Clearly there ought to be some relationship between the standard reduction potential E_0 and the standard change in free energy of oxidation, since both are measures of the tendency of an oxidative event to occur. That relationship is given by the equation

$$\Delta G^0 = -nF\Delta E_0 \tag{5-9}$$

where ΔG^0 = standard free-energy change for given electron transfer

n = number of electrons transferred

F = faraday (23,040 cal/volt)

ΔE_0 = difference in standard reduction potential of two carriers involved in transfer

In any series of reversibly oxidizable species such as shown in Figure 5-7, electrons will tend to flow spontaneously from the reduced form of the more electronegative pairs (i.e., those with the most negative E_0 values) to the oxidized form of the more electropositive compounds (those having the more positive E_0 values). Thus by arranging the electron carriers of the respiratory chain in a thermo-dynamic sequence, a stepwise flow of electrons to successively more positive acceptors is possible, until the electrons finally reach oxygen, the most electropositive acceptor of all, which is thereby reduced to water.

For each step in the sequence, then, the free energy released in the transfer of electrons from donor to acceptor is directly pro-portional to the difference in the reduction potential values for the two intermediates. If one looks at Figure 5-7 for particularly large changes in free energy (i.e., for large differences in standard reduction potentials between successive carriers) that might be used to generate ATP, three transfers are found for which the energy change is especially large:

Transfer Step	ΔE_0 (volts)	ΔG^0 (kcal/mole)
NAD to FMN	0.27	-12.4
Cytochrome b to cytochrome c_1	0.20	-9.2
Cytochrome a/a_3 to O_2	0.54	-24.8

Oxidative Phosphorylation

The existence of three transfers characterized by large energy changes is in good agreement with an observation made more than twenty years ago: when a pair of electrons travels along the electron transport chain from $NADH_2$ to O_2, three molecules of ATP are formed by phosphorylation of ADP. Since the generation of ATP requires at least 7.3 kcal of free energy per mole under standard conditions (con-siderably more under actual cellular conditions), these are the only three transfers that yield enough energy to couple to ATP synthesis. As Figure 5-7 indicates, it is in fact at each of these three sites that the transfer of electrons down the chain is coupled to the synthesis of one ATP. Since these ATPs are formed by phosphorylation reactions coupled to oxidative electron transport, the process is called *oxidative phosphorylation*.

We can now rewrite Equation 5-7 for the oxidation of $NADH_2$ via the electron transport chain to include the three phosphorylations—bearing in mind, of course, that the actual transfer of electrons occurs stepwise, with ATP synthesis coupled to the three most energy-rich transfers:

$$NADH_2 + \tfrac{1}{2}O_2 + 3ADP + 3P_i \rightarrow NAD + H_2O + 3ATP \qquad (5\text{-}10)$$

And since there are 10 such $NADH_2$ molecules generated upon the complete oxidation of glucose during glycolysis and the TCA cycle, together they account for the synthesis of 30 more ATPs per glucose.

The two molecules of $FADH_2$, on the other hand, are considerably lower in energy yield than $NADH_2$, the standard reduction potential for the $FAD/FADH_2$ couple being only -0.05 volt. As can be seen from the position of FAD in the electron transport scheme of Figure 5-7, its electrons enter the chain only after the first ATP-generating site, such that each $FADH_2$ upon oxidation by the electron transport chain generates only two ATPs instead of three. Equation 5-8 can therefore be rewritten as follows:

$$FADH_2 + \tfrac{1}{2}O_2 + 2ADP + 2P_i \rightarrow FAD + H_2O + 2ATP \qquad (5\text{-}11)$$

The two molecules of $FADH_2$ generated per glucose during the TCA cycle therefore contribute a total of four ATPs to the overall yield.

To summarize electron transport and oxidative phosphorylation, then, we note that, per glucose oxidized, there are 12 coenzymes to be oxidized: 2 $FADH_2$'s (one from each turn of the TCA cycle) and 10 $NADH_2$'s (two from glycolysis, two from the gateway step, and six from the TCA cycle). The total ATP yield from oxidative phosphorylation is therefore 34 ATPs per glucose, obtained by multiplying Equation 5-10 by 10 and Equation 5-11 by 2:

10 × Equation 5-10: $10NADH_2 + 5O_2 + 30ADP + 30P_i \rightarrow$
$10NAD + 10H_2O + 30ATP$

2 × Equation 5-11: $2FADH_2 + O_2 + 4ADP + 4P_i \rightarrow$
$2FAD + 2H_2O + 4ATP$

Overall: $10NADH_2 + 2FADH_2 + 6O_2 + 34ADP + 34P_i \rightarrow$
$10NAD + 2FAD + 12H_2O + 34ATP \qquad (5\text{-}12)$

Like the TCA cycle, all the reactions involved in electron transport and oxidative phosphorylation are located in the mitochondria of eukaryotic cells. But while the TCA-cycle enzymes are thought to be present in the soluble matrix of the mitochondrion, the proteins, enzymes, and electron carriers of the transport system are all embedded in the inner mitochondrial membrane, which is infolded into *cristae* to increase the surface area for the respiratory apparatus. The details of mitochondrial membrane structure are illustrated in Figure 5-8. In prokaryotes the electron transport chain is also membrane-bound, but the association is with the cellular membrane.

The electron-transferring proteins and carriers are believed to be

Structural Organization of the Electron Transport Chain

organized within the membrane into discrete clusters or complexes called *respiratory assemblies*. In at least some mitochondrial preparations, the inner membrane has been shown to be lined with knoblike structures that protrude into the surrounding matrix, as shown in Figure 5-8b and c. Each such protruding knob (or "lollipop," as the

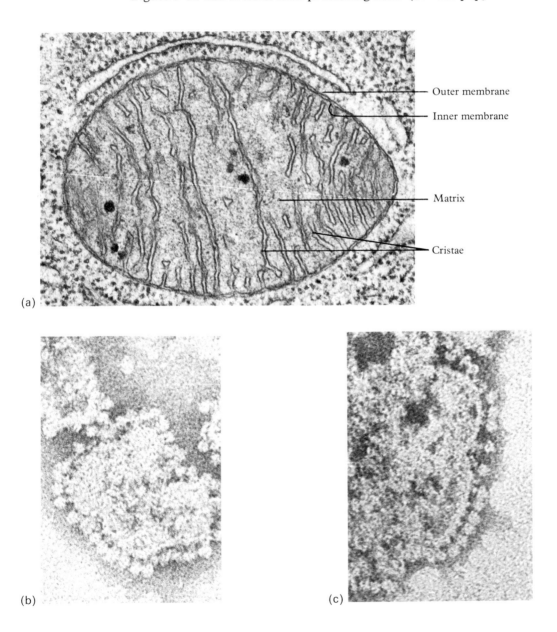

(a)

(b)   (c)

structure is sometimes fancifully called) is a unit of F_1 (for *factor 1*), an enzyme known to have ATPase (ATP-hydrolyzing) activity in the test tube but thought to be involved in the reverse reaction (ATP generation) in the cell. Mitochondrial membrane fragments that lack the F_1 protein cannot couple electron transport to ATP synthesis and do not have the characteristic protruding knobs.

The Coupling of ATP Synthesis to Electron Transport

Under normal physiological conditions the coupling of electron transport to phosphorylation is tight and obligatory. Not only is it impossible to get ATP generation without electron transfer, but the tightness of coupling also means that no electrons will flow unless ATP can be synthesized. This is an important cellular regulatory mechanism, since the ATP/ADP ratio plays an important role in regulating the activity of the electron transport scheme (and, via accumulation or depletion of reduced coenzymes, that of the entire respiratory apparatus). When the ATP/ADP ratio is low, phosphorylation is favored both thermodynamically and kinetically, so that ATP synthesis proceeds, electrons are transported, coenzymes are recycled, and substrates are oxidized. When the ATP/ADP ratio is high, there is a shortage of ADP (and sometimes inorganic phosphate as well) for phosphorylation, electron transport is slowed or halted, coenzymes build up in the reduced form, oxidized coenzymes are not available as electron acceptors in the TCA cycle and glycolytic pathway, and the whole process is slowed or stopped. In addition to this kind of kinetic control of energy metabolism, the ATP/ADP ratio also plays a critical role in allosteric control of key regulatory enzymes since, as discussed below, both ATP and ADP serve as allosteric effectors in the regulation of both glycolysis and the TCA cycle.

Although the obligatory nature of the coupling of ATP synthesis to electron transport is well established, the actual mechanism whereby the transfer of electrons from one intermediate in the chain to the next

Figure 5-8   Structure and organization of the mitochondrion. Shown in (a) is an intact mitochondrion from the pancreas of a bat (68,000X). Both the outer membrane and the inner membrane can be distinguished. The cristae represent infoldings of the inner membrane into the internal volume of the organelle, which is otherwise occupied by the semifluid matrix. By careful isolation of submitochondrial particles, the F_1 projections on the inner membranes are clearly visible with the electron microscope, as seen at high magnification (320,000X) in (b) and (c). (a) Courtesy of K. R. Porter; (b) and (c) courtesy of E. Racker.

can drive the phosphorylation of ADP remains obscure, despite intensive efforts during the past 25 years. Three alternative hypotheses have been proposed to explain this conservation of energy during electron transport. As is so often the case in cellular physiology, they have formidable names, but the concepts are fairly straightforward. In the chronological order in which they were proposed, these are (1) chemical coupling, (2) chemiosmotic coupling, and (3) conformational coupling.

According to the *chemical coupling hypothesis*, oxidative phosphorylation occurs in a manner similar to the phosphorylation reactions that occur during glycolysis (Reactions Gly-6 and Gly-7) and in the TCA cycle (Reactions TCA-4 and TCA-5), involving the formation of a high-energy phosphorylated intermediate, followed by phosphate transfer to ADP. A model sequence for such a process might look as follows, using as an example the transfer of electrons from $NADH_2$ to FMN at the first step in the transport chain (similar reaction sequences would presumably occur at the other two steps in the chain that couple to a phosphorylation event):

$$NADH_2 + FMN + X + P_i \rightarrow NAD + FMNH_2 + X \sim P \qquad (5\text{-}13)$$

$$\frac{X \sim P + ADP \rightarrow X + ATP \qquad\qquad\qquad\qquad (5\text{-}14)}{}$$

$$NADH_2 + FMN + ADP + P_i \rightarrow NAD + FMNH_2 + ATP \qquad (5\text{-}15)$$

The compound labeled X in this scheme represents a hypothetical high-energy coupling intermediate, and the structure $X \sim P$ is intended to denote a high-energy phosphate bond. Numerous variations on this theme have been advanced, but they all involve one or more such hypothetical high-energy intermediates, and they all therefore share in the uncertainty generated by the inability to isolate such intermediates despite years of intense effort.

An attractive alternative to the chemical coupling model is the *chemiosmotic hypothesis* advanced in the early 1960s by P. Mitchell, a British biochemist. This model, which has gained substantial support in recent years, proposes that electron transport is accompanied by a net translocation of H^+ ions (protons) from the inside to the outside of the inner mitochondrial membrane. The resulting proton gradient then provides the driving force for ATP synthesis. To explain how a gradient of protons is established and maintained across the membrane, the theory assumes that the proteins involved in the hydrogen-transferring steps of electron transport are positioned in the membrane in such a way that H^+ ions are always obtained from the inside of the mitochondrion and released to the outside.

Three such steps are shown in Figure 5-9. Consider, for example, the step in the electron transport chain at which reduced coenzyme Q ($CoQH_2$) is oxidized and the ion of cytochrome b is reduced. The step involves a net release of H^+ (Reaction 5-16) that is complemented in the last step of electron transport by the uptake of H^+ in the reduction of oxygen (Reaction 5-17):

$$CoQH_2 + 2Fe^{+++}_{(cytochrome\ b)} \rightarrow CoQ + 2Fe^{++}_{(cytochrome\ b)} + 2H^+ \qquad (5\text{-}16)$$

$$2Fe^{++}_{(cytochrome\ a/a_3)} + \tfrac{1}{2}O_2 + 2H^+ \rightarrow 2Fe^{+++}_{(cytochrome\ a/a_3)} + H_2O \qquad (5\text{-}17)$$

A net movement of H^+ ions across the membrane can be achieved by positioning the enzymes responsible for these reactions such that the H^+ ions liberated by Reaction 5-16 are released to the outside while those required by Reaction 5-17 are drawn from the inside. To understand the other proton-pumping mechanism shown in Figure 5-9, recall that the reduced form of NAD, though usually represented by

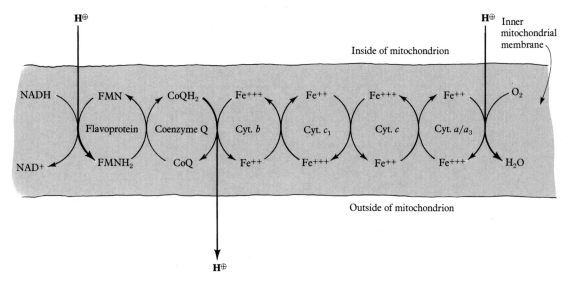

Figure 5-9 Establishment of a hydrogen-ion concentration gradient across the inner membrane of the mitochondrion, as postulated by the chemiosmotic theory for coupling of oxidative phosphorylation to electron transport. According to the model, the gradient results from the specific orientation of electron transport proteins within the membrane such that protons are drawn from the inside of the mitochondrion but released to the outside. The energy represented by the proton (or pH) gradient is then used to drive ATP synthesis. The outer mitochondrial membrane does not pose a permeability barrier and is therefore not shown in the figure.

the shorthand form $NADH_2$, ought really to be written as $NADH +$ H^+ (see Figure 4-5). This means that the reoxidation of reduced NAD by FMN in the first step of the electron transport chain involves a net uptake of H^+:

$$NADH + H^+ + FMN_{\text{(flavoprotein)}} \rightarrow NAD^+ + FMNH_{2\,\text{(flavoprotein)}} \quad (5\text{-}18)$$

Again it is a postulation of the chemiosmotic theory that the flavo-protein responsible for this step derives its protons specifically from the inside of the membrane.

By coupling the transfer of electrons to a directional pumping of protons in this way, some of the free energy released in the transport process is stored as a gradient of pH and charge that can be used in turn to drive ATP synthesis. To explain this connection between a proton gradient and ATP synthesis, the chemiosmotic theory suggests that the F_1 enzyme, in catalyzing Reaction 5-19, removes the elements of water as H^+ and OH^- ions, the former pulled from the active site by the high internal pH (low internal H^+ concentration) and the latter drawn to the outside by the lower pH prevailing there:

$$ADP + P_i \rightarrow ATP + H^+ + OH^- \qquad (5\text{-}19)$$

↓   ↓
to   to
inside outside

The chemiosmotic theory therefore dispenses with the need for a high-energy chemical intermediate and depends upon a proton gradient as the link between electron transport and ATP formation. This means that ATP synthesis may not be coupled directly to specific electron-transferring steps with large free-energy changes, as shown in Figure 5-7, but may depend instead upon those steps characterized by a net uptake or release of protons. Still lacking, however, is a definitive experimental criterion to distinguish between chemical coupling and the chemiosmotic model.

The *conformational coupling hypothesis* is the most recent contribution to what must at present be regarded as a problem in theoretical biology. This model envisions macromolecular components within the membrane that, rather like allosteric enzymes, have two different conformational configurations, one representing a high-energy form. Electron transport is considered to result in a conversion of the low-energy configuration to the high-energy form, while the reverse process drives the phosphorylation of ADP by an enzyme included in the conformational unit. Conformational changes consistent with such a model can be observed in mitochondria under differing states of electron transport chain activity.

Thus three basically different schemes (and several variations of each) are presently available to explain the coupling of oxidative phosphorylation. Each has its passionate devotees and its skeptics. Each appears to be the best explanation for some data while remaining in apparent conflict with others. The definitive data necessary to decide which (if any) of these models is correct presumably lie in experiments as yet undone.

To summarize the whole process of respiratory metabolism, it is necessary only to add the summary equation for glycolysis + TCA cycle (Equation 5-6) to that for electron transport and oxidative phosphorylation (Equation 5-12): **Summary of Respiratory Metabolism**

$$\text{Glucose} + 6H_2O + 10NAD + 2FAD + 4ADP + 4P_i \rightarrow$$
$$6CO_2 + 10NADH_2 + 2FADH_2 + 4ATP$$

$$10NADH_2 + 2FADH_2 + 6O_2 + 34ADP + 34P_i \rightarrow$$
$$10NAD + 2FAD + 12H_2O + 34ATP$$

$$\text{Glucose} + 6O_2 + 38ADP + 38P_i \rightarrow 6CO_2 + 6H_2O + 38ATP \quad (5\text{-}20)$$

Thus we return in the end to a summary form of the equation that should be familiar to you. We have, in a sense, spent Chapters 4 and 5 simply quantitating the yield of energy in what we knew at the outset (see Equation 4-7) to be an exergonic reaction. Now, however, you should be in a much better position to appreciate the intricacies of chemistry, thermodynamics, and cellular structure this equation seeks to summarize.

To determine the efficiency of respiratory metabolism, recall that the complete oxidation of glucose to CO_2 and H_2O occurs with a standard free-energy change of 686 kcal/mole. Since approximately the same free-energy change characterizes this overall reaction under actual cellular conditions (that is, $\Delta G \simeq \Delta G^0$), we can use this value as the basis for our calculations. Assuming a ΔG value of -12 kcal/mole for the hydrolysis of ATP under cellular conditions, the 38 ATPs produced per glucose correspond to about 456 kcal of energy conserved per mole of glucose oxidized. The efficiency of the process is therefore about 65 percent, far in excess of that obtainable with the most efficient of artificial machines. **Efficiency of Respiratory Metabolism**

We conclude our discussion of respiratory metabolism with a brief view of the ways in which the cell regulates these reaction sequences to keep its ATP supplies finely tuned to its actual needs. That a **Regulation of Respiratory Metabolism** process such as aerobic energy metabolism is subject to regulation should come as no surprise after the initial discussion of metabolic control in Chapter 3. After all, a cell possesses the metabolic machinery for glucose catabolism not simply to oxidize indiscriminately large quantities of glucose (or other substrate) to carbon dioxide and water, but rather to generate ATP in just the right amounts and at just the right rates to meet cellular energy needs on a moment-by-moment basis.

There are, of course, many ways by which cells can regulate their activities. *Genetic regulation*, for example, entails the control of the synthesis of specific proteins; regulation of gene readout therefore determines both the kinds and, in conjunction with protein turnover, the amounts of enzymes in the cell, as well as the times, conditions, or stages under which specific enzymes appear, disappear, or change in abundance. Induction or repression of protein synthesis in response to environmental signals or developmental programs is a regulatory mechanism of great significance to the cell biologist. It is, however, not so important a consideration for the regulation of energy metabolism as for other metabolic processes—respiratory metabolism is such a key process to virtually all cells that the enzymes of the glycolytic pathway, TCA cycle, and oxidative phosphorylation are usually *constitutive*, meaning that they are generally present in most cells under almost all conditions.

Regulation by *compartmentation* is another control mechanism. This involves both the localization of specific enzymes and metabolic capabilities within different organelles or other subcellular structures and also the detailed spatial arrangement within such structures. Respiratory metabolism illustrates both these points in eukaryotic cells. The entire glycolytic sequence occurs in the soluble portion of the cytoplasm (the *cytosol*), while the enzymes responsible for further oxidation of pyruvate are localized in the mitochondrion, as are the electron carriers and enzymes of electron transport and oxidative phosphorylation. Within the mitochondrion there is further localization of TCA-cycle enzymes in the soluble matrix and of the electron carriers and ATP-generating system in the respiratory assemblies on the inner membrane. Compartmentation not only allows locally high concentrations of specific enzymes and metabolic intermediates but also enables the cell to regulate specific processes by controlling the movement of key intermediates (such as pyruvate, ADP, and ATP) into and out of compartments.

The major control mechanism of interest to us in the present context, however, is the *allosteric regulation* of key enzymes, a topic already introduced in Chapter 3 and raised again here because of the illustrative examples from the glycolytic and TCA pathways, as shown in Figure 5-10. Allosteric regulation, you will recall, is the means whereby the activity of an enzyme can be inhibited (or in some cases enhanced) by a specific substance that is usually neither a substrate nor a product of that enzyme and often, in fact, represents the end product of the pathway initiated by the regulatory enzyme.

To appreciate fully the means whereby a cell regulates its respiratory metabolism, it is important to keep in mind that we are dealing here with *energy* metabolism and that the metabolic pathways we have been considering exist first and foremost to generate ATP. To the cell, then, the most important end product of respiratory metabolism is not lactate, not CO_2, and not reduced coenzymes. It is ATP. With that perspective it should come as no surprise that the most important factor in the control of cellular energy metabolism is the level of ATP. We have already seen how the coupling of ATP synthesis to electron transport renders the activity of the electron transport chain (and indirectly, therefore, that of the entire respiratory process) sensitive to the ATP/ADP ratio of the cell. In addition to this kind of mass action control of electron transport by ADP availability, however, the ATP/ADP balance in the cell also plays a critical role in the allosteric regulation of both glycolysis and the TCA cycle.

Control of the TCA cycle by the ATP/ADP ratio is effected at the step involving the oxidative conversion of isocitrate to α-ketoglutarate and CO_2 (Reaction TCA-3). Since this is a cyclic pathway, it is not possible to identify an "initial step" corresponding to the first reaction in a linear sequence. Isocitrate dehydrogenase, the enzyme that catalyzes this reaction, is doubly sensitive to the ATP/ADP ratio: it is both inhibited by ATP and activated by ADP. This, of course, requires that the enzyme have at least two separate effector-binding sites. (In fact the situation with isocitrate dehydrogenase is even more complex, because it is also allosterically sensitive to both inhibition by $NADH_2$ and stimulation by NAD.)

When the cellular need for ATP suddenly decreases and ATP begins accumulating (with a corresponding decrease in ADP), the enzyme is inhibited by the elevated ATP concentration and TCA-cycle activity decreases to keep the rate of ATP generation in tune with ATP consumption. Similarly, should the cell suddenly begin using larger quantities of ATP, the decrease in cellular ATP levels relieves the inhibition of isocitric dehydrogenase, and the concomitant rise in ADP concentration has a stimulatory effect on the activity of this enzyme and consequently on the whole TCA cycle. Isocitrate dehydrogenase can therefore be regarded as the "pacemaker" enzyme of the TCA cycle, rendering the whole process exquisitely sensitive to the

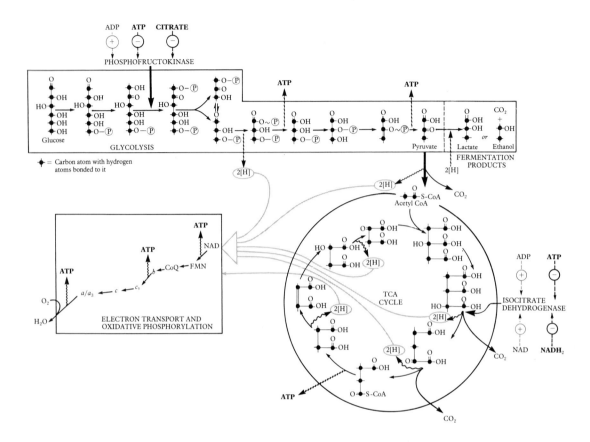

Figure 5-10 Allosteric regulation of respiratory metabolism. Regulation of glycolysis is achieved primarily by feedback inhibition of the enzyme phosphofructokinase by both ATP and citrate. Phosphofructokinase is also subject to allosteric stimulation by ADP, rendering the enzyme doubly sensitive to the ATP/ADP status of the cell. Regulation of the TCA cycle focuses on the enzyme isocitrate dehydrogenase, which is sensitive to feedback inhibition by both ATP and $NADH_2$ and to stimulation by ADP and NAD.

energy status of the cell, as represented by the ATP/ADP (and $NADH_2$/NAD) ratio. These effects are illustrated in Figure 5-10.

The corresponding allosterically regulated pacemaker enzyme in the glycolytic sequence is phosphofructokinase, the enzyme responsible for the phosphorylation of fructose-6-phosphate to generate fructose-1,6-diphosphate (Reaction Gly-3). The choice of this enzyme for allosteric regulation seems to violate the general principle that feedback inhibition by the product of a pathway usually affects the first enzyme in the sequence, since the glycolytic pathway is generally considered to begin with the phosphorylation of glucose (Reaction Gly-1). A possible rationale is that glucose, glucose-6-phosphate, and fructose-6-phosphate are common intermediates involved in a variety of metabolic pathways such that the cell could ill afford to have their levels dictated solely by the ATP status of the cell. The further phosphorylation of fructose-6-phosphate represents the real commitment to glycolysis, since the fructose-1,6-diphosphate that results has only one metabolic fate, that of cleavage and oxidation by the glycolytic sequence. From a regulatory point of view, then, we might do well to regard fructose-6-phosphate as a generally available cellular intermediate from which the actual sugar-splitting pathway departs. In this sense, allosteric regulation of phosphofructokinase seems eminently reasonable.

In any case, the reaction catalyzed by phosphofructokinase is the rate-limiting step in glycolysis, and this enzyme, like isocitrate dehydrogenase, is inhibited by ATP and stimulated by ADP (or possibly AMP). Thus glycolysis shares with the TCA cycle a great sensitivity to the ATP/ADP level of the cell. Note, however, that in the case of phosphofructokinase we are dealing with an allosteric enzyme for which a single substance, ATP, plays a role in both the catalytic activity and the allosteric inhibition of the enzyme. The contradictory effects of ATP on enzyme activity this appears to imply are resolved by differences in the ATP affinities of the active and allosteric sites, as explored further in Problem 5-7.

In addition to its sensitivity to ATP and ADP, phosphofructokinase is also subject to allosteric inhibition by citrate. This sensitivity serves as a crucial regulatory link between glycolysis and the TCA cycle to ensure that the rates of these two processes are carefully synchronized with each other. Thus when the level of citrate rises in the cell due to overproduction of pyruvate, acetyl CoA, and citrate, the activity of the glycolytic pathway is effectively throttled as citrate binds to its effector site on phosphofructokinase, converting the enzyme reversibly to its inactive form.

These allosteric effects of ATP, ADP, and other intermediates on the enzymes phosphofructokinase and isocitrate dehydrogenase are summarized in Figure 5-10. Together they represent the principal (though not the only) means by which the component parts of respiratory metabolism are integrated and continuously adjusted to meet cellular energy needs with great fidelity and efficiency. In its control, as in its chemistry, respiratory metabolism can only be regarded as a marvel of design and engineering. No transistors, no mechanical parts, no noise, no pollution—and all done in units of organization that require an electron microscope to visualize. Yet the process goes on routinely and continuously in almost every living cell with a degree of integration, efficiency, fidelity, and control that we can scarcely understand well enough to appreciate, let alone aspire to reproduce in our test tubes.

Practice Problems   5-1. Biological Oxidation: The process of biological oxidation occurs in three separate phases. Name each phase, indicate where in a eukaryotic cell that phase occurs, state briefly the overall purpose of each phase, indicate the major molecular inputs and outputs, and write an overall equation for each phase.

5-2. Glucose Catabolism: Suppose you feed glucose-6-^{14}C (glucose labeled radioactively on carbon atom number 6) to an aerobic culture of bacteria and extract and separate glycolytic and TCA-cycle intermediates shortly thereafter. Which carbon atom(s) would you expect to find labeled first with ^{14}C in the following molecules? (Give number of carbon atom(s) or place an asterisk next to radioactive atom(s) in the structural formula.)
(a) Fructose-1,6-diphosphate   (d) Acetyl coenzyme A
(b) Glycerate-2-phosphate   (e) Isocitrate
(c) Lactate   (f) Succinate

5-3. Energy Storage: Energy can be transferred and stored in cells in a variety of forms, of which the most immediately available is ATP. In order of decreasing immediacy of availability, several important energy transfer and storage forms commonly used by cells are ATP, NADH$_2$, acetyl CoA, pyruvate, and glucose.
(a) Calculate the number of ATP equivalents represented by each of these molecules, assuming complete oxidation under aerobic conditions (example: 1 glucose = 38 ATPs).
(b) Assuming the ΔG^0 for the hydrolysis of ATP to be -7.3 kcal/mole under physiological conditions, calculate the weight in grams of each compound that would be required to obtain 1 kcal of usable (free) energy. (Molecular weights for ATP, NADH$_2$, and acetyl CoA are 507, 663, and 809, respectively.) Can you explain why energy storage in cells usually involves carbohydrates?
(c) Following are catalog prices quoted recently by a chemical company for these intermediates:

Compound	Current Price
ATP	$5.90/g
NADH$_2$	$25/g
Acetyl CoA	$191/100 mg
Pyruvate	$1.95/5 g
Glucose	$2/kg

Calculate for each the cost of purchasing 1 kcal of free energy (energy of hydrolysis of ATP equivalents) in each form.

5-4. Coenzymes: All the oxidative steps in glycolysis and the TCA cycle use NAD as the electron acceptor except for the oxidation of succinate to fumarate, which uses FAD instead:

$$\text{Succinate} + \text{FAD} \rightleftharpoons \text{fumarate} + \text{FADH}_2 \qquad \Delta G^0 \simeq 0 \text{ kcal/mole}$$

(a) What is there about the *chemistry* of this reaction that makes it different from the other oxidative events of glycolysis and the TCA cycle?

(b) What is there about the *thermodynamics* of this reaction that makes it different from other oxidative events of glycolysis and the TCA cycle?

5-5. Electron Transport: Rotenone is an extremely potent insecticide and fish poison. At the molecular level, its mode of action is to block electron transfer from FMN to coenzyme Q.

(a) Why do fish and insects die after digesting rotenone?

(b) Would you expect the use of rotenone as an insecticide to be a potential hazard to other forms of animal life (people, for example)? Explain.

(c) Would you expect the use of rotenone as a fish poison to be a potential hazard to aquatic plants that might be exposed to the compound? Explain.

5-6. Glutamate Metabolism: Glutamate is a five-carbon amino acid that can be oxidatively deaminated as shown below to yield α-ketoglutarate, an intermediate in the TCA cycle. Suppose that an aerobic bacterial culture uses glutamate as its sole energy source and that the cells excrete 1 mole of ammonia (NH_3) and 1 mole of oxaloacetate back into the medium for every mole of glutamate used.

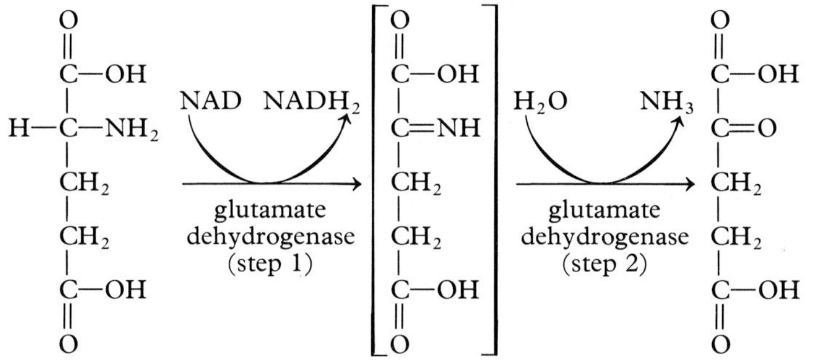

Glutamate   Enzyme-bound α-Ketoglutarate
intermediate

(5-21)

(a) Calculate the number of ATP molecules synthesized per molecule of glutamate metabolized.

(b) Write an overall balanced equation for the energy metabolism these bacteria are carrying out.

‡5-7. Allosteric Regulation: Phosphofructokinase is an especially interesting allosteric enzyme because a single substance, ATP, serves as both substrate and allosteric inhibitor for the enzyme. This fact creates an apparent contradiction of effects, since increasing levels of substrate should increase the rate of an

enzyme-catalyzed reaction while increasing levels of an allosteric inhibitor should render the enzyme inactive.

(a) In what sense does phosphofructokinase sensitivity to ATP violate a general principle of allosteric regulation?

(b) Why is it logical that a reaction which consumes ATP should be inhibited by ATP if that reaction is part of a catabolic pathway like glycolysis?

(c) Can you explain how an enzyme can function adequately when the same compound serves as both substrate and allosteric inhibitor?

(d) How can an allosteric inhibitor like ATP regulate the rate of a reaction when a given enzyme molecule has only two alternative configurations, such that it is either in the completely inactive form (ATP bound to effector site) or the fully active form (ATP not bound)?

‡5-8. Uncoupling of Oxidative Phosphorylation: Oxidative phosphorylation is normally tightly coupled to electron transport in the sense that no electrons will flow if ATP synthesis cannot occur concomitantly (due, for example, to low levels of either ADP or P_i). However, chemical agents are known that uncouple ATP synthesis from electron transport, allowing electron flow to occur in the absence of ATP generation. One such *uncoupling agent* is 2,4-dinitrophenol, which is highly toxic to humans, causing a marked increase in metabolism and temperature, profuse sweating, collapse, and death. For a brief period in the 1950s, however, sublethal doses of dinitrophenol were actually prescribed as a means of weight reduction in humans.

(a) Why would an uncoupling agent like 2,4-dinitrophenol be expected to cause an increase in metabolism (as evidenced by consumption of oxygen and catabolism of foodstuffs)?

(b) Based on what you know about allosteric regulation of the glycolytic pathway and the TCA cycle, why would an uncoupling agent like dinitrophenol be likely to have greater, more far-reaching effects upon respiratory metabolism than might be predicted by the simple lack of control of the rate of electron transport?

(c) Why would consumption of dinitrophenol lead to an increase in temperature and profuse sweating?

(d) Assuming the chemical coupling hypothesis, postulate a mode of action for dinitrophenol.

(e) Dinitrophenol has been shown to carry protons across biological membranes. How could this observation be used to explain its uncoupling action?

(f) Why would dinitrophenol have been considered as a drug for weight reduction? Can you guess why it was abandoned as a reducing aid?

$$
\begin{array}{l}
\text{O} \\
\parallel \\
\text{C—COOH} \\
\mid \\
\text{C—COOH} \\
\parallel \\
\text{CH—COOH}
\end{array}
$$

Aconitate

‡5-9. Thermodynamics of Aconitase Equilibrium: The conversion of citrate to isocitrate in the TCA cycle actually occurs by a dehydration/rehydration reaction with *aconitate* as an isolatable intermediate. A single enzyme, aconitase, catalyzes the conversion of citrate to aconitate and aconitate to isocitrate. An equilibrium mixture of citrate, aconitate, and isocitrate contains about 90, 4, and 6 percent of the three acids, respectively.

(a) Why is the conversion of citrate to isocitrate necessary in the TCA cycle?

(b) What is the K^0 and the ΔG^0 (at $25°C$) for each of the two steps (citrate $\rightarrow$ aconitate; aconitate $\rightarrow$ isocitrate)? For the overall process?

(c) Would the TCA cycle proceed under standard conditions? Why or why not?

(d) How can the TCA cycle proceed under cellular conditions?

References

Goodwin, T. W. (ed.), *The Metabolic Roles of Citrate* (New York: Academic Press, 1968).

Krebs, H. A., "The History of the Tricarboxylic Acid Cycle," *Perspectives in Biology and Medicine* (1970), *14*: 154.

Lowenstein, J. M. (ed.), *Citric Acid Cycle: Control and Compartmentation* (New York: Marcell Dekker, 1969).

Mitchell, P., "Chemiosmotic Coupling in Energy Transduction: A Logical Development of Biochemical Knowledge," *J. Bioenergetics* (1972), *3*: 5.

Munn, E. A., *The Structure of Mitochondria* (New York: Academic Press, 1974).

Packer, L., and A. Gómez-Puyou, *Mitochondria: Bioenergetics, Biogenesis and Membrane Structure* (New York: Academic Press, 1976).

Racker, E., *A New Look at Mechanisms in Bioenergetics* (New York: Academic Press, 1976).

Slater, E. C., "The Coupling between Energy-Yielding and Energy-Utilizing Reactions in Mitochondria," *Quarterly Rev. Biophys.* (1971), *4*: 35.

Slater, E. C., "The Mechanism of Energy Conservation in the Mitochondrial Respiratory Chain," *Harvey Lectures* (1972), *66*: 19.

Tandler, B., and C. L. Hoppel, *Mitochondria* (New York: Academic Press, 1972).

Fats and Proteins as Alternative Energy Sources

6

We have seen in the preceding two chapters how the monosaccharide glucose is oxidized to carbon dioxide and water, with the trapping of energy as ATP. We considered glucose as the starting material in our discussion of respiratory metabolism because it is the single most important energy source for most cells. Cells do not depend exclusively upon glucose to meet their energy needs, however. Indeed there are many organisms, especially among the microorganisms, that routinely utilize other foodstuffs. Even human cells, which are normally nourished by blood glucose, can oxidize other carbohydrates as well and can also metabolize fats and proteins. The general rule in all cases, however, seems to be the same: to convert all such alternative substrates as quickly and as efficiently as possible into intermediates in the mainstream glycolytic and TCA pathways so that a minimum number of additional enzymes are needed to effect complete metabolism. Indeed, as pointed out in Chapter 5, it is a unifying feature of cellular energy metabolism that the glycolytic pathway and the TCA cycle serve as a common framework around which nearly all other catabolic pathways are constructed.

We shall be concerned in this chapter with the two major classes of alternative energy sources—fats and proteins. In general, fats serve storage and insulating roles in the body, and proteins play functional and structural roles unrelated to energy metabolism. Under appropriate conditions, however, both fats and proteins may be oxidized to obtain energy and indeed serve as routine energy sources for some cells. We shall not be nearly so concerned with the details of chemistry and regulation as we were for glucose catabolism but shall simply sketch in broad outline the sequence of reactions necessary to bring these substrates into the mainstream of energy metabolism.

The Oxidative Metabolism of Fats

Carbohydrates are the most abundant and, in general, the most important substrates for cellular energy metabolism, but fats are also commonly used as fuel. Both plants and animals depend upon fat as the major means of storing energy, not only because of its high caloric value but also because fat can be stored in an unhydrated, concentrated form. Fat is a common foodstuff stored in the seeds of plants and, as many of us are all too aware, is also the storage form of choice in higher animals, including ourselves. The oxidation of stored fat can account for as much as 50 percent of the ATP production in vertebrate tissues such as the liver, kidneys, and heart muscle, and it is virtually the only source of energy in organisms as diverse as hibernating bears, migrating birds, and germinating sunflower seedlings.

We shall be concerned here only with the *neutral fats*, which are esters of *glycerol* and *fatty acids*. Since glycerol has three hydroxyl groups, it can be triply esterified with fatty acids to yield a *triglyceride*, as shown in Figure 6-1. Fatty acids, in turn, are long-chain hydrocarbons with a carboxylic acid group on one end. The general formula is shown in Figure 6-1; n is usually an even number, and the total number of carbon atoms is usually somewhere between 12 and 20, with the 16- and 18-carbon fatty acids being especially prominent in higher plants and animals. Further structural variety is provided by the introduction of one or more double bonds to create the unsaturated or polyunsaturated fatty acids so characteristic of vegetable oils (an *oil* is simply a neutral fat that is liquid at room temperature).

The catabolic metabolism of fats begins with the cleavage of triglycerides into the glycerol and fatty-acid constituents, using enzymes called *lipases* (Reaction 6-1). The glycerol is then converted to an intermediate of the glycolytic pathway (dihydroxyacetone phosphate; see Figure 4-4) by a phosphorylation/dehydrogenation sequence, with glycerol-3-phosphate as the intermediate (Reaction 6-2).

$$
\begin{array}{llll}
\text{CH}_2\text{—O—}\overset{\displaystyle\text{O}}{\overset{\|}{\text{C}}}\text{—R}_1 & & \text{CH}_2\text{—OH} & \text{HO—}\overset{\displaystyle\text{O}}{\overset{\|}{\text{C}}}\text{—R}_1 \\[2ex]
\text{CH—O—}\overset{\displaystyle\text{O}}{\overset{\|}{\text{C}}}\text{—R}_2 \;+\; 3\text{H}_2\text{O} \xrightarrow{\;\text{lipase}\;} & & \text{CH—OH} \;+\; \text{HO—}\overset{\displaystyle\text{O}}{\overset{\|}{\text{C}}}\text{—R}_2 \quad (6\text{-}1) \\[2ex]
\text{CH}_2\text{—O—}\overset{\displaystyle\text{O}}{\overset{\|}{\text{C}}}\text{—R}_3 & & \text{CH}_2\text{—OH} & \text{HO—}\overset{\displaystyle\text{O}}{\overset{\|}{\text{C}}}\text{—R}_3 \\[1ex]
\quad\text{Triglyceride} & & \quad\text{Glycerol} & \quad\text{3 Fatty acids}
\end{array}
$$

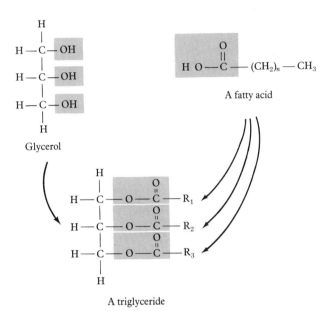

Figure 6-1 The structures of glycerol, a fatty acid, and a triglyceride. The three hydroxyl groups of glycerol and the carboxyl group of the fatty acid are shaded to emphasize their involvement in formation of the three ester bonds (also shaded) of the triglyceride. The common fatty acids range in length from 12 to 20 carbon atoms and may have one or more sites of unsaturation. The three fatty acids of a given triglyceride need not be identical; R_1, R_2, and R_3 may differ both in length of the carbon chain and in occurrence of carbon–carbon double bonds.

$$
\begin{array}{ccc}
\underset{\text{Glycerol}}{\begin{array}{l} CH_2-OH \\ | \\ CH-OH \\ | \\ CH_2-OH \end{array}}
\xrightarrow[\substack{\text{glycerol} \\ \text{kinase}}]{ATP \qquad ADP}
\underset{\substack{\text{Glycerol-3-} \\ \text{phosphate}}}{\begin{array}{l} CH_2-OH \\ | \\ CH-OH \\ | \\ CH_2-O-\textcircled{P} \end{array}}
\xrightarrow[\substack{\text{glycerol-3-phosphate} \\ \text{dehydrogenase}}]{NAD \qquad NADH_2}
\underset{\substack{\text{Dihydroxy-} \\ \text{acetone} \\ \text{phosphate}}}{\begin{array}{l} CH_2-OH \\ | \\ C=O \\ | \\ CH_2-O-\textcircled{P} \end{array}}
\end{array}
$$

(6-2)

Fatty acids also follow the general rule of catabolism to an inter-mediate of the main respiratory pathway, but the process is somewhat more complex, involving successive removals of two-carbon acetate units that enter the TCA cycle as acetyl coenzyme A. The sequential degradation of fatty acids into two-carbon fragments is called β-oxidation because, as we shall see, the initial oxidative event occurs on the carbon atom in the β position to (i.e., the second carbon from) the carboxylic acid group.

β-Oxidation of Fatty Acids

The process of β-oxidation (Figure 6-2) involves several successive cycles of enzymatic attack on the long-chain fatty acid, each cycle leading to the release of an acetate group as acetyl CoA and leaving the fatty acid shortened by two carbons but ready for another round of reactions. As you examine the β-oxidation cycle, notice carefully the extent to which it draws upon reactions and coenzymes already familiar to you from Chapters 4 and 5.

The process begins with an activation step to form an energy-

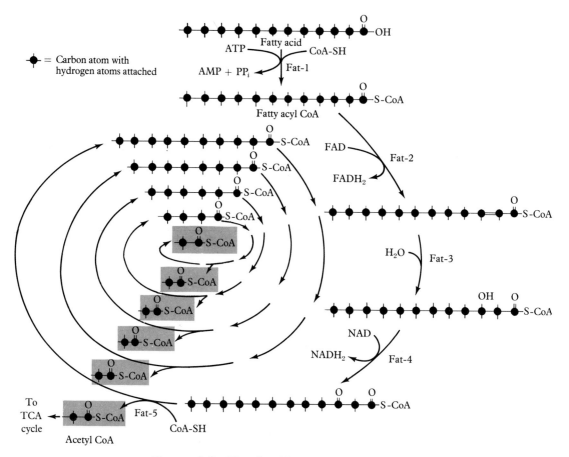

Figure 6-2 The β-oxidation of a fatty acid. The spiral nature of the process is illustrated, with the reaction numbers and chemical details shown for the first cycle of the spiral only. Since the particular fatty acid shown (lauric acid or *laurate*) has 12 carbon atoms, five cycles of β-oxidation are required to degrade it to six molecules of acetyl coenzyme A, each of which is further metabolized via the TCA cycle.

rich thioester with the sulfhydryl group of coenzyme A (see Figure 5-1). This activation step occurs only once per fatty-acid molecule, as the initial reaction in the whole sequence. A molecule of ATP provides the required driving force:[*]

$$CH_3-(CH_3)_n-\overset{\displaystyle O}{\overset{\displaystyle \|}{C}}-OH + CoA-SH + ATP \xrightarrow[\text{thiokinase}]{}$$

$$CH_3-(CH_2)_n-\overset{\displaystyle O}{\overset{\displaystyle \|}{C}}-S-CoA + H_2O + AMP + PP_i \quad \text{(Fat-1)}$$

The activated fatty acid, called a *fatty acyl CoA*, is then dehydrogenated as follows, using the familiar coenzyme FAD as the hydrogen acceptor and forming the corresponding α,β-unsaturated acyl CoA compound:

$$CH_3-(CH_2)_{n-2}-CH_2-CH_2-\overset{\displaystyle O}{\overset{\displaystyle \|}{C}}-S-CoA + FAD \xrightarrow[\text{dehydrogenase}]{}$$

$$CH_3-(CH_2)_{n-2}-CH=CH-\overset{\displaystyle O}{\overset{\displaystyle \|}{C}}-S-CoA + FADH_2 \quad \text{(Fat-2)}$$

Water is then added across the double bond to yield a β-hydroxylated acid:

$$CH_3-(CH_2)_{n-2}-CH=CH-\overset{\displaystyle O}{\overset{\displaystyle \|}{C}}-S-CoA + H_2O \xrightarrow[\text{hydrolyase}]{}$$

$$CH_3-(CH_2)_{n-2}-\overset{\displaystyle OH}{\overset{\displaystyle |}{C}H}-CH_2-\overset{\displaystyle O}{\overset{\displaystyle \|}{C}}-S-CoA \quad \text{(Fat-3)}$$

---

[*]To streamline the equations for fatty-acid oxidation, the enzyme names have in each case been simplified to indicate just the class or kind of enzyme involved. To satisfy those with a special interest in enzyme nomenclature, the complete enzyme names are as follows:

Reaction	Enzyme Name
Fat-1	Fatty-acid thiokinase
Fat-2	Fatty-acyl CoA dehydrogenase
Fat-3	3-Hydroxy fatty-acyl CoA hydrolyase
Fat-4	3-Hydroxy fatty-acyl CoA dehydrogenase
Fat-5	β-Ketothiolase

For at least three of these reactions (Fat-1, Fat-2, and Fat-5), there are two or more forms of the enzyme, each specific for fatty acids of different chain lengths.

In the next step the hydroxyl group is oxidized to the keto group with NAD as the hydrogen acceptor:

$$CH_3-(CH_2)_{n-2}-\overset{\overset{\displaystyle OH}{|}}{C}H-CH_2-\overset{\overset{\displaystyle O}{\|}}{C}-S-CoA + NAD \xrightarrow[\text{dehydrogenase}]{}$$

$$CH_3-(CH_2)_{n-2}-\overset{\overset{\displaystyle O}{\|}}{C}-CH_2-\overset{\overset{\displaystyle O}{\|}}{C}-S-CoA + NADH_2 \qquad \text{(Fat-4)}$$

The β-keto acid generated by Reaction Fat-4 is not very stable. It is cleaved with uptake of another molecule of coenzyme A into acetyl CoA and an acyl CoA compound with two fewer carbon atoms than the original compound:

$$CH_3-(CH_2)_{n-2}-\overset{\overset{\displaystyle O}{\|}}{C}-CH_2-\overset{\overset{\displaystyle O}{\|}}{C}-S-CoA + CoA-SH \xrightarrow[\text{thiolase}]{}$$

$$CH_3-(CH_2)_{n-2}-\overset{\overset{\displaystyle O}{\|}}{C}-S-CoA + CH_3-\overset{\overset{\displaystyle O}{\|}}{C}-S-CoA \qquad \text{(Fat-5)}$$

The acetyl CoA produced in this step can be fed directly into the TCA cycle, since both β-oxidation and the TCA cycle are located in the mitochondrion of the eukaryotic cell. (An exception to this generalization occurs in germinating seeds of fat-storing plant species like castor bean or sunflower, in which the acetyl CoA from fat degradation is used for sugar synthesis rather than energy metabolism, and the enzymes of β-oxidation are packaged along with those required for sugar synthesis in membrane-bounded organelles called *glyoxysomes*.)

Note that the newly formed acyl CoA compound of Reaction Fat-5 is identical to the activated starting compound used in Reaction Fat-2 except that it is two carbons shorter. This means that the new acyl CoA compound can again undergo the series of reactions represented by Reactions Fat-2 through Fat-5 to remove another two-carbon unit. This establishes the spiral nature of the process, which can continue down the fatty-acid backbone, shortening the chain by two carbon atoms at a time. The complete β-oxidation of a fatty acid containing n carbon atoms would give rise eventually to $n/2$ molecules of acetyl CoA, each of which can be metabolized via the Krebs cycle to CO_2 and H_2O. (Actually the process gets more complicated when the molecule gets down to the four-carbon level, but such details need not concern us here; we shall simply assume that β-oxidation runs to completion, with all carbons of the fatty-acid chain appearing eventually as acetyl CoA.) The spiral nature of β-oxidation is illustrated in Figure 6-2 for a fatty acid with 12 carbon atoms.

The process of β-oxidation is a particularly striking example of the conservatism with which cells conduct their metabolism. Faced with the task of metabolizing a long-chain fatty acid with a structure quite unlike any of the intermediates in the glycolytic and TCA pathways, the cell rises magnificently to the challenge: the complete oxidation of a fatty acid requires the addition of only five new enzymes to the cellular repertoire, three of which in fact catalyze reactions analogous to those of an already-existing sequence in the TCA cycle. (Compare Reactions Fat-2, Fat-3, and Fat-4 above with Reactions TCA-6, TCA-7, and TCA-8 of Chapter 5.)

We do not usually think of proteins primarily as an energy source for cells, since they clearly contribute more fundamentally to the ongoing functioning of the cell in their roles as enzymes, structural proteins, chromosomal components, contractile elements, transport proteins, and antibodies. Yet proteins can, in fact, be used for energy, and under certain conditions (starvation, for example) they can contribute significantly to the energy economy of the cell or organism. The proteins catabolized for energy yield come from three different sources: dietary protein, storage protein, and metabolic turnover of cellular components. Catabolism of dietary protein is obviously a characteristic of higher animals, while the use of storage protein for energy is best illustrated by the germination of protein-storing seeds like beans or peas. All cells undergo metabolic turnover of most proteins and protein-containing structures, and it is not surprising that the cell, with its expected efficiency, ensures that the energy available from such protein catabolism is conserved as ATP.

The problem of protein catabolism is really one of amino acid oxidation, since proteins are simply chains of α-amino acids. An α-amino acid, in turn, is a carboxylic acid with an amino group, $-NH_2$, on the carbon atom immediately adjacent to the acid group, as shown in Figure 6-3. Successive amino acid units in proteins are linked by covalent *peptide bonds*, also illustrated in Figure 6-3. Protein degradation therefore begins with enzymatic hydrolysis of the peptide bonds to yield free amino acids. Thereafter the problem becomes one of tracing the catabolism of the resulting amino acids.

Enzymes that hydrolyze proteins to amino acids are called *proteases*. Proteases are common to all cells but have been particularly well studied in the digestive tract of vertebrates, since dietary proteins

The Oxidative Metabolism of Proteins

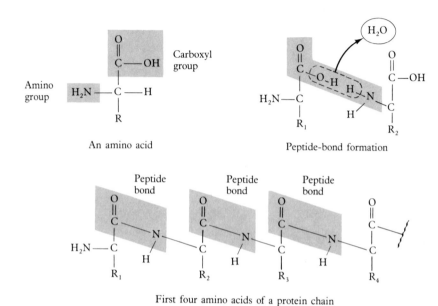

An amino acid   Peptide-bond formation

First four amino acids of a protein chain

Figure 6-3   The structure of an amino acid and the chemistry of peptide-bond formation. The amino and carboxyl groups of the amino acid are shaded to emphasize their involvement in formation of the peptide bonds (also shaded) linking successive amino acid units in a protein chain. The "R" group may be any of the different side chains characteristic of the 20 different amino acids found in proteins. The actual order of the amino acids in a specific protein is genetically determined.

cannot be absorbed directly whereas free amino acids can. Essentially, protease function can be summarized as in Reaction 6-3, but keep in mind that each peptide bond is attacked and hydrolyzed separately, and the hydrolysis of successive peptide bonds in a protein often occurs sequentially.

Oxidation of Amino Acids   The pathways for amino acid catabolism illustrate again the general principle that alternative substrates are converted in as few steps as possible to mainstream intermediates. However, proteins are made up of 20 different kinds of amino acids, many of which require separate degradative pathways. At first glance this diversity of amino acids seems a needless extravagance that violates the concept of the cell as a highly conservative organizational unit. And if proteins were used primarily as a source of energy, it would indeed be needlessly extravagant to construct them of 20 different kinds of monomers, each with a specific degradative pathway. But proteins serve other, much

$$\begin{array}{cccccc}
\overset{\displaystyle O}{\underset{\displaystyle C}{\|}} & \overset{\displaystyle O}{\underset{\displaystyle C}{\|}} & \overset{\displaystyle O}{\underset{\displaystyle C}{\|}} & \cdots & \overset{\displaystyle O}{\underset{\displaystyle C}{\|}} & \\
\end{array}$$

Protein with n amino acid groups

$+ (n-1)H_2O \longrightarrow \longrightarrow \longrightarrow \longrightarrow$

stepwise
hydrolysis of
peptide bonds
by protease

(6-3)

$$H_2N \quad \underset{\substack{CH \\ | \\ R_1}}{\overset{\displaystyle O}{\overset{\|}{C}}-OH} \quad + \quad H_2N \quad \underset{\substack{CH \\ | \\ R_2}}{\overset{\displaystyle O}{\overset{\|}{C}}-OH} \quad + \quad H_2N \quad \underset{\substack{CH \\ | \\ R_3}}{\overset{\displaystyle O}{\overset{\|}{C}}-OH} \quad + \cdots + \quad H_2N \quad \underset{\substack{CH \\ | \\ R_n}}{\overset{\displaystyle O}{\overset{\|}{C}}-OH}$$

n Separate amino acids

more fundamental functions in the cell, and it is not surprising that the performance of so many different tasks requires many proteins varying in size, shape, and charge, as well as in function. This great diversity of function and structure is possible precisely because of the chemical diversity of the amino acids from which proteins are constructed. In a real sense, then, the complexity of the metabolic pathways necessary for amino acid catabolism is the price a cell must pay for the structural and functional diversity it requires of its proteins.

A detailed examination of the various multienzyme sequences needed for the oxidation of 20 different amino acids is neither appropriate nor necessary for our purposes. Instead it will suffice to note several general principles and to look at one or two illustrative examples. Despite their number and complexity, all the pathways for amino acid catabolism converge eventually into a few terminal pathways leading to such intermediates in the TCA cycle as acetyl CoA, oxaloacetate, fumarate, α-ketoglutarate, and succinyl CoA. (Not all the carbon atoms of each of the 20 amino acids actually enter the TCA cycle, however, since decarboxylation reactions result in the loss of some carbon atoms as CO_2 along the way.) We shall look at just two pathways—one (for alanine) involves the direct conversion of an amino acid to a TCA-cycle intermediate; the other (for phenylalanine) is a more complicated sequence.

The catabolic metabolism of amino acids always includes the removal of the amino group somewhere in the sequence, often near the beginning. This removal is accomplished either by a process called *transamination* or by direct *oxidative deamination*. The resulting compound in both cases is an α-keto acid, which either is itself an intermediate in the glycolysis/TCA cycle scheme or is convertible into such an intermediate. We shall look at transamination here, leaving the equally important process of oxidative deamination for Problem 6-4. Transamination really just involves the transfer of the amino group from an amino acid to an α-keto acid acceptor, such that the *deamination* of the amino acid is accompanied by the *amination* of the α-keto acid. This is a convenient, much-used mechanism for shifting amino groups between carbon skeletons and is used by the cell not only in degradative pathways but also in synthetic routes. The general transamination reaction is shown as Reaction 6-4, while Reaction 6-5 depicts a specific case—the conversion of alanine to pyruvate, using as the amino acceptor the compound α-ketoglutarate already familiar to us.

$$
\begin{array}{cc}
\underset{\substack{| \\ R_1 \\ \alpha\text{-Amino} \\ \text{acid}}}{\overset{\overset{\displaystyle O}{\|}}{\underset{|}{C}\text{--OH}}} &
\underset{\substack{| \\ R_2 \\ \alpha\text{-Keto} \\ \text{acid}}}{\overset{\overset{\displaystyle O}{\|}}{\underset{|}{C}\text{--OH}}}
\end{array}
\qquad \xrightarrow{\text{transamination}} \qquad
\begin{array}{cc}
\underset{\substack{| \\ R_1 \\ \alpha\text{-Keto} \\ \text{acid}}}{\overset{\overset{\displaystyle O}{\|}}{\underset{|}{C}\text{--OH}}} &
\underset{\substack{| \\ R_2 \\ \alpha\text{-Amino} \\ \text{acid}}}{\overset{\overset{\displaystyle O}{\|}}{\underset{|}{C}\text{--OH}}}
\end{array}
$$

Left structure: H$_2$N—C—H with R$_1$ (α-Amino acid) + O=C with R$_2$ (α-Keto acid) → O=C with R$_1$ (α-Keto acid) + H$_2$N—C—H with R$_2$ (α-Amino acid), each bearing C—OH topped by O=.

(6-4)

Left structure: H$_2$N—C—H with CH$_3$ (Alanine) + O=C with CH$_2$—CH$_2$—COOH (α-Ketoglutarate) → O=C with CH$_3$ (Pyruvate) + H$_2$N—C—H with CH$_2$—CH$_2$—COOH (Glutamate), each bearing C—OH topped by O=.

(6-5)

Note, then, that in a single transamination step, alanine has been converted into pyruvate, which can be oxidized directly to acetyl CoA (Reaction 5-1, the gateway to the TCA cycle), thereby bringing alanine quickly into the mainstream of respiratory metabolism:

$$\text{Alanine} \xrightarrow[\text{amination}]{\text{trans-}} \text{pyruvate} \xrightarrow[\substack{\text{decarboxylation} \\ \text{(gateway step)}}]{\text{oxidative}} \text{acetyl CoA} \qquad (6\text{-}6)$$
$$\downarrow$$
$$\text{TCA cycle}$$

Although the conversion of alanine into a mainstream intermediate is a direct, single-step process, other (indeed most) amino acids require considerably more complicated pathways with many intermediates. In fact some of the catabolic pathways for amino acids appear at first glance to be almost unnecessarily long and complicated for what is accomplished. Often, however, the intermediates in such pathways have other functions in the cell, as starting points for biosynthetic routes to other essential compounds. As an example of a relatively complex pathway, consider the catabolism of the aromatic amino acids phenylalanine and tyrosine, as shown in Figure 6-4. This pathway, though quite involved, is worth our attention here as an especially rich example of a pathway with intermediates essential for the synthesis of other cellular constituents.

Included among the compounds synthesized by anabolic pathways that diverge from the degradative pathway for phenylalanine and tyrosine are cellular constituents as diverse as the melanin pigments that shield our skin from the sun, the hormones epinephrine (also called adrenalin), norepinephrine, and thyroxine, and the alkaloid drugs morphine and codeine. Since there are no known alternative synthetic routes for these compounds, organisms that require these substances are ultimately dependent for their synthesis upon the degradation of phenylalanine or tyrosine. And since most higher animals cannot synthesize these aromatic amino acids, the starting point for such compounds is really the phenylalanine or tyrosine present in the diet.

Heritable Metabolic Defects in Phenylalanine Catabolism

In addition to illustrating a complicated metabolic sequence for amino acid degradation from which a variety of synthetic routes diverge, the pathway shown in Figure 6-4 also lends itself well to an interesting digression on *heritable metabolic defects*. Such defects, or "inborn errors of metabolism," as they are also called, result from the inability of the cells of an organism to carry out a specific step in a metabolic sequence. In many cases it has been possible to demonstrate that such defects are, in fact, caused by a deficiency or complete lack of a specific enzymatic activity. The pathway for phenylalanine and tyrosine catabolism provides us with three such examples: albinism, phenylketonuria, and alcaptonuria.

Albinism is the lack, partial or complete, of normal pigmentation

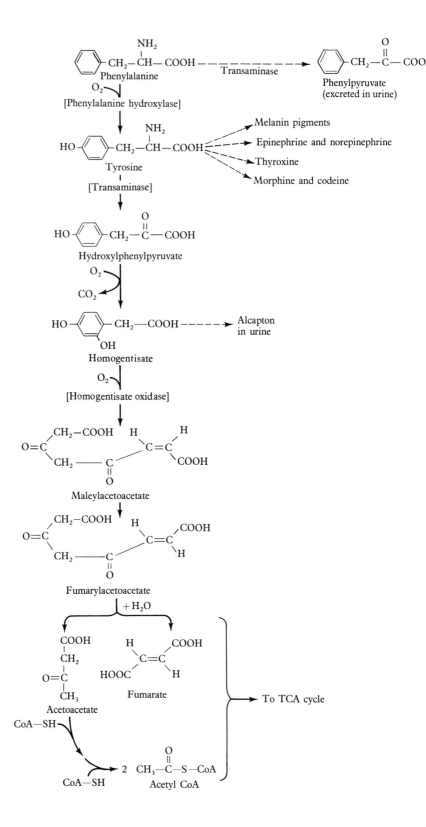

Figure 6-4 Pathway for the catabolism of phenylalanine and tyrosine. Both these aromatic amino acids are degraded by a sequence of reactions involving transamination, oxidative opening of the aromatic ring, and eventual hydrolytic cleavage to two 4-carbon compounds, acetoacetate and fumarate. The acetoacetate is activated by coenzyme A and further cleaved to acetyl CoA. The fumarate and acetyl CoA that enter the TCA cycle represent eight of the nine carbon atoms of the original aromatic amino acids, the other carbon atom having been lost as CO_2 in step 3. Each step in this sequence is, of course, catalyzed by a separate, specific enzyme even though the figure includes the names of only those enzymes of interest to the text discussion. Pathways shown in dashed lines diverge from the main catabolic pathway, either to provide for the synthesis of other useful molecules (such as melanin, epinephrine, norepinephrine, thyroxine, morphine, and codeine) or to dispose of the excessively high levels of specific substances like phenylalanine or homogentisate that accumulate in the genetic absence of particular enzymes in the catabolic pathway.

of the skin, hair, and eyes. As you might well guess, it is caused by an inability to synthesize the melanin pigments mentioned earlier. Melanin synthesis requires the activity of an enzyme (tyrosinase) that oxidizes tyrosine, and albino organisms can be shown to have little or no tyrosinase activity.

Phenylketonuria and alcaptonuria can also be traced to enzymatic deficiencies in phenylalanine metabolism, but they are characterized by the accumulation of harmful substances rather than by the lack of a beneficial substance. *Phenylketonuria*, or PKU, is caused by the absence of the enzymatic activity (phenylalanine hydroxylase) that carries out the initial step in the sequence shown in Figure 6-4. Unable to be converted to tyrosine, phenylalanine is instead transaminated to phenyl-pyruvate, which can be further metabolized to other phenylketo compounds:

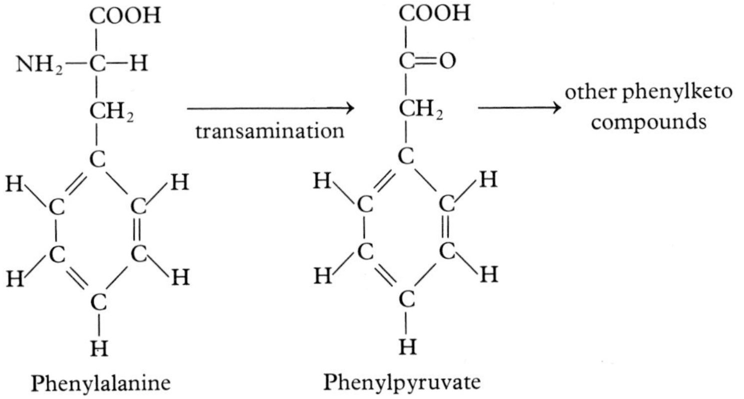

Phenylalanine → (transamination) → Phenylpyruvate → other phenylketo compounds (6-7)

These compounds build up in the blood and are, as the name of the disorder suggests, excreted in the urine. The brain is severely affected by the phenylketo compounds in the blood, causing serious mental retardation in children if the condition is not detected and treated promptly. Fortunately chemical tests for PKU are now in widespread usage with newborn babies, and the symptoms of the disease can usually be prevented provided that the intake of dietary phenylalanine is restricted from birth by a diet limited to foods naturally low in phenylalanine.

Alcaptonuria is a similar disorder in the sense that an abnormal metabolite builds up in the body and is excreted in the urine. In this case the missing or defective enzyme (homogentisate oxidase) is the one normally responsible for the fourth step of the sequence shown in Figure 6-4. In the absence of that enzyme homogentisate appears in the urine, where it darkens oxidatively to the pigment *alcapton*. Other than darkening of the urine, persons afflicted with this disorder show no further symptoms in early life, but they are susceptible to later pigmentation of connective tissue, along with arthritis.

Chemotrophy Behind, Phototrophy Ahead

Following that brief interlude, we return to the main theme of chemotrophic energy metabolism just long enough to summarize the discussion of Chapters 4, 5, and 6 and anticipate the topic of Chapter 7. We have come to grips in these last three chapters with the metabolic means whereby a molecule of glucose (or related sugar) is progressively oxidized, leading under aerobic conditions to the eventual release of each carbon atom in its fully oxidized form, CO_2. In the process 12 pairs of electrons are removed, picked up by coenzymes, and passed stepwise to molecular oxygen, O_2, with the storage of a significant portion of the free-energy yield as "high-energy" phosphate bonds of ATP. In the absence of O_2, the metabolism of glucose is limited to the fermentative production of lactate (or ethanol and CO_2). The ATP yield is much more modest in this case (2 ATPs per glucose by fermentation versus 38 by respiration) but is nonetheless adequate to meet the needs of species and tissues that function, whether obligately or facultatively, under anaerobic conditions. We have also seen the central position that the glycolytic pathway, the TCA cycle, and the electron transport system occupy in chemotrophic metabolism. Into this mainstream pathway flow a variety of tributary sequences, accounting for the catabolism of alternative energy sources such as fats and amino acids.

These, then, are the main pathways by which chemotrophs obtain needed energy from the more important organic foodstuffs, and these are pathways common to almost all cells, whether plant, animal, or microbial. To be sure, many organic substrates and pathways of a more specialized nature have gone unmentioned, and we have neglected almost entirely those microorganisms that derive their energy by oxidation of inorganic compounds. But in the pathways of these three chapters we have encountered the main metabolic processes upon which most of the chemotrophic world depends for its energy needs.

Throughout this discussion of chemotrophic metabolism we have assumed the availability of oxidizable food molecules—carbohydrates, fats, and proteins—with energy packaged into their bonds. Having seen how the chemotrophs get that energy out, it is now time to ask how that energy gets in. For that, we proceed to Chapter 7 and the sunlit world of the phototrophs.

Practice Problems

6-1. True or False: Answer each of the following as true (T) or false (F).
(a) The total number of enzymes required to degrade neutral fats under aerobic conditions is about the same as the total number of enzymes required for aerobic protein catabolism.
(b) Fatty-acyl CoA is to a fatty acid what acetyl CoA is to acetate.
(c) To degrade a fatty acid with $2n$ carbon atoms completely to acetyl CoA requires n cycles of β-oxidation.
(d) The β-oxidation of a fatty acid with an odd number of carbon atoms probably terminates in propionyl CoA instead of acetyl CoA.
(e) The amino acid aspartate can be converted to a TCA-cycle intermediate in a single step.

$$CH_3-CH_2-\overset{\overset{\displaystyle O}{\|}}{C}-S-CoA$$
Propionyl CoA

6-2. Fatty-acid Oxidation: Palmitate is a 16-carbon, saturated fatty acid $(C_{16}H_{32}O_2)$ that can be oxidized completely to CO_2 and H_2O by a combination of β-oxidation and the TCA cycle.
(a) Calculate the net number of ATP molecules generated by the complete catabolism of palmitate to the two-carbon (acetyl CoA) level, assuming that all the reduced coenzymes generated in the process of β-oxidation are re-oxidized by the electron transport chain. (*Hint*: See answer to Problem 6-1c first.)
(b) Next calculate the number of ATP molecules generated by the further oxidation of the eight resulting acetyl CoA molecules to CO_2 and H_2O (again assuming complete regeneration of the coenzymes by electron transport).
(c) Now calculate the total number of ATP molecules generated by the complete oxidation of one molecule of palmitate to CO_2 and H_2O. Write a balanced overall equation for this process.

$$HO-\overset{\overset{\displaystyle O}{\|}}{C}-CH_2-\overset{\overset{\displaystyle H_2N}{|}}{C}H-\overset{\overset{\displaystyle O}{\|}}{C}-OH$$
Aspartate

6-3. Comparative Energy Content of Carbohydrates and Fats: Higher plants and animals "prefer" to store energy reserves as fat rather than as carbohydrate because fat has a higher energy content per unit weight. The calculations specified here are designed to quantitate this difference.

(a) Consider first the utilization of glucose ($C_6H_{12}O_6$; molecular weight = 180) as an energy source and calculate the moles of ATP generated during the complete oxidative metabolism of 1 g of glucose. Assuming that the ΔG value for the hydrolysis of ATP under physiological conditions is -12 kcal/mole, how much energy is obtained from 1 g of glucose?

(b) Now consider instead the utilization of palmitate ($C_{16}H_{32}O_2$; molecular weight = 256) and calculate the moles of ATP generated upon complete oxidation of 1 g of palmitate (see Problem 6-2c).

(c) How much more efficient on a per-gram basis is fat as a form of energy storage compared to carbohydrate, assuming the values for glucose and palmitate to be representative of carbohydrates and fats in general? Why do organisms as diverse as castor beans and humans prefer fat as a means of storing energy reserves?

‡(d) Bearing in mind that respiratory metabolism is essentially an oxidative process, how might you explain the difference in the energy content of carbohydrate and fat on a per-gram basis?

6-4. **Ammonia Liberation and Excretion:** Transamination provides a convenient mechanism for transferring amino groups between carbon skeletons. But it does not accomplish the net removal of nitrogen from the pool of organic molecules, as clearly must occur if net catabolism of amino acids is to occur in the cell. Transamination does, however, allow for the collection of amino groups on a common carbon skeleton, that of the five-carbon amino acid glutamate, from which the amino group is liberated as free ammonia by *oxidative deamination*:

$$
\begin{array}{c}
\text{COOH} \\
|\\
\text{H}_2\text{N}-\text{C}-\text{H} \\
|\\
\text{CH}_2 + \text{NAD} + \text{H}_2\text{O} \\
|\\
\text{CH}_2 \\
|\\
\text{COOH} \\
\text{Glutamate}
\end{array}
\quad
\xrightarrow[\text{dehydrogenase}]{\text{glutamate}}
\quad
\begin{array}{c}
\text{COOH} \\
|\\
\text{O}=\text{C} \\
|\\
\text{CH}_2 + \text{NADH}_2 + \text{NH}_3 \\
|\\
\text{CH}_2 \\
|\\
\text{COOH} \\
\alpha\text{-Ketoglutarate}
\end{array}
$$

(6-8)

In general, glutamate is the only amino acid that has a specific deaminating enzyme. Thus the amino groups collected from the various amino acids by transamination reactions ultimately appear on glutamate and are liberated as ammonia (or ammonium ion, NH_4^+) via Reaction 6-8. The α-ketoglutarate formed in the process can either be metabolized further via the TCA cycle or be used as an amino acceptor in further transamination reactions. The ammonia liberated as shown above is either excreted as such (as is the case for most microorganisms and many aquatic animals, such as the bony fishes) or is first "packaged" into the organic molecule urea (or, in some species, into uric acid) for eventual excretion in the urine (as occurs in most terrestrial vertebrates). Ammonia is highly soluble in water but is toxic to cells. Urea is less soluble but nontoxic.

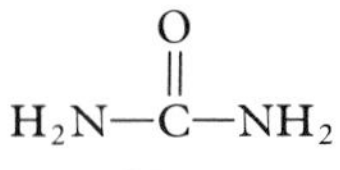

$$\underset{\text{Urea}}{H_2N-\overset{\overset{\displaystyle O}{\|}}{C}-NH_2}$$

(a) What do you suppose is the rationale behind collection of amino groups on a single carbon skeleton for deamination via Reaction 6-8?

(b) The following scheme provides for the continued deamination and oxidative degradation of a variety of amino acids, with glutamate as the sole collector of nitrogen for deamination. Fill in the names of the missing compounds in the diagram.

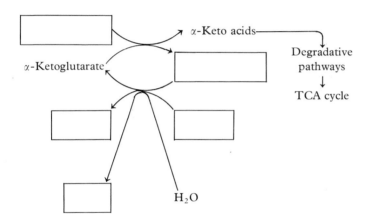

(c) Why do terrestrial vertebrates tend to excrete nitrogen as urea or uric acid, whereas aquatic animals frequently eliminate nitrogen directly as ammonia or ammonium ion?

(d) Why is metamorphosis in the frog accompanied by the appearance in the liver of the enzymes necessary to synthesize urea?

(e) Why do plants have no means for excreting nitrogenous wastes? (Have you ever seen a plant urinate?)

6-5. Isoleucine Degradation: The following scheme shows the degradation pathway for the six-carbon amino acid *isoleucine*, which you probably have not encountered previously. As indicated in the scheme, the reaction sequence leads eventually to an intermediate that can be cleaved to acetyl CoA and propionyl CoA. The structures designated *P, Q, R, S,* and *T* are the five missing intermediates in this pathway, though not in that order. The acetyl CoA that results from this process can enter the TCA cycle directly, whereas the propionyl CoA must be first carboxylated (in an ATP-requiring process) to succinyl CoA.

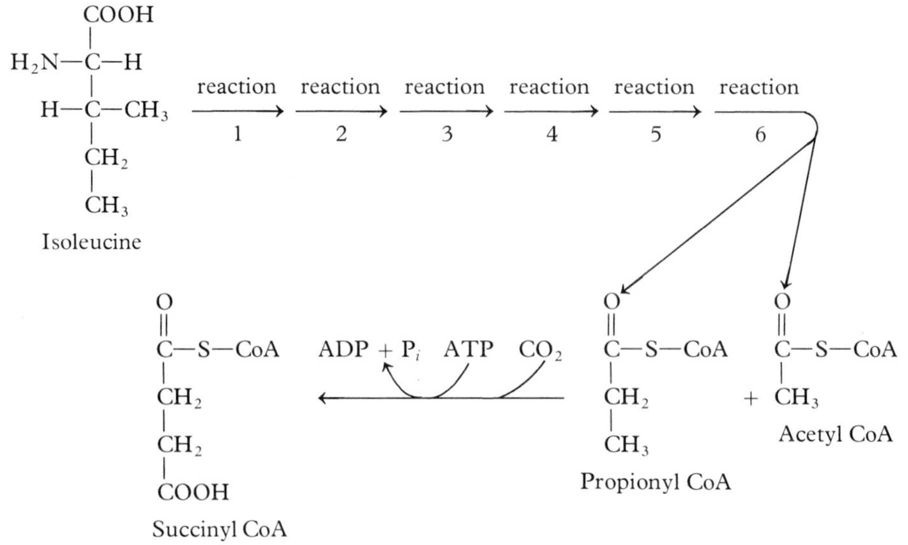

COOH
|
H₂N—C—H
|
H—C—CH₃   reaction   reaction   reaction   reaction   reaction   reaction
|
CH₂   →   →   →   →   →
| 1   2   3   4   5   6
CH₃

Isoleucine

(6-9)

O
‖
C—S—CoA   ADP + P_i ATP CO₂ C—S—CoA   C—S—CoA
| | |
CH₂   ←   CH₂   + CH₃
| | Acetyl CoA
CH₂   CH₃
|
COOH   Propionyl CoA

Succinyl CoA

O   O   O
‖   ‖   ‖
C—S—CoA   C—S—CoA   C—S—CoA
| | |
H—C—CH₃   H—C—CH₃   H—C—CH₃
| | |
HO—C—H   CH₂   O=C
| | |
CH₃   CH₃   CH₃

P   *Q*   *R*

O   COOH
‖ |
C—S—CoA   C=O
| |
C—CH₃   H—C—CH₃
‖ |
C—H   CH₂
| |
CH₃   CH₃

S   *T*

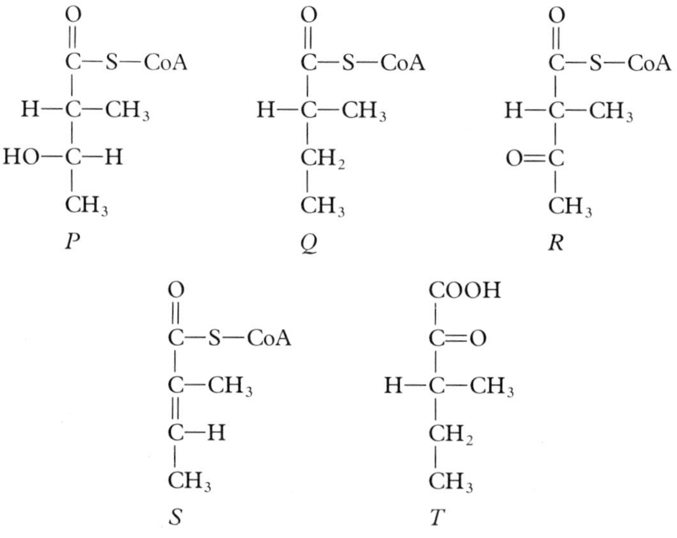

(a) Drawing upon your knowledge of metabolic reactions already encountered in this and previous chapters, order the structures P, Q, R, S, and T into a metabolic sequence that will convert isoleucine into acetyl CoA and propionyl CoA.

(b) For each of the six steps in this pathway, indicate a similar reaction from an already familiar pathway that serves as a useful prototype. (*Hint*: All are in Chapter 6 except Reaction 2, which has its cousin in Chapter 5.)

(c) In which carbon atoms of the resulting acetyl CoA or succinyl CoA would a ^{14}C label appear if a bacterial culture capable of degrading isoleucine were fed ^{14}C-isoleucine with the radioactive label

(i) In position 2?   (iii) In position 1?

(ii) In position 5?   (iv) Uniformly in all positions?

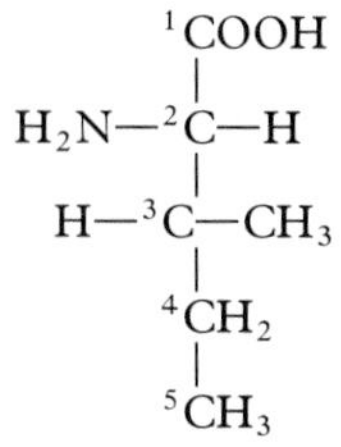

‡6-6. Maple Syrup Urine Disease: A rare genetic defect that causes severe mental retardation in infants is called maple syrup urine disease because of the peculiar odor of the urine. The odor is caused by the presence in the urine of the three keto acids shown below. (Note that keto acid A below is intermediate *T* from Problem 6-5.) Infants suffering from this defect invariably are found to have all three of these keto acids present in the urine:

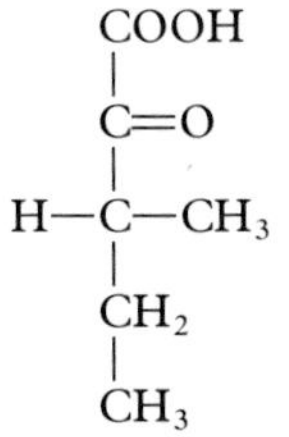

$$^1COOH$$
$$H_2N-^2C-H$$
$$H-^3C-CH_3$$
$$^4CH_2$$
$$^5CH_3$$

Isoleucine

```
   COOH            COOH            COOH
    |               |               |
   C=O             C=O             C=O
    |               |               |
 H—C—CH3           CH              CH2
    |             /   \             |
   CH2         CH3     CH3          CH
    |                               |
   CH3                            /   \
                                CH3    CH3

Keto acid A       Keto acid B       Keto acid C
```

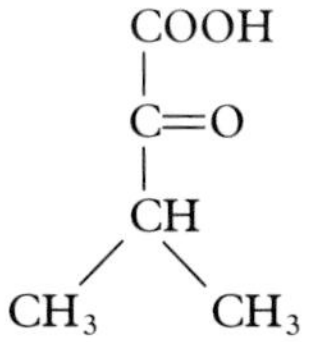

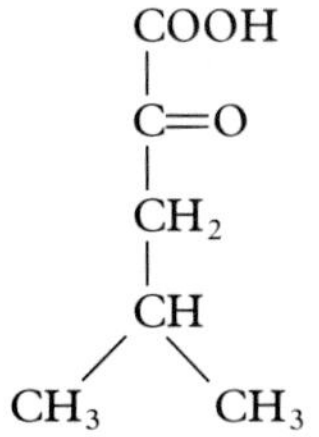

(a) What common features does this genetic defect share with phenylketonuria?

(b) Based on the pathway deduced in Problem 6-4, postulate an explanation for the appearance of keto acid A in the urine of infants suffering from this defect.

(c) What explanation can you offer for the presence of keto acids B and C in the urine?

(d) What do you conclude from the observation that all three keto acids invariably occur together in maple syrup urine disease?

(e) What course of action would you suggest to prevent mental retardation in infants afflicted with this condition? Do you think this treatment would be more or less difficult than the corresponding course of action for phenyl-ketonuria?

References

Brady, R. O., "Hereditary Fat-Metabolism Diseases," *Sci. Amer.* (Aug. 1973), *229*: 88.

Greville, G. D., and P. K. Tubbs, "The Catabolism of Long-Chain Fatty Acids in Mammalian Tissues," *Essays in Biochemistry*, vol. 4 (New York: Academic Press, 1968).

Lehninger, A. L., *Biochemistry*, 2nd ed. (New York: Worth, 1975).

Stanburgy, J. O., J. B. Wyngaarden, and D. S. Frederickson (eds.), *The Metabolic Basis of Inherited Disease*, 2nd ed. (New York: McGraw-Hill, 1966).

Stumpf, P. K., "Metabolism of Fatty Acids," *Ann. Rev. Biochem.* (1969), *38*: 159.

Phototrophic Energy Metabolism

7

In the first three chapters of Part 2 we have been preoccupied with one of the two major solutions to the universal problem of meeting the energy needs of living cells. The general question can be phrased in terms of ATP, the common energy currency of the cell, to inquire into the external sources of energy used by the cell for the generation of ATP. The chemotrophic answer, of course, is from the bond energies of preformed, oxidizable organic molecules that serve as food for the organism. Thus the starting point for the energy metabolism discussed in Chapters 4, 5, and 6 was in all cases an oxidizable organic molecule— be it a sugar like glucose, a fatty acid like palmitate, or an amino acid like alanine—available to the organism from its environment.

We are ready now to turn to the second major solution to the problem of meeting energy needs—that of the phototrophs. These are the organisms capable of harnessing the energy of solar radiation for ATP synthesis. All phototrophic cells and organisms are also capable of chemotrophic metabolism and depend upon that capability whenever light is not available. In fact a phototroph is really best regarded as a chemotroph that can also use light as a source of energy and does so preferentially whenever possible. Such organisms are usually also capable of using CO_2 to meet all their carbon needs. We shall use the term "phototroph" to refer throughout this discussion to all organisms that can use sunlight for energy and CO_2 for carbon needs. Our consideration of phototrophic energy metabolism will therefore deal with two general questions: how solar energy is trapped and converted to chemical free energy that can be used for biosynthesis and how CO_2

is used as the sole carbon source in that biosynthesis. Both these aspects are inherent in the term *photosynthesis*, which thus becomes the dominant topic of this chapter.

Photosynthesis Defined

Respiration, broadly defined, is really the *oxidative decarboxylation* of organic substrates, carried out by chemotrophic cells to obtain the energy necessary for maintenance of life processes. During respiration, for example, the glucose molecule is subjected to a series of enzyme-catalyzed conversions designed to oxidize all six carbon atoms fully, liberating each as CO_2 and conserving a portion of the energy of oxidation as ATP.

In similarly broad terms, photosynthesis can be defined as the *reductive carboxylation* of organic substrates (viewed alternatively as the *fixation of CO_2* into organic form) carried out by chlorophyll-containing phototrophic cells at the expense of solar energy. Fully oxidized carbon atoms (as CO_2) are fixed (covalently linked) to organic acceptor molecules and are subsequently reduced and rearranged into molecules of sugar (or other needed compounds), with sunlight supplying the energy necessary to drive both the fixation and the reduction.

There is, then, both a "photo" and a "synthesis" component to our definition of photosynthesis. In terms of energy transactions (the "photo" component), we are dealing with the photochemical process whereby the electromagnetic energy of visible light is converted in a biological system into chemical energy. As we shall see, this chemical energy appears initially as a state of electronic excitation in a chlorophyll molecule, then as the high-energy bonds of ATP and the reduced coenzyme $NADPH_2$, and eventually as the chemical bond energies of organic molecules. In terms of carbon metabolism (the "synthesis" component), photosynthesis can be thought of as a collection of non-photochemical, ATP-driven events whereby carbon dioxide is fixed into organic compounds, followed by its reduction to the oxidation level of sugars. The reducing energy necessary for this purpose comes from the light-driven oxidation of a suitable electron (hydrogen) donor. In plants this donor is water, but among the photosynthetic bacteria a variety of alternative donors are used, such as hydrogen gas (H_2), hydrogen sulfide (H_2S), or even oxidizable organic molecules.

The Scope and Significance of Photosynthesis

Few phenomena in nature can compare with photosynthesis in sheer scope and grandeur. Ours is a planet dominated by the color green, a pleasant distinction owed to the prominence of chlorophyll-containing organisms that have covered every bit of the earth's surface that is the least bit hospitable to the maintenance of phototrophic life. The amount of energy trapped and carbon converted into organic form by this photosynthetic network is mind-boggling. Each year more than 100 billion tons of organic matter are produced by photosynthetic organisms on a worldwide basis, an amount that has been estimated to equal the weight of all the buildings in New York City!

The importance of photosynthesis goes beyond the sheer weight of its aggregate product, however. Photosynthesis is, in a very real sense, the single most vital metabolic process underlying all life on this planet, since all forms of life, whatever their immediate sources of energy, are dependent ultimately upon the light energy of the sun. The transduction of solar energy into chemical energy that occurs in photosynthetic prokaryotes and in the chloroplasts of plants is crucial to the continued existence of the entire biological world, for the chemical energy released in cells upon oxidation of foodstuffs represents the energy of sunlight, originally entrapped within the molecules of organic compounds during photosynthesis. And not only is the energy expended by living organisms derived from photosynthetic capital, but most of the materials that enter into cellular structure come also from this source. From the products of photosynthesis and from a few simple inorganic compounds obtained from the environment, there are built up in living organisms all the multifarious and complex molecules that constitute the cellular structure of living organisms or are otherwise essential to their existence.

In its overall formulation, photosynthesis can be viewed as an oxidation-reduction process: CO_2 is *reduced* to the oxidation level of the carbon atoms in a sugar molecule, and an appropriate hydrogen donor is *oxidized*. In these terms an appropriate general equation for photosynthesis can be written as follows:

Photosynthesis as an Oxidation-Reduction Process

$$CO_2 + 2H_2A \xrightarrow{\text{light}} [CH_2O] + 2A + H_2O \qquad (7\text{-}1)$$

where H_2A is a general electron (hydrogen) donor, A is the oxidized form of H_2A, and $[CH_2O]$ is cellular material in which carbon is at the oxidation level of carbohydrate.

The casting of photosynthesis in the framework of Equation 7-1 is a deliberate attempt to avoid perpetuating the hackneyed idea that the electron donor of photosynthesis is inevitably water. This is true for the green plants and eukaryotic algae but not for the photosynthetic bacteria which, in the absence of the photochemical machinery necessary to extract electrons from water, oxidize a variety of other substrates including H_2, H_2S, $H_2S_2O_3$, and even organic compounds such as lactate or succinate. Although much of this chapter in fact deals with photosynthesis in green plants with water as the electron donor (thereby rendering the present coverage just as hackneyed as that of most other introductions to photosynthesis), the proper context for the discussion nonetheless remains the broader biological perspective, which admits a variety of donors.

Photosynthesis in Green Plants

For the case of the green plant, with water as the electron donor, we can rewrite Equation 7-1 in the more specific (and for most readers probably the more familiar) form of Equation 7-2:

$$6CO_2 + 6H_2O \xrightarrow{\text{light}} C_6H_{12}O_6 + 6O_2 \qquad (7\text{-}2)$$

The actual chemistry and photochemistry are far more complex than suggested by either equation, however. It will be our goal in this chapter to look at some of the component parts of photosynthesis and see how each contributes to these two overall equations. Our discussion begins with the "synthesis" component—the nonphotochemical fixation, reduction, and metabolism of carbon that appears to be common to all photosynthetic organisms. We shall then move on to the "photo" component—the light-dependent events by which high-energy phosphate bonds (ATP) and reducing power ($NADPH_2$) are generated by oxidation of water (or, in the photosynthetic bacteria, other electron donors).

Photosynthetic Carbon Metabolism

The process of photosynthetic carbon fixation and metabolism may be considered in four steps: (1) the initial fixation of CO_2 into organic form by carboxylation of an acceptor molecule; (2) the reduction of the fixed carbon from the oxidation level of an acid to that of an aldehyde as found in sugars; (3) the subsequent synthesis of carbohydrate or other cellular components from the fixed and reduced carbon; and (4) the eventual regeneration of the acceptor molecule. Each of these steps is discussed here; taken together they constitute the *Calvin cycle* for photosynthetic carbon metabolism, named after Melvin Calvin, who received the Nobel prize for his work on the elucidation of these pathways. The Calvin cycle is illustrated in Figure 7-1.

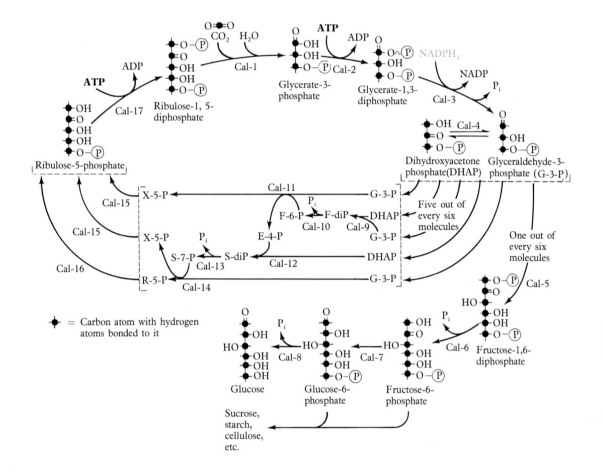

Figure 7-1 The Calvin cycle for photosynthetic carbon metabolism. Carbon as CO_2 is fixed into organic form, then reduced to the oxidation level characteristic of carbon in sugar molecules. For net synthesis of one glucose molecule, six molecules of CO_2 must be fixed, resulting in the formation of 12 molecules of glyceraldehyde-3-phosphate (G-3-P). Only one out of every six molecules of G-3-P (two per glucose synthesized) is actually used for sugar synthesis, since the other five out of every six molecules (ten per glucose) are reutilized within the cycle to regenerate the initial five-carbon acceptor molecule, ribulose-1,5-diphosphate. Abbreviations used in the regeneration sequence shown in brackets are as follows: G, glyceraldehyde; DHAP, dihydroxyacetone phosphate; F, fructose; E, erythrose; S, sedoheptulose; X, xylulose; and R, ribose. For a more detailed representation of the regeneration sequence, see Figure 7-2.

Carbon Fixation   Photosynthetic carbon metabolism begins with the actual fixation of CO_2 into organic form. For most photosynthetic organisms, this process involves the carboxylation of (addition of CO_2 to) a five-carbon sugar called *ribulose-1,5-diphosphate* to form two molecules of glycerate-3-phosphate, which is also called 3-phosphoglyceric acid, or PGA:[*]

$$
\begin{array}{c}
\text{CH}_2\text{—O—}\textcircled{P} \\
| \\
\text{C}=\text{O} \\
| \\
\text{H—C—OH} \\
| \\
\text{H—C—OH} \\
| \\
\text{CH}_2\text{—O—}\textcircled{P} \\
\text{Ribulose-1,5-diphosphate}
\end{array}
\;+\; CO_2 + H_2O
\xrightarrow[\substack{\text{ribulose-1,5-} \\ \text{diphosphate} \\ \text{carboxylase}}]{}
\begin{array}{c}
\text{CH}_2\text{—O—}\textcircled{P} \\
| \\
\text{H—C—OH} \\
| \\
\text{C}=\text{O} \\
| \\
\text{OH} \\
\\
\text{OH} \\
| \\
\text{C}=\text{O} \\
| \\
\text{H—C—OH} \\
| \\
\text{CH}_2\text{—O—}\textcircled{P} \\
\text{2 Molecules of} \\
\text{glycerate-3-phosphate}
\end{array}
\;+\;
\qquad \text{(Cal-1)}
$$

As in the case of the glycolytic pathway, the subsequent split of one larger molecule (a hexose in the case of glycolysis, a pentose in the Calvin cycle) into two smaller molecules is anticipated by the use of a doubly phosphorylated species, such that each of the eventual cleavage products has a phosphate group. Glycerate-3-phosphate is generally regarded as the first detectable product of carbon fixation by the Calvin cycle, since the six-carbon intermediate that is postulated to be the immediate product of the carboxylation reaction has never been isolated. As indicated by the shaded letters in Reaction Cal-1, the newly fixed carbon atom appears as the carboxyl group on one of the two resulting molecules of glycerate-3-phosphate.

---

[*] Note that the symbol $\textcircled{P}$ is being used for organically linked phosphate. Recall that each phosphate group, if written out in detail, would appear as follows:

$$
\begin{array}{c}
\text{O} \\
\| \\
\text{—O—P—OH} \\
| \\
\text{OH}
\end{array}
$$

Remember, too, that the phosphate group exists predominantly in the ionized form at near-neutral pH and that the structure is resonance-stabilized, with the extra electron pair delocalized over each of the three free P—O bonds.

The glycerate-3-phosphate formed upon fixation of CO_2 is reduced to glyceraldehyde-3-phosphate in a reaction sequence that is the exact reversal of the oxidative sequence in glycolysis (Reactions Gly-6 and Gly-7) except that the coenzyme involved is not NAD but a molecular relative called *nicotinamide adenine dinucleotide phosphate*, NADP:

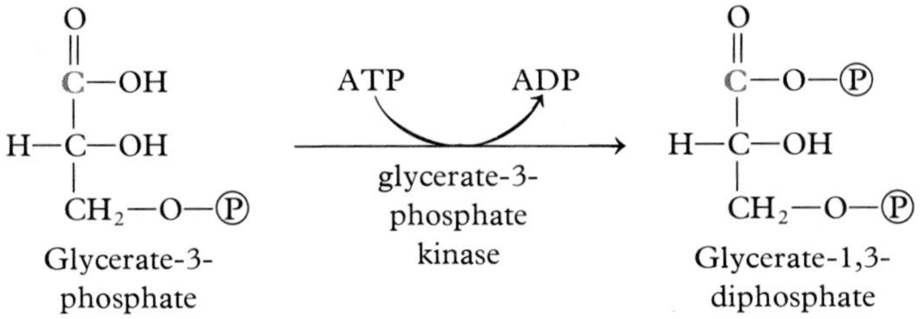

Reduction of Glycerate-3-phosphate

$$
\begin{array}{c}
\text{O} \\
\parallel \\
\text{C—OH} \\
\mid \\
\text{H—C—OH} \\
\mid \\
\text{CH}_2\text{—O—}\textcircled{P}
\end{array}
\qquad
\overset{\text{ATP} \quad \text{ADP}}{\underset{\substack{\text{glycerate-3-} \\ \text{phosphate} \\ \text{kinase}}}{\xrightarrow{\hspace{2cm}}}}
\qquad
\begin{array}{c}
\text{O} \\
\parallel \\
\text{C—O—}\textcircled{P} \\
\mid \\
\text{H—C—OH} \\
\mid \\
\text{CH}_2\text{—O—}\textcircled{P}
\end{array}
\qquad \text{(Cal-2)}
$$

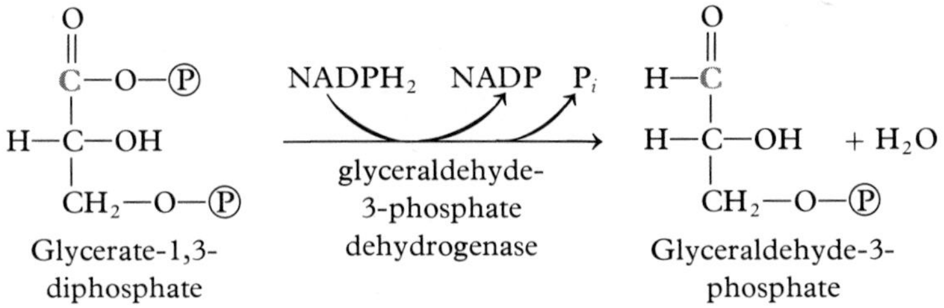

Glycerate-3-phosphate   Glycerate-1,3-diphosphate

$$
\begin{array}{c}
\text{O} \\
\parallel \\
\text{C—O—}\textcircled{P} \\
\mid \\
\text{H—C—OH} \\
\mid \\
\text{CH}_2\text{—O—}\textcircled{P}
\end{array}
\qquad
\overset{\text{NADPH}_2 \quad \text{NADP} \quad \text{P}_i}{\underset{\substack{\text{glyceraldehyde-} \\ \text{3-phosphate} \\ \text{dehydrogenase}}}{\xrightarrow{\hspace{2cm}}}}
\qquad
\begin{array}{c}
\text{O} \\
\parallel \\
\text{H—C} \\
\mid \\
\text{H—C—OH} \quad + \text{H}_2\text{O} \\
\mid \\
\text{CH}_2\text{—O—}\textcircled{P}
\end{array}
\qquad \text{(Cal-3)}
$$

Glycerate-1,3-diphosphate   Glyceraldehyde-3-phosphate

NADP is the coenzyme of choice in a large number of synthetic pathways in biological systems. It is a derivative of the NAD molecule with which we are already familiar (see Figure 4-5); the two coenzymes differ only by the presence in NADP of a third phosphate group, attached to the hydroxyl group on carbon atom 3 of the ribose in the adenosine half of the molecule.

As Reactions Cal-2 and Cal-3 indicate, this reductive sequence requires both ATP (to activate the acid for reduction) and $NADPH_2$ (as the source of reducing power). It is therefore at this point (as well as in the eventual regeneration of ribulose-1,5-diphosphate in Reaction Cal-17) that the Calvin cycle couples to the light-dependent steps of photosynthesis.

Formation of carbohydrate from the glyceraldehyde-3-phosphate of Reaction Cal-3 ought to be straightforward for the veteran of Chapter 4. The pathway involved is just the reversal of the initial steps of glycolysis (Reactions Gly-1 through Gly-5) except that the phosphate groups contributed by ATP in the glycolytic direction are removed by simple hydrolysis in the Calvin cycle. The reaction sequence is as follows:

Carbohydrate Synthesis

$$\text{Glyceraldehyde-3-phosphate} \underset{\text{triose isomerase}}{\overset{\longrightarrow}{\rightleftharpoons}}$$

$$\text{dihydroxyacetone phosphate} \qquad \text{(Cal-4)}$$

$$\text{Glyceraldehyde-3-phosphate} + \text{dihydroxyacetone phosphate} \underset{\text{aldolase}}{\overset{\longrightarrow}{\rightleftharpoons}}$$

$$\text{fructose-1,6-diphosphate} \qquad \text{(Cal-5)}$$

$$\text{Fructose-1,6-diphosphate} \underset{\text{phosphatase}}{\longrightarrow} \text{fructose-6-phosphate} + P_i \quad \text{(Cal-6)}$$

$$\text{Fructose-6-phosphate} \underset{\text{isomerase}}{\rightleftharpoons} \text{glucose-6-phosphate} \quad \text{(Cal-7)}$$

$$\text{Glucose-6-phosphate} \underset{\text{phosphatase}}{\longrightarrow} \text{glucose} + P_i \qquad \text{(Cal-8)}$$

By summing this sequence (Cal-4 through Cal-8), it can be shown that two molecules of glyceraldehyde-3-phosphate give rise to one molecule of glucose. Glucose is often represented in this way as the end product of photosynthetic carbon metabolism. This, however, is more a definition of convenience than of fact, since the designation of an end product for the Calvin cycle is really rather arbitrary once you get past fructose-1,6-diphosphate, the form in which carbon exits from the actual cycle (Figure 7-1). In a sense you could even regard the whole phototrophic organism—plant, algal cell, or bacterium—as the "end product" of photosynthesis, since the Calvin cycle is the sole source of all the fixed carbon from which every molecule and structure in the organism is elaborated. Even in a much more immediate sense, glucose is at best only a formal end product of photosynthesis, since relatively little carbon actually appears as free glucose. Most of it is either converted into transport forms like sucrose or into storage carbohydrates like starch.

Sucrose, as you may recall from Chapter 4, is a disaccharide formed by linking glucose and fructose together. It is the most common form in which carbon is translocated in higher plants. As shown in Figure 7-1, sucrose synthesis involves the phosphorylated forms of glucose and fructose as intermediates. *Starch*, on the other hand, is a polysaccharide consisting of many glucose units linked together in a process that again depends upon phosphorylated rather than free glucose as an intermediate. Starch is the storage polysaccharide of higher plants, but most photosynthetic organisms have comparable storage molecules made up of units of glucose or other sugars or both. Thus glucose may be regarded formally as the end product of photosynthesis for ease of balancing equations and emphasizing the complementary relationship between photosynthesis and respiration, but in actual fact it is the phosphorylated sugars and compounds derived from them that represent the real fate and flow of fixed carbon in autotrophic cells.

At first glance it may seem that the metabolic sequence represented by Reactions Cal-1 through Cal-8 is the extent of carbon chemistry necessary to understand photosynthesis, since we have, after all, observed both the reductive fixation of CO_2 and the synthesis of glucose. So far, however, we have simply added one carbon atom to a five-carbon sugar and converted the products eventually to a six-carbon sugar. The sequence from Reaction Cal-1 to Cal-8, in other words, accounts for the synthesis of a six-carbon sugar, but with the fixation of only a single carbon atom. Missing, of course, is the regeneration of the starting compound, without which the sequence could not continue beyond the time it takes to exhaust the initial supply of that starting compound. It is the regeneration of the ribulose-1,5-diphosphate which is accomplished in the remainder of the Calvin cycle (and which, in fact, makes it a cycle). On the average, only one out of every six molecules of glyceraldehyde-3-phosphate can be used to synthesize glucose by this route; the other five are required within the cycle for the regeneration of the starting compound. This means that, on the average, six molecules of CO_2 must be fixed for the net synthesis of one glucose molecule, which is in turn consistent with the carbon balance necessary if the system is to be self-sustaining.

Regeneration of Ribulose-1,5-diphosphate

To see the stoichiometry clearly, consider Equation 7-3, which summarizes Reactions Cal-1 through Cal-3 with the coefficients necessary to generate six molecules of glyceraldehyde-3-phosphate:

3 Ribulose-1,5-diphosphate + 3CO_2 + 6NADPH$_2$ + 6ATP →

6 glyceraldehyde-3-phosphate + 6NADP + 6ADP + 6P$_i$ + 3H$_2$O   (7-3)

On the average, one out of these six molecules of glyceraldehyde-3-phosphate will be used for the synthesis of glucose (one-half molecule with the present stoichiometry) by the sequence of Reactions Cal-4 through Cal-8, summarized as follows:

1 Glyceraldehyde-3-phosphate → $\frac{1}{2}$ glucose + P$_i$ (7-4)

The remaining five molecules of glyceraldehyde-3-phosphate serve then as the starting point for the regeneration of the three molecules of ribulose-1,5-diphosphate with which the cycle commenced. The problem is thus one of regenerating three 5-carbon molecules from five 3-carbon molecules (with the loss of the two extra phosphates not needed when you rearrange carbons from five to three molecules). The key to understanding the chemical rearrangements by which this is accomplished lies in three kinds of reactions, each of which occurs twice in the regeneration sequence. The first involves the *condensation* of one 3-carbon unit onto a second 3- (or 4-) carbon unit to yield a single 6- (or 7-) carbon compound; the second is a *phosphate-splitting* reaction to get rid of the second phosphate group that results when two monophosphorylated sugars are so condensed; and the third reaction (called *transketolation*) involves the transfer of a two-carbon unit from a keto compound to an aldo compound.

Now we can put together a simple sequence of reactions that converts five 3-carbon molecules into three 5-carbon molecules. As shown in Figure 7-2 (and in lesser detail in Figure 7-1), this is accomplished by two series of reactions, Cal-9 to Cal-11 and Cal-12 to Cal-14, each consisting of a condensation reaction (labeled A), a phosphate-splitting step (B), and a transketolation (C), in that order. The five-carbon compounds formed by this sequence of reactions (Cal-9 through Cal-14) are xylulose-5-phosphate (two molecules) and

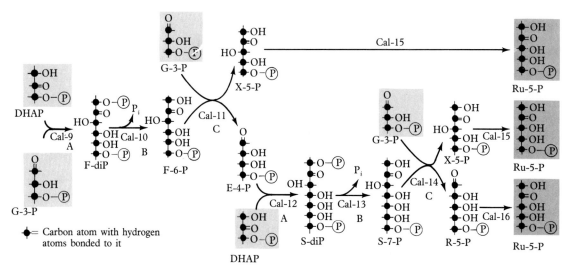

Figure 7-2   The regeneration sequence of the Calvin cycle, in which five 3-carbon compounds (glyceraldehyde-3-phosphate or dihydroxyacetone phosphate) are converted into three 5-carbon compounds (ribose-5-phosphate or xylulose-5-phosphate, both convertible into ribulose-5-phosphate). The sequence (shown in brackets in Figure 7-1) involves two series of reactions, each consisting of a condensation step (A), a phosphate-cleaving step (B), and a transketolation step (C) that shifts a two-carbon fragment from one sugar to another. Each of the five initial 3-carbon molecules is shaded in light grey, and each of the resulting 5-carbon compounds is shaded in dark grey. For abbreviations of compound names see the legend for Figure 7-1.

ribose-5-phosphate (one molecule). These compounds are readily converted into ribulose-5-phosphate by Reactions Cal-15 and Cal-16, respectively. We can therefore summarize the regeneration sequence of Figure 7-2 by writing the following summary sequence:[*]

$$5 \text{ Glyceraldehyde-3-phosphate} \rightarrow 3 \text{ ribulose-5-phosphate} + 2P_i \quad (7\text{-}5)$$

Finally there remains but to convert the three resulting molecules of ribulose-5-phosphate back to the diphosphate form required as the CO_2 acceptor in the initial step. This conversion is accomplished by a phosphorylation reaction, using as the phosphate donor ATP derived from the photochemical reactions we shall encounter shortly:

$$3 \text{ Ribulose-5-phosphate} + 3ATP \rightarrow 3 \text{ ribulose-1,5-diphosphate} + 3ADP$$

$$(\text{Cal-17})$$

We return thus to the initial acceptor molecule from which the process departed, and we are now in a position to sum up the carbon metabolism of the Calvin cycle. This we can do by writing an overall equation for each of the various parts of the cycle (with coefficients appropriate for the formation and utilization of six molecules of glyceraldehyde-3-phosphate) and then adding them to arrive at a net overall equation:

$$3 \text{ Ribulose-1,5-diphosphate} + 3CO_2 + 3H_2O \xrightarrow{\text{Cal-1}} 6 \text{ glycerate-3-phosphate}$$

$$6 \text{ Glycerate-3-phosphate} + 6NADPH_2 + 6ATP \xrightarrow[\text{and Cal-3}]{\text{Cal-2}}$$

$$6 \text{ glyceraldehyde-3-phosphate} + 6NADP + 6ADP + 6P_i + 6H_2O$$

$$1 \text{ Glyceraldehyde-3-phosphate} \xrightarrow{\text{Cal-4 through Cal-8}} \tfrac{1}{2} \text{ glucose} + P_i$$

$$5 \text{ Glyceraldehyde-3-phosphate} \xrightarrow{\text{Cal-9 through Cal-16}}$$

$$3 \text{ ribulose-5-phosphate} + 2P_i$$

$$3 \text{ Ribulose-5-phosphate} + 3ATP \xrightarrow{\text{Cal-17}} 3 \text{ ribulose-1,5-diphosphate} + 3ADP$$

Net: $$3CO_2 + 6NADPH_2 + 9ATP \longrightarrow$$
$$\tfrac{1}{2} \text{ glucose} + 6NADP + 9ADP + 9P_i + 3H_2O \quad (7\text{-}6)$$

[*]Actually the regeneration sequence of Figure 7-2 requires three molecules of glyceraldehyde-3-phosphate and two molecules of dihydroxyacetone phosphate. But these two 3-carbon compounds are readily interconvertible (Reaction Cal-4) and are adequately represented as five molecules of glyceraldehyde-3-phosphate in Equation 7-5, since the purpose of that equation is to summarize the overall sequence for the regeneration of the five-carbon sugar ribulose.

Or, expressed on a per-glucose basis,

$$6CO_2 + 12NADPH_2 + 18ATP \longrightarrow$$
$$1 \text{ glucose} + 12NADP + 18ADP + 18P_i + 6H_2O \qquad (7\text{-}7)$$

Some Do It Differently: The C₄ Plants and the Hatch-Slack Pathway

The Calvin cycle, summarized in Equation 7-7, is used universally in all photosynthetic organisms for the reduction and processing of fixed carbon. However, the reaction described as initiating the entire cycle (Cal-1) is actually only one of two possible carbon-fixing mechanisms. It is, to be sure, the more common mechanism and the sole means of photosynthetic carbon fixation in the preponderance of plant species. Nevertheless some monocotyledonous species, especially the tropical grasses (including agronomically important species such as corn, sorghum, and sugarcane) preface the Calvin cycle with a short fixation sequence usually referred to as the *Hatch-Slack pathway* after two Australians who played key roles in its elucidation. The Hatch-Slack pathway is shown in Figure 7-3, which also illustrates the relationship of this pathway to the Calvin cycle.

The initial fixation step in the Hatch-Slack (HS) pathway involves the carboxylation of the three-carbon acceptor, phosphoenol pyruvate, to form oxaloacetate:

$$
\begin{array}{l}
\text{COOH} \\
| \\
\text{C}-\text{O}-\textcircled{P} + CO_2 \\
|| \\
\text{CH}_2
\end{array}
\xrightarrow[\text{pyruvate carboxylase}]{\text{phosphoenol}}
\begin{array}{l}
\text{COOH} \\
| \\
\text{C}=\text{O} + P_i \\
| \\
\text{CH}_2 \\
| \\
\text{COOH}
\end{array}
\qquad \text{(HS-1)}
$$

Phosphoenol pyruvate

Oxaloacetate

As in the Calvin cycle, the shading is used here and in subsequent equations to indicate the position of the newly fixed carbon atom. Phosphoenol pyruvate should already be familiar to you as a high-energy intermediate in the glycolytic sequence, and oxaloacetate as an intermediate in the TCA cycle. In fact the interconversion of phosphoenol pyruvate and oxaloacetate is also possible in most chemotrophic cells, but it is catalyzed by a different enzyme than that used in this alternative means of carbon fixation and it usually functions in the reverse direction, converting oxaloacetate to phosphoenol pyruvate in the synthesis of glucose (see Chapter 8).

Oxaloacetate, although thought to be the immediate product of carboxylation in such species, is rapidly converted into either malate

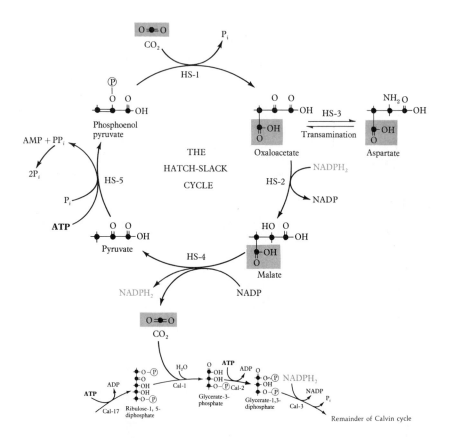

Figure 7-3 The Hatch-Slack cycle of carbon fixation in C_4 plants and its relationship to the Calvin cycle common to all photosynthetic cells. To aid in tracing carbon through the cycle, the incoming carbon (from the newly fixed CO_2 molecule) is shaded in each molecule of the Hatch-Slack cycle in which it appears. Note that the carbon atom released as CO_2 in step HS-4 is the same as that fixed as CO_2 in step HS-1. The particular sequence shown is that thought to be operative in those species of C_4 plants that release CO_2 from the Hatch-Slack cycle by oxidative decarboxylation of malate. A similar sequence can be written for the species in which carbon flow is through aspartate (via Reaction HS-3) instead of malate, except that the aspartate is apparently reconverted to oxaloacetate first and the oxalo-acetate is then decarboxylated to yield pyruvate and CO_2 directly. For the spatial distribution of the Hatch-Slack and Calvin cycles within the leaves of C_4 plants, see Figure 7-4.

(by NADP-mediated reduction; Reaction HS-2) or aspartate (by trans-amination; Reaction HS-3), both of which become highly and rapidly labeled when short-term pulses of radioactively labeled CO_2 are administered to plants with the Hatch-Slack pathway.

$$
\begin{array}{ccc}
\begin{array}{c} \text{COOH} \\ | \\ \text{C}{=}\text{O} \\ | \\ \text{CH}_2 \\ | \\ \text{COOH} \\ \text{Oxaloacetate} \end{array}
& + \text{ NADPH}_2 \underset{\substack{\text{malate} \\ \text{dehydrogenase}}}{\rightleftharpoons}
& \begin{array}{c} \text{COOH} \\ | \\ \text{H}{-}\text{C}{-}\text{OH} \\ | \\ \text{CH}_2 \\ | \\ \text{COOH} \\ \text{Malate} \end{array} + \text{NADP}
\end{array}
$$

(HS-2)

$$
\begin{array}{c} \text{COOH} \\ | \\ \text{C}{=}\text{O} \\ | \\ \text{CH}_2 \\ | \\ \text{COOH} \\ \text{Oxaloacetate} \end{array}
\quad
\underset{\text{transaminase}}{\xrightarrow{\substack{\text{H}_2\text{N} \quad\quad\quad \text{O} \\ | \quad\quad\quad\quad || \\ \text{R}{-}\text{C}{-}\text{COOH} \quad \text{R}{-}\text{C}{-}\text{COOH}}}}
\quad
\begin{array}{c} \text{COOH} \\ | \\ \text{H}{-}\text{C}{-}\text{NH}_2 \\ | \\ \text{CH}_2 \\ | \\ \text{COOH} \\ \text{Aspartate} \end{array}
\quad \text{(HS-3)}
$$

We have already encountered the malate dehydrogenase reaction (HS-2) as a step in the TCA cycle (see Reaction TCA-8 of Chapter 5), where it functions in the reverse direction, oxidizing malate to oxaloacetate. Reaction HS-3 should also be familiar, since it is a further example of the transamination reaction discussed in Chapter 6 (see Reaction 6-4).

Because the immediate products of carboxylation in plants with the Hatch-Slack pathway are four-carbon compounds (oxaloacetate, malate, and aspartate), such plants are called "C_4 plants" to distinguish them from "C_3 plants" in which the immediate product of fixation is the three-carbon compound glycerate-3-phosphate (Reaction Cal-1). Species of C_4 plants appear to depend upon either Reaction HS-2 or HS-3 for further metabolism of newly formed oxaloacetate and can be classified accordingly as "malate formers" or "aspartate formers."

Although the newly fixed carbon of C_4 plants appears initially in oxaloacetate and then in malate or aspartate, it eventually finds its way into all the intermediates of the Calvin cycle. This is because the initial fixation into four-carbon compounds is followed by a decarboxylation, and the liberated CO_2 (the *same* carbon atom as that fixed initially; see Figure 7-3) is then refixed by the carboxylating enzyme of the Calvin cycle (Reaction Cal-1). Using malate as our example (a similar reaction could be written for aspartate), the decarboxylation/carboxylation sequence is as follows:

$$
\begin{array}{c}
\text{COOH} \\
| \\
\text{H—C—OH} \\
| \\
\text{CH}_2 \\
| \\
\text{COOH}
\end{array}
\;+\; \text{NADP} \xrightarrow[\text{malic enzyme}]{}
\begin{array}{c}
\text{COOH} \\
| \\
\text{C=O} \\
| \\
\text{CH}_3
\end{array}
\;+\; CO_2 \;+\; \text{NADPH}_2
$$

Malate   Pyruvate

(HS-4)

$$
\text{Ribulose-1,5-diphosphate} + H_2O + CO_2 \xrightarrow[\substack{\text{ribulose-1,5-} \\ \text{diphosphate} \\ \text{carboxylase}}]{}
$$

2 glycerate-3-phosphate   (Cal-1)

Remaining after the decarboxylation of Reaction HS-4 is the three-carbon compound pyruvate, which can be phosphorylated to regenerate the initial acceptor molecule, phosphoenol pyruvate:[*]

$$
\begin{array}{c}
\text{COOH} \\
| \\
\text{C=O} \\
| \\
\text{CH}_3
\end{array}
\;+\; P_i
\;\;\xrightarrow[\substack{\text{pyruvate, } P_i \\ \text{dikinase}}]{\text{ATP}\;\;\;\;\text{AMP} + \text{PP}_i}\;\;
\begin{array}{c}
\text{COOH} \\
| \\
\text{C—O—}\textcircled{P} \\
|| \\
\text{CH}_2
\end{array}
\;\;\; (\text{HS-5})
$$

Pyruvate   Phosphoenol pyruvate

The overall process (HS-1 through HS-5) is therefore cyclic, and the net result is a "feeder system" that provides for the initial entrapment of CO_2 by an alternative to the carboxylation of ribulose-1,5-diphosphate used in the Calvin cycle.

---

[*] Note that the phosphorylation of pyruvate in the regeneration step of the Hatch-Slack pathway (Reaction HS-5) is not achieved by a simple transfer of phosphate from ATP. Such a reaction would be thermodynamically very unfavorable; indeed the highly exergonic nature of the reverse process as it occurs in the glycolytic pathway (Reaction Gly-10) renders that an essentially irreversible process. Instead the phosphorylation of pyruvate in the Hatch-Slack pathway is coupled to the cleavage of ATP to AMP and inorganic pyrophosphate, PP_i. This device was encountered in the previous chapter (see Reaction Fat-1 of Chapter 6) and will appear again in the next chapter as a distinguishing feature of many biosynthetic reactions with an especially high energy requirement. The utility of pyrophosphate cleavage is discussed further in Chapter 8; for the present, it will suffice to note that when accompanied by the further cleavage of pyrophosphate to inorganic phosphate,

$$
PP_i + H_2O \xrightarrow[\text{pyrophosphatase}]{} 2P_i
$$

this becomes a mechanism for releasing the energy of *both* high-energy anhydride bonds of ATP, thereby exerting a much greater thermodynamic "pull" and making possible reactions such as the phosphorylation of pyruvate that would not be feasible otherwise.

The Hatch-Slack or C_4 pathway is therefore not a substitute for the Calvin cycle but simply an initial carbon-fixing process that precedes the Calvin cycle in certain species. In other words, C_4 plants possess the Hatch-Slack pathway in addition to rather than in place of the Calvin cycle.

Since the CO_2 trapped by the Hatch-Slack cycle is passed ultimately to the Calvin cycle anyway, one might well wonder what advantage is afforded to the C_4 plant that offsets the disadvantages of carrying around the extra enzymatic baggage and using the extra ATPs required for operation of the Hatch-Slack feeder system. To appreciate fully the Hatch-Slack cycle and the competitive advantages it confers upon C_4 plants would require a more extended discussion of plant physiology than is appropriate here. Suffice it to say that unlike the C_3 plants, the C_4 species have two distinctively different types of photosynthetic leaf cells and that the two component cycles of Figure 7-3 are actually located in different cell types. The initial fixation of CO_2 (via the Hatch-Slack cycle) occurs in *mesophyll cells*, which occupy more external locations in the leaf, with ready access to atmospheric CO_2. Subsequent carbon metabolism (by the Calvin cycle), on the other hand, takes place in internal *bundle sheath cells* found in close proximity to the translocation system of the leaf. Thus the outer cells can collect CO_2 from the air and move it (as four-carbon compounds) to the inner cells, where it can be passed to the Calvin cycle with minimal danger of its return to the environment. These features of C_4 leaf design are illustrated in Figure 7-4.

Figure 7-4 Spatial localization of the Hatch-Slack and Calvin cycles within the leaf of a C_4 plant. Carbon dioxide entering the leaf through a stomate is fixed into organic form by the Hatch-Slack pathway localized within the *mesophyll* cells. The immediate product of phosphoenol pyruvate carboxylation is oxaloacetate, but this is quickly converted to malate (by reduction; R = OH) or aspartate (by transamination; R = NH_2), depending upon the particular species of plant. Carbon moves into the *bundle sheath* cells in the leaf interior as the four-carbon acid, which is then decarboxylated. The CO_2 thus released is refixed by the Calvin cycle, yielding ultimately starch and sucrose. The sucrose passes into the adjacent vascular tissue for translocation to other parts of the plant. Meanwhile the pyruvate remaining from the decarboxylation reaction returns to the mesophyll cells for regeneration of phosphoenol pyruvate. The Hatch-Slack pathway operating in the mesophyll cells can be viewed as a pumping mechanism designed to maintain a locally high concentration of CO_2 for the carboxylating enzyme of the Calvin cycle operating in the bundle sheath cells. The particular C_4 leaf shown here is that of the grass *Panicum mileaceum*. (Courtesy of M. D. Hatch.)

To appreciate fully the merits of this arrangement, it is necessary to realize that Calvin cycle activity is always accompanied by *photo-respiration*, a light-dependent *release* of CO_2 and *consumption* of O_2, which obviously detracts from the efficiency of overall photosynthetic carbon fixation. Photorespiration, it turns out, involves a competition between CO_2 and O_2 for ribulose-1,5-diphosphate carboxylase, the CO_2-fixing enzyme of the Calvin cycle. Under conditions of high CO_2

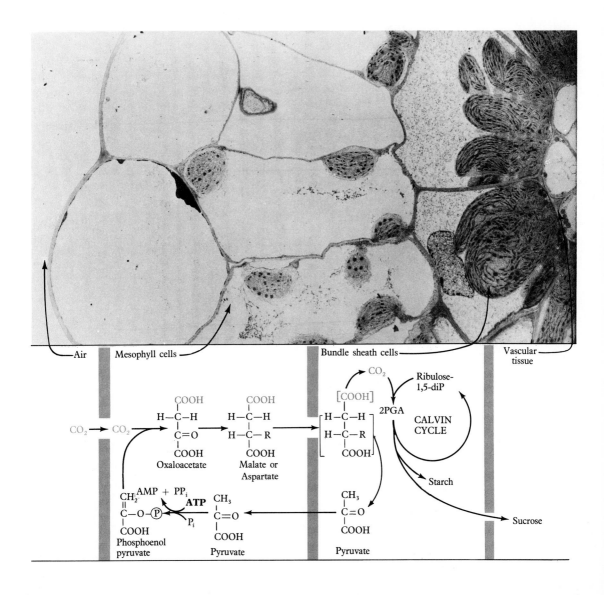

and low O_2, this enzyme functions as a carboxylase, fixing carbon as shown in Reaction Cal-1. At low CO_2 and high O_2, however, the same enzyme displays an alternative *oxygenase* function that, without going into detail, leads to the oxygen consumption and CO_2 evolution of photorespiration.

A critical function of the Hatch-Slack pathway, therefore, is to maintain the CO_2 concentration at a high level in the bundle sheath cells that carry out the Calvin cycle by essentially "pumping" CO_2 into these cells from the atmosphere surrounding the leaf. In addition, with the Calvin cycle and its associated photorespiration localized only in the inner cells of the bundle sheath, any CO_2 released by photorespiration must diffuse through the surrounding mesophyll cells before escaping from the leaf. This virtually ensures that the released CO_2 will be refixed by the Hatch-Slack pathway operative in the mesophyll cells, thereby preventing its loss from the leaf. As a result C_4 plants such as corn and sugarcane are characterized by net photosynthetic rates often two to three times those of C_3 plants such as the cereal grains. The agricultural significance of such differences can hardly be overestimated.

The Light Reactions of Photosynthesis

So far we have seen no involvement of light in any of the foregoing; as long as ATP and $NADPH_2$ are available, carbon is fixed and sugar is synthesized, with no direct requirement for light. In fact the nonphotochemical reactions of the Calvin cycle (with or without the initial C_4 fixation sequence) are frequently referred to as the "dark reactions" of photosynthesis to emphasize that light is not involved. The term "dark reactions" can be somewhat misleading, however; it simply indicates that light is not directly involved in these reactions and must not be misconstrued to mean that these reactions occur only in the dark. Actually the Calvin cycle sequence normally operates only in the light because of its ongoing requirement for concomitant regeneration of ATP and $NADPH_2$, both light-requiring processes.

Not only have we failed to discover the involvement of light as yet, but there is not even anything particularly unique about the ability of phototrophic organisms to fix carbon into organic compounds. Certainly CO_2 fixation processes (carboxylation reactions) occur in animals and other nonphotosynthetic organisms as well—somewhere in your tissues, you are undoubtedly fixing carbon right now. What *is* unique about autotrophs is their ability to carry out sustained *net fixation* of carbon, driven by the energy of solar radiation. Only photosynthetic organisms can utilize solar energy to extract electrons from poor donors like water and use them to reduce organic compounds.

Which brings us, therefore, to the photochemical reactions of photosynthesis—to the reason, in other words, that the "synthesis" is in fact "photo."

We can get no further in our discussion without mentioning *chlorophyll*, because this is the pigment that serves as the direct interface between sunlight and the biosphere—between photons of light and the chemistry of carbon fixation. In all photosynthetic organisms from bacteria to higher plants, it is the chlorophyll molecule that absorbs the energy of photons of light and uses that energy to excite electrons from their usual energy levels to higher energy states. This is, in itself, not a unique property of chlorophyll; many molecules absorb light energy and undergo electron excitation. But only chlorophyll can pass the energized electron (via several intermediates) to a molecule of the coenzyme NADP to reduce it to $NADPH_2$. Photosynthetic cells may contain a variety of other *accessory pigments* (especially the *carotenoids* and the *phycobilins* that frequently account for the coloration of plant parts and algal cells) which also absorb light energy that can be used to drive carbon fixation, but the energy must first be transferred from such pigments to chlorophyll. No other pigment molecule can substitute for chlorophyll as a source of electrons for NADP reduction.

Shown in Figure 7-5 is the molecular structure for *chlorophyll a*, the form of chlorophyll common to all oxygen-producing (water-oxidizing) phototrophs. In addition to chlorophyll *a*, all these organisms except the blue-green algae also contain a second kind of chlorophyll. This may be chlorophyll *b* (in green plants), chlorophyll *c* (in brown algae, diatoms, and dinoflagellates), or chlorophyll *d* (in red algae). Each of these substances is a chemical variant of the basic structure shown in Figure 7-5; chlorophyll *b*, for example, has the CH_3 group on ring II replaced by a CHO group. The photosynthetic bacteria, which cannot extract electrons from water and consequently do not evolve oxygen, contain only a single type of chlorophyll, called *bacteriochlorophyll*. Like chlorophylls *b*, *c*, and *d*, bacteriochlorophyll is simply a chemical variation of the basic structure shown in Figure 7-5. One form of bacteriochlorophyll, for example, differs from chlorophyll *a* in having all the carbon-carbon bonds of ring II saturated (fully hydrogenated).

The light reactions of photosynthesis can be considered conveniently as two component processes: (1) the actual light-dependent reduction of NADP (photoreduction) accompanied by the oxidation of an electron donor (water for the oxygen-evolving phototrophs); and (2) the light-dependent generation of ATP (photophosphorylation). We shall deal with each in turn.

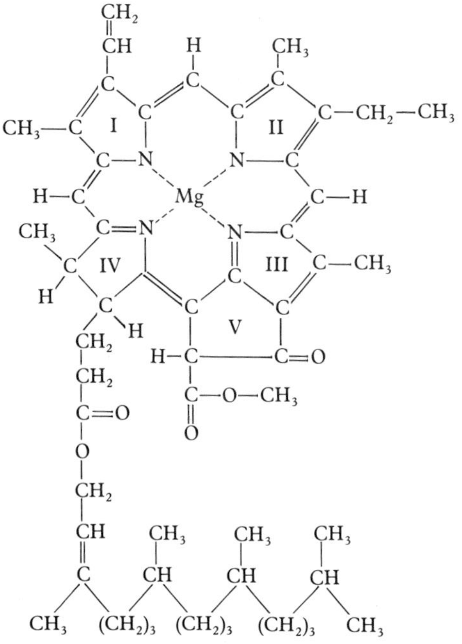

Figure 7-5 The structure for chlorophyll *a*.

Photoreduction (NADPH₂ Generation) When chlorophyll (or any other light-sensitive pigment, for that matter) absorbs a photon of light, the energy excites an electron to a higher energy level. The added energy can then be either reemitted by the excited molecule (as heat or light) or it can be passed on to another molecule to initiate a photochemical reduction. If the energy has not been used for this latter purpose in less than a nanosecond (10^{-9} sec), it has surely been lost as heat or light (usually the latter, as a fluorescent emission). In the case of chlorophyll, the energized electron can be passed, via several intermediates, to the coenzyme NADP, resulting in the reduction of the coenzyme and the oxidation of the chlorophyll. This light-driven reduction of NADP is called *photoreduction*. Like most biological reductions, this one requires two electrons per molecule of coenzyme reduced, so the reaction can be written as follows:

$$NADP + 2\ chlorophyll + 2H^+ + light \rightarrow NADPH_2 + 2\ chlorophyll^+ \quad (7\text{-}8)$$

If NADP is to be reduced continuously in this manner, then clearly something somewhere must be concomitantly oxidized, since the chlorophyll must regain the electron it lost to NADP in order to function catalytically in photoreduction. As mentioned previously, the photosynthetic bacteria use inorganic compounds such as H_2S or even organic molecules as electron donors, but for eukaryotic phototrophs (algae and higher plants) the universal source of electrons is water. Our discussion will center primarily on the eukaryotic case.

In an ill-understood reaction (or, more likely, series of reactions), an electron-deficient chlorophyll molecule is able to extract electrons from the water molecule, oxidizing the water to oxygen. This light-dependent oxidative splitting of water is called *photolysis*. Again the electrons move in pairs, so the reaction for the oxidation of water is as follows:

$$H_2O + 2\,\text{chlorophyll}^+ + \text{light} \rightarrow \tfrac{1}{2}O_2 + 2\,\text{chlorophyll} + 2H^+ \qquad (7\text{-}9)$$

By summing Reactions 7-8 and 7-9, we arrive at an overall equation that accounts for the use of water, the evolution of oxygen, and the generation of $NADPH_2$ during photosynthesis:

$$NADP + H_2O + \text{light} \xrightarrow[\text{chlorophyll}]{} NADPH_2 + \tfrac{1}{2}O_2 \qquad (7\text{-}10)$$

So far we have been concerned only with the overall process of water oxidation and NADP reduction, as summarized by Reaction 7-10. The actual mechanism is more complex than the summary equation might suggest, however. This becomes clear when we realize that the chlorophyll molecules which donate electrons to NADP are not the same molecules which extract electrons from water. Clearly, if electrons are being picked up from water by one system of chlorophyll molecules but are being donated to NADP by another system, there must be some sort of transport system moving electrons from one system to the other. Further evidence for the complexity of the process comes from the known participation in the light reactions of a variety of compounds that are capable of being reversibly oxidized and reduced. These include the *cytochromes* (iron-containing compounds similar to those involved in electron transport in the mitochondrion), the *plastoquinones*, *plastocyanins*, and *ferredoxin*. The exact ordering of these components to provide for the efficient transfer of electrons from water to NADP via two chlorophyll systems is not yet completely clear, but a likely model is that shown in Figure 7-6.

Inherent in this and all similar schemes is the involvement of two separate light-trapping systems to provide the energy input necessary for spanning the electrochemical gradient represented by NADP/$NADPH_2$ ($E_0 = -0.32$ volt) and O_2/H_2O ($E_0 = +0.82$ volt). One light-trapping system, called *photosystem I*, uses light to accomplish the chemistry summarized in Reaction 7-8: electrons of chlorophyll are energized and used to reduce a highly electronegative but as yet unknown compound called "Z" (sometimes also called "ferredoxin

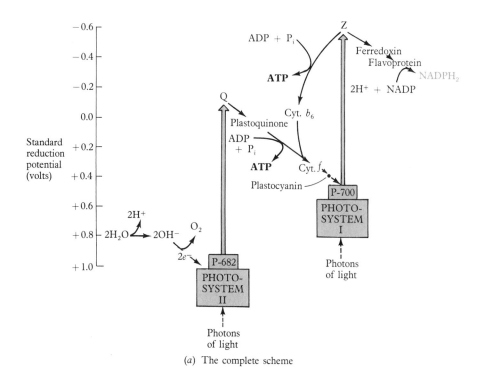

(a) The complete scheme

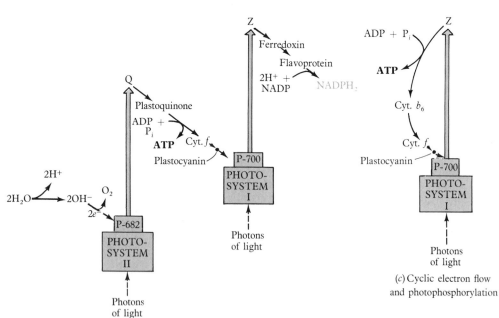

(b) Noncyclic electron flow and photophosphorylation

(c) Cyclic electron flow and photophosphorylation

reducing substance," FRS), which in turn reduces ferredoxin. From ferredoxin the electrons are transferred "downhill" (via a flavoprotein called *NADP reductase*) to NADP, thereby generating the reduced coenzyme required by the dark reactions. Meanwhile the second light-trapping system, *photosystem II*, uses a separate grouping of chlorophyll molecules to accomplish the chemistry summarized in Reaction 7-9: electrons are extracted from water, energized, and passed to a more electronegative receptor designated as "Q." Net electron flow from water to NADP is possible because of the electron transport chain connecting photosystems I and II. Electrons passed to the carrier Q by photosystem II flow down a chain of carriers consisting, at a minimum, of plastoquinone, cytochrome *f*, and plastocyanin. Photosystem I accepts these electrons, energizes them further, and passes them eventually to NADP.

The essential features of the complete system are (1) the photochemical system (II) that uses the energy of absorbed light to elevate an electron from the level of water ($+0.82$ volt) to that of the carrier Q at the "top" of the transport chain (-0.10 volt); (2) a second photochemical system (I) that uses additional light energy to elevate the electrons from the level of the plastocyanin at the "bottom" of the transport chain ($+0.40$ volt) to that of the compound Z at the most electronegative end of the scheme (-0.60 volt); (3) the enzymes and carriers necessary to mediate the actual extraction of electrons from water (or, more specifically, from OH^- ions derived from water), with liberation of O_2; (4) the enzymes and carriers necessary to mediate the actual transfer of electrons from Z via ferredoxin to NADP; and (5) the electron carriers that connect the two photosystems and allow net, unidirectional transport of electrons from photosystem II to I (and thereby from water to NADP).

Figure 7-6 The light reactions of photosynthesis. The major reactions involved in the light-dependent transfer of electrons from water to NADP and in the coupled photophosphorylation of ADP to ATP are shown in part (*a*) with the intermediates in electron transport arranged according to the reduction potentials. The particular scheme shown here should be regarded only as a likely model, since many of the details of photosynthetic electron transport are still unclear, and a variety of such schemes have been presented. The separation of the composite scheme (*a*) into its noncyclic (*b*) and cyclic (*c*) components is intended primarily as an aid in understanding the two modes of electron flow; in the intact chloroplast, the composite scheme is presumed to function.

The two separate photosystems of Figure 7-6 are apparently required because the energy gradient from water to NADP is too great to be spanned by a single photochemical event. The best evidence for the actual existence of two separate photochemical reactions is the so-called *Emerson enhancement effect*, discovered by R. Emerson and co-workers in 1957. Their observation was that a greater photosynthetic yield can be achieved with red light of two slightly different wavelengths than is possible by summing the yields obtained with either wavelength separately. This effect can be explained in terms of the sensitivity of photosystem I to red light of slightly longer wavelengths (690–720 nm) than those to which photosystem II is maximally sensitive (below 690 nm).

Photophosphory-
lation (ATP
Generation)

Thus far we have seen how electrons are extracted from a thermodynamically reluctant donor (water) and transferred, via a chain of reversibly reducible intermediates, to a reluctant acceptor (NADP) by using the energy of light to overcome the energy gradient. This accounts for the generation of $NADPH_2$, one of the two necessary Calvin cycle reactants for which the light reactions are responsible. We come now to a consideration of the means by which the second light-dependent compound, ATP, is generated.

As we already know from the discussion of oxidative phosphorylation in Chapter 5, ADP can be phosphorylated to ATP as electrons are transported down an electrochemical gradient, as in the mitochondrial electron transport chain. The makings of a similar chain can be seen in Figure 7-6a, as electrons flow downhill between the "top" of photosystem II and the "bottom" of photosystem I via a series of electron carriers. And ATP synthesis is coupled, probably at a single point, to this "downhill" flow of electrons just as it is to the analogous flow in the electron transport chain of respiratory metabolism. The only difference is that the electrochemical gradient in that case is created by oxidative metabolism and here by photochemical events. Hence the ATP-generating process of respiratory metabolism is called *oxidative phosphorylation* and that of photosynthesis is termed *photophosphorylation*.

In a sense, then, ATP formation or photophosphorylation is a by-product of the two-photosystem transfer of electrons from water to NADP, but it is probably the need for concomitant ATP generation that explains the existence of two separate photosystems with the opportunity for phosphorylation during the downhill flow of electrons between them. This overall process, as illustrated in Figure 7-6b, is termed *noncyclic photophosphorylation* inasmuch as ATP generation is accomplished by a unidirectional flow of electrons from H_2O to NADP.

To summarize noncyclic electron flow, we can rewrite Equation 7-10 as follows (based on the passage of a pair of electrons through the system):

$$\text{NADP} + \text{H}_2\text{O} + \text{ADP} + \text{P}_i + \text{light} \xrightarrow[\text{noncyclic photophosphorylation}]{}$$
$$\text{NADPH}_2 + \tfrac{1}{2}\text{O}_2 + \text{ATP} \qquad (7\text{-}11)$$

By multiplying this equation by 12 (equivalent to allowing 12 electron pairs to flow from H_2O to NADP), we can account for the 12 NADPH_2 molecules needed to synthesize one glucose molecule by the Calvin cycle (compare Equations 7-7 and 7-12):

$$12\text{NADP} + 12\text{H}_2\text{O} + 12\text{ATP} + 12\text{P}_i + \text{light} \xrightarrow[\text{noncyclic photophosphorylation}]{}$$
$$12\text{NADPH}_2 + 6\text{O}_2 + 12\text{ATP} \qquad (7\text{-}12)$$

Having seen how both ATP and NADPH_2 are generated in light-dependent reactions, our discussion is nearly complete, since these are the only two needs for the continued fixation of CO_2 by the Calvin cycle. A comparison of Reaction 7-12 with the summary equation for the Calvin cycle (Reaction 7-7) reveals one final problem, however. Equation 7-12 suggests that NADPH_2 and ATP are obligatorily generated in equimolar amounts, such that the production of the 12 NADPH_2 required by the stoichiometry of the Calvin cycle is accompanied by the generation of 12 ATP, whereas in fact the Calvin cycle requires proportionately more ATP (18 ATP versus 12 NADPH_2 per glucose). Where, we may ask, do the extra ATPs come from?

The answer is provided by a chemical option that makes it possible for photosystem I to generate ATP *without* concomitant NADP reduction. As shown in Figure 7-6c, this is accomplished by allowing the electrons from compound Z to pass down an electron transport chain and return to chlorophyll P-700 rather than being used to reduce NADP. As might be predicted, this flow of electrons is coupled at one point at least to the generation of ATP. By this means ATP is produced, but no water is split, no oxygen is evolved, and no NADPH_2 is formed. Light is simply used to cycle electrons around photosystem I and through a transport chain, generating ATP and returning the electrons to the chlorophyll from whence they came. For obvious reasons this process is termed *cyclic photophosphorylation*. The summary equation for this process can be represented (per electron pair) as follows:

$$\text{ADP} + \text{P}_i + \text{light} \xrightarrow[\text{cyclic photophosphorylation}]{} \text{ATP} \qquad (7\text{-}13)$$

To meet the stoichiometric needs of the Calvin cycle (Equation 7-7), we need only specify the flow of 12 pairs of electrons through the noncyclic pathway (12 × Equation 7-11) and 6 pairs of electrons through the cyclic option (6 × Equation 7-13). We can then write the following summary equations:

Noncyclic: $12NADP + 12H_2O + 12ADP + 12P_i \xrightarrow{\text{light}}$
$$12NADPH_2 + 6O_2 + 12ATP$$

Cyclic: $6ADP + 6P_i \xrightarrow{\text{light}} 6ATP$

Sum: $12NADP + 12H_2O + 18ADP + 18P_i \xrightarrow[\text{photosystems I and II}]{\text{light absorbed by}}$
$$12NADPH_2 + 6O_2 + 18ATP \qquad (7\text{-}14)$$

Summary of Photosynthesis

Now we are in a position to write an overall equation for photosynthesis (as carried out by C_3 plants) by summing the light and dark reactions. If we add the summary equation for the Calvin cycle (Equation 7-7) to the summary expression for the light reactions (Equation 7-14), we arrive at an overall expression as follows:

Calvin cycle: $6CO_2 + 12NADPH_2 + 18ATP \longrightarrow$
$$C_6H_{12}O_6 + 12NADP + 18ADP + 18P_i + 6H_2O$$

Light reactions: $12NADP + 12H_2O + 18ADP + 18P_i \xrightarrow{\text{light}}$
$$12NADPH_2 + 6O_2 + 18ATP$$

Overall: $6CO_2 + 6H_2O \xrightarrow{\text{light}} C_6H_{12}O_6 + 6O_2 \qquad (7\text{-}15)$

This, then, brings us back to where we started at Equation 7-2. But you should now be in a much better position to understand some of the metabolism and photochemical complexity behind what might otherwise appear to be a misleadingly simple reaction.

Energetics and Efficiency of Photosynthesis

Thus far solar energy has been invoked as the driving force behind photosynthesis, and chlorophyll has been introduced as the pigment uniquely responsible for absorbing sunlight and converting it into chemical energy. Nothing, however, has yet been said about either the quality (wavelength) or quantity (number of photons) of light required by chlorophyll in the photosynthetic process.

Light, of course, is electromagnetic radiation, and visible light is restricted to the rather narrow wavelength range from about 400 to about 700 nm. The relationship between the wavelength λ and the energy content e of a photon (quantum packet) of light is given by the equation

$$e = \frac{hc}{\lambda} \qquad (7\text{-}16)$$

where e = energy content of one photon or quantum of light
  λ = wavelength of that photon in nm (10^{-9} m)
  h = Planck's constant = 1.58×10^{-37} kcal-sec/photon
  c = velocity of light = 3.0×10^{17} nm/sec

Like chemical molecules, light can be usefully considered in units of moles. A mole (6.023×10^{23} quanta) of light is called an *einstein*, and its energy content E can be obtained by simply multiplying the preceding expression by Avogadro's number, N:

$$E = \frac{Nhc}{\lambda} = \frac{(6.023 \times 10^{23})(1.58 \times 10^{-37})(3.0 \times 10^{17})}{\lambda} \qquad (7\text{-}17)$$

$$= \frac{2.86 \times 10^4}{\lambda} \text{ kcal/einstein}$$

Note that the energy of light is inversely proportional to wavelength, such that light at the violet end of the visible spectrum is highest in energy while that at the red end is lowest. Table 7-1 summarizes the energy content and color of light at several points along the visible spectrum.

TABLE 7-1 Energy Content of Light within the Visible Spectrum

Wavelength (nm)	Color	Energy Content (kcal/einstein)
400	Violet	72
500	Blue-green	57
600	Orange	48
700	Infrared	41

Clearly the energy content of the photons of light used for a specific photochemical process is dependent upon the wavelengths of light actually absorbed by the pigment involved. For photosynthesis, the primary pigments are obviously the chlorophylls, although the accessory pigments may also account for significant light absorption.

Figure 7-7 shows the absorption spectra for chlorophylls *a* and *b*. The absorption spectrum shown in Figure 7-8, on the other hand, is not that of an isolated pigment but that of an intact photosynthetic tissue, the thallus of the green alga *Ulva taeniata*. Also presented in Figure 7-8 is the action spectrum for photosynthesis in the thallus. An *action spectrum* depicts the dependence of an actual physiological process on wavelength; an *absorption spectrum* indicates the relative extent to which light of different wavelengths is absorbed. Determination of the action spectrum for a light-dependent process is often one of the first steps toward identifying the pigment involved in the process.

Note the agreement between the action spectrum and the absorption spectrum of Figure 7-8 and the extent to which both reflect the absorption characteristics of isolated chlorophyll, as shown in Figure 7-7. It is this strong absorption in the blue and red spectral regions with an absorption minimum around 550 nm that gives much of the photosynthetic world, including the *Ulva* thallus, its characteristic green color. Even a casual comparison of Figures 7-7 and 7-8 indicates, however, that the absorption spectrum of the algal thallus only approximates that of isolated chlorophyll. In particular, the thallus tissue absorbs and utilizes light in the wavelength range 480–600 nm more effectively than would be predicted from the absorption profile for chlorophyll alone. This is indicative of absorption by

Figure 7-7 The absorption spectra for ether extracts of chlorophylls *a* and *b*.

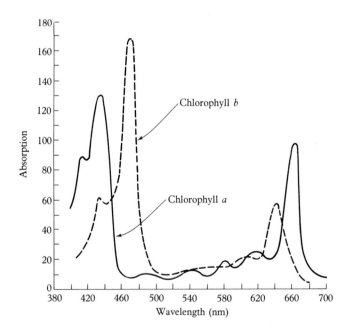

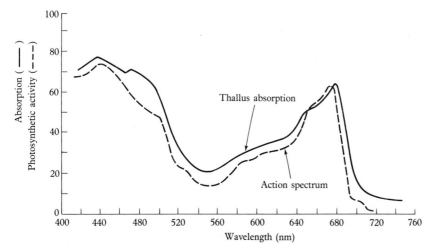

Figure 7-8 The action spectrum and absorption spectrum for the photosynthetic thallus of the green alga *Ulva taeniata.* Note the agreement between the absorption by the intact tissue shown here and the absorption spectra of isolated chlorophylls as shown in Figure 7-7. (The slightly higher wavelengths for the absorption maxima of the intact tissue are the result of the aqueous environment rather than the ether solvent of Figure 7-7.)

accessory pigments, which often modify the absorption and action spectra of photosynthetic tissues.

Since the electrons of light-activated chlorophyll are excited through a potential gradient of about -1.0 volt (see Figure 7-6), we can readily calculate (from Equation 5-9) the free-energy change this represents:

$$\Delta G^0 = -nF\Delta E_0 = (-1)(23.04)(-1.0) = 23 \text{ kcal/mole} \qquad (7\text{-}18)$$

Clearly, then, visible light of any wavelength is sufficiently energetic to drive the one-electron transfer involved in the chlorophyll photosystems. At 670 nm, the approximate absorption maximum for chlorophyll a in most cells, light has an energy content of about 43 kcal/einstein. Assuming that one photon of light is required to excite one electron, the excitation of electrons by red light occurs with an efficiency of energy transduction of about 54 percent.

If we assume that the scheme shown in Figure 7-6 and summarized in Equation 7-12 is essentially correct, we can also calculate the overall efficiency of photosynthesis from sunlight to sugar molecule. According to the scheme of Figure 7-6, the synthesis of one molecule of glucose requires that 12 pairs of electrons traverse the noncyclic route, with two excitations per electron. Additionally, six pairs of electrons must cycle in photosystem I to generate the extra ATP required, and this photosystem involves a single excitation per electron. The number of excitation events per glucose is therefore 48 from noncyclic electron flow (24 electrons × 2 excitations per electron) and 12 from cyclic transport (12 electrons × 1 excitation per electron), for a total of 60 excitation events per sugar. This is equivalent to ten photons absorbed per CO_2 fixed, a figure consistent with the experimental estimates of eight to ten photons per CO_2 obtained by many investigators in the highly controversial research area of determining photosynthetic quantum requirements. For red (670-nm) light, 60 photons represents about 60 × 43 = 2580 kcal of light absorbed per mole of glucose, and since glucose differs in free energy from CO_2 and H_2O by 686 kcal/mole, the overall efficiency of photosynthetic energy transduction is about 27 percent. Although efficiencies of this order can be observed experimentally in the laboratory with algal cells or isolated chloroplasts, photosynthesis under field conditions occurs with a much lower overall efficiency, such that only 1 to 2 percent of the sun's energy falling on a green plant during the growing season is actually used for photosynthesis, and only about 0.2 to 0.4 percent of the total energy of incident solar radiation is stored by photosynthesizing plants annually. Problem 7-2 will provide you with an opportunity to quantitate the practical efficiency of photosynthesis for an actual agricultural crop and to consider some of the reasons that account for the apparent discrepancy between theoretical laboratory values and practical agricultural experience.

Chloroplast Structure and the Photosynthetic Apparatus

In Chapter 5 we saw that in the eukaryotic cell all the reactions of respiratory metabolism beyond the pyruvate molecule are localized in the mitochondrion. Analogously most of the events of photosynthesis occur within the *chloroplast* in the eukaryotic cell.* Both these promi-

---

*All the reactions of the Calvin cycle shown in Figure 7-1 (Reactions Cal-1 through Cal-17) occur within the chloroplast, as does the sequence of reactions necessary to synthesize starch from phosphorylated glucose. This is clear from the observable accumulation of starch within chloroplasts during photosynthesis. Also clear, however, is that the inner membrane of the chloroplast is impermeable to sucrose, the major form in which carbon is translocated in higher plants, and that the final steps in sucrose biosynthesis occur not in the chloroplast but in the cytoplasm. Specifically, it appears that carbon flows out of the chloroplast not as a six-carbon sugar but as the three-carbon compound dihydroxyacetone phosphate, and that the subsequent metabolism required to synthesize sucrose from the three-carbon level occurs in the cytoplasm. This includes steps Cal-4 to Cal-7, as well as the reactions necessary to synthesize sucrose from glucose-6-phosphate and fructose-6-phosphate. Thus the sequence from dihydroxyacetone phosphate to glucose-6-phosphate apparently occurs both in the chloroplast (for starch synthesis) and in the cytoplasm (for sucrose synthesis). Note also that, both for the mitochondrion and for the chloroplast, carbon flow across the inner membrane occurs as a three-carbon compound, pyruvate in one case and dihydroxyacetone phosphate in the other.

nent eukaryotic organelles can be seen clearly in the algal cell shown in Figure 7-9. Prokaryotes, on the other hand, possess neither mitochondria nor chloroplasts and therefore do not compartmentalize respiratory or photosynthetic processes in the same sense that eukaryotes do.

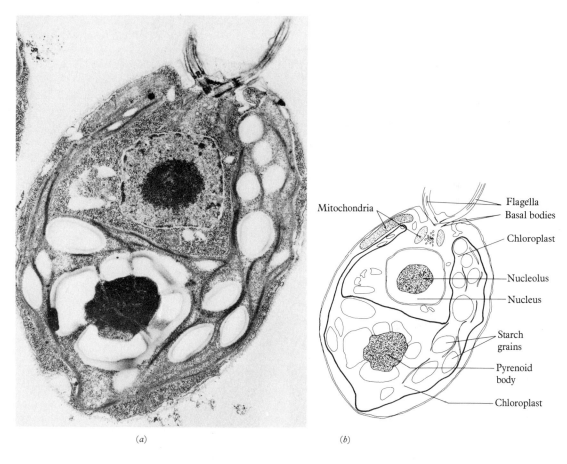

(a) (b)

Figure 7-9 Internal structure of a eukaryotic algal cell. Although leaf cells of higher plants usually contain numerous chloroplasts, algal cells generally have only one or a few. (a) The flagellated green alga *Chlamydomonas reinhardi* shown here (10,750×) has a single chloroplast that occupies much of the cell. The outline of the chloroplast is indicated by the dark line in the accompanying sketch (b). Structures within the chloroplast include the starch grains and the pyrenoid body. Other cellular features labeled on the sketch include the nucleus, nucleolus, mitochondria, and flagella with basal bodies. (Photo (a) courtesy of D. L. Ringo.)

Shown in Figure 7-10 is a typical chloroplast as visualized in the electron microscope. Like the mitochondrion, the chloroplast has both an outer and an inner membrane, with the inner membrane giving rise to an intricate network of specialized internal membranous structure. In the case of the chloroplast, the internal network of membranes is characterized by many flattened *vesicles* (hollow, saclike structures) called *thylakoid disks*, and these in turn are organized into stacks called

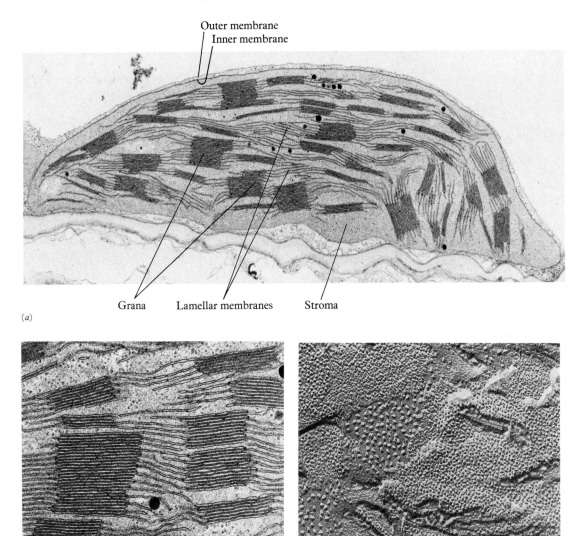

Outer membrane
Inner membrane

Grana   Lamellar membranes   Stroma

(a)

(b)   (c)

grana (singular: granum). The grana are interconnected by membranous strands called lamellae running longitudinally through the semifluid matrix or *stroma* of the chloroplast.

TABLE 7-2 Localization of Processes within Mitochondria and Chloroplasts

		Carbon Metabolism		Electron Transport	
Process	Organelle	Pathway	Localization	Process	Localization
Respiratory metabolism	Mitochondrion	TCA cycle, β-oxidation	Matrix	Oxidative phosphorylation	Cristae
Photosyn- thesis	Chloroplast	Calvin cycle	Stroma (matrix)	Photoreduction and photophos- phorylation	Grana and lamellae

As indicated in Table 7-2, the similarity between mitochondria and chloroplasts extends to the localization of the component parts of the respiratory and photosynthetic processes within their respective organelles. In both cases the reactions involved in actual carbon metabolism (the TCA cycle and β-oxidation for respiration, the Calvin cycle for photosynthesis) occur in the soluble matrix of the organelle, whereas the components of electron transport and phosphorylation are localized on specialized structures derived from the inner membrane. For the chloroplast this means that the grana contain essentially all the photosynthetic pigments, the enzymes required for the light reactions, the carriers involved in electron transport, and the factors which couple that transport to phosphorylation. In photosynthetic prokaryotes (both bacteria and blue-green algae), all these components are also membrane-associated, but with the cellular rather than an organellar membrane. A photosynthetic bacterium is shown in Figure 7-11.

Figure 7-10 Structure and organization of the chloroplast. Shown here (*a*) at 21,960× is an intact chloroplast from the leaf of timothy grass, clearly illustrating the outer and inner membranes, the grana, and the lamellar membranes running through the stroma. At higher magnification (36,783×), the individual thylakoid disks of grana (*b*) can be seen better. In (*c*) is a thylakoid membrane (73,200×), as viewed by the freeze-etching technique described in the text. The particles seen embedded within the membrane are thought to be quantasomes, the structural units corresponding to the photosystems involved in the light reactions. See Figure 7-12 for a model of the possible organization of these particles within the membrane. (Photos (*a*) and (*b*) courtesy of W. P. Wergin and E. H. Newcomb; (*c*) courtesy of D. Branton.)

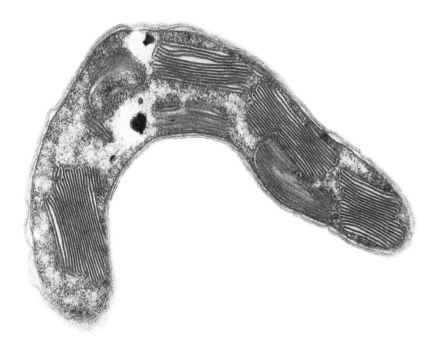

Figure 7-11 Structure and organization of a photosynthetic bacterium. Photosynthetic prokaryotes do not contain chloroplasts but instead have the pigments, electron carriers, and enzymes of the light reactions localized on membranes derived from the cellular membrane. Shown here (37,500×) is the bacterium *Ectothiorhodospira mobilis.* (Courtesy of S. W. Watson.)

A further similarity between mitochondrial and chloroplast structure is evident in the molecular organization of electron transport and phosphorylation. The photosystems of Figure 7-6 are not just biochemical abstractions to explain light absorption and electronic excitation. Like the respiratory assemblies of the mitochondrion, they are thought to represent discrete units of organization embedded in the inner membrane. Each of the photosystems (or *photosynthetic units,* as they are often called) is a characteristic assembly of light-absorbing pigments, consisting of about 200 to 300 chlorophyll molecules plus the accessory pigments appropriate to the particular species of plant and the specific photosystem.

Although all the pigment molecules within a photosynthetic unit are capable of absorbing photons of light, only one molecule in the cluster can actually use that energy in the photochemical reaction whereby an electron is passed to an acceptor molecule to initiate electron flow. That special pigment molecule is called the *reaction center* (or "energy trap") of the photosystem and is in both photosystems a long-wavelength form of chlorophyll *a*. (Each photosystem contains several varieties of chlorophyll *a*, which can be distinguished from one another by the characteristic wavelengths at which they absorb maximally.) In photosystem I the reaction center is a form of chlorophyll *a* called P-700. (The "P" stands for pigment and the "700" designates the absorption peak in nanometers.) Thus it is P-700 that is shown as the actual source of the electrons for NADP reduction in Figure 7-6. The corresponding chlorophyll molecule at the reaction center of photosystem II has an absorption maximum at 682 nm and is accordingly designated in Figure 7-6 as P-682. A photosystem, then, can be thought of as a unit of organization within the chloroplast inner membrane, consisting of an antennalike network of light-gathering pigments with a special photochemically active chlorophyll molecule at its center. Light energy absorbed by a pigment molecule anywhere in the network is transferred to the reaction center, where it can be used by the P-700 or P-682 molecule to initiate photochemical electron flow.

If photosystems are real structural entities (rather than, for example, statistical ratios between pigment molecules and reaction centers within the membrane), then what the physiologist calls a photosynthetic unit might be expected to manifest itself morphologically as well. In fact electron microscopy of thylakoid membranes often reveals regular repeating structures that have been termed *quantasomes*. Quantasomes were originally reported to be structural units with dimensions of about 200 × 100 Å (20 × 10 nm), but it is now possible to distinguish two kinds of particles, differing in both size and localization within the membrane. This distinction was made possible by use of a technique called *freeze-etching*.

For freeze-etching, a biological sample (the thylakoid membrane in this case) is frozen and fractured with a sharp blow. A platinum-carbon etching or replica is then made of the fractured sample. When the etching is examined in the electron microscope, what is in essence being observed is the fracture plane of the sample. Since the fracture plane apparently runs along the interior of the membrane (a point on which there is still some disagreement), it is possible by this technique in effect to "open up" a membrane and observe its inner components. Applied to the thylakoid membrane, freeze-etching has revealed two different kinds of particles, which are separated from each other by the fracture plane, presumably because they are located on opposing sides of the thylakoid membrane. Particles with a diameter of about 110 Å are thought to be the structural units of photosystem I, while those with a diameter of about 175 Å correspond to photosystem II (Figure 7-10c). A model to illustrate the possible organization of these particles within the thylakoid membrane is presented in Figure 7-12.

In addition to the particles firmly embedded within the thylakoid membrane, a third type of particle with a diameter of about 100 Å is found more loosely attached to the outer surface of the membrane. These particles appear as knoblike bumps when the surface of the thylakoid membrane is examined by high-resolution electron microscopy and are therefore assigned a more superficial location on the model of Figure 7-12. These 100-Å particles can be isolated and shown to have ATPase activity. They are generally regarded as analogous to the mitochondrial coupling factor (F_1) discussed in Chapter 5 and are therefore called CF_1 particles (for chloroplast coupling factor). Though characterized by ATP-cleaving activity in the test tube, CF_1, like the mitochondrial F_1, is thought to function in an ATP-generating capacity in the cell. Thus CF_1 is likely to be the enzyme responsible for the ATP synthesis that accompanies light-driven electron transport in both the cyclic and noncyclic modes.

The Coupling of ATP Synthesis to Photosynthetic Electron Transport

The phosphorylation of ADP to ATP as it occurs during the light reactions of photosynthesis is strikingly similar in overall design to the oxidative phosphorylation of respiratory metabolism. In both cases the flow of electrons down a potential gradient provides the driving force for the coupled phosphorylation events. Both processes also

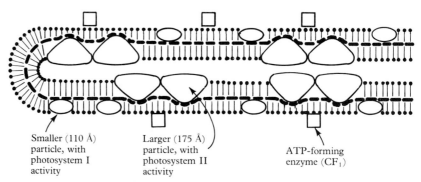

Smaller (110 Å) particle, with photosystem I activity

Larger (175 Å) particle, with photosystem II activity

ATP-forming enzyme (CF_1)

Figure 7-12 A model for the organization of the thylakoid membranes. The particles thought to correspond to units of photosystems I and II are apparently located within the membrane that makes up each thylakoid disk, while the coupling factor CF_1 responsible for ATP generation is located closer to the surface. The dotted line indicates the fracture plane that would be necessary to separate the 110-Å and 175-Å particles from each other, as is observed by the freeze-etching technique described in the text.

share a similar structural basis in that the units of electron transport are embedded in the inner membrane of the respective organelle, while the actual phosphorylation event involves an enzyme localized in a particle on the membrane surface. It is probably not surprising, then, that the same three mechanisms proposed to explain oxidative phosphorylation—chemical, chemiosmotic, and conformational coupling—have also been proposed to explain photophosphorylation. Nor is it likely to come as a surprise that, as in the mitochondrial case, the available data do not yet allow a definitive choice between these models.

Studies with chloroplast phosphorylation have provided especially strong supportive evidence for the chemiosmotic hypothesis, which depends, as you will recall, upon an H^+ gradient induced across the membrane as electron transport proceeds. Illuminated chloroplasts can be shown to take up H^+ ions from their surroundings and to release K^+ and Mg^{++} ions. These ion movements are strikingly similar to those seen in the mitochondrion but with the distinction that the *direction* is reversed: H^+ ions are ejected during mitochondrial electron transport but are taken up in the chloroplast case. In an important series of experiments involving rapid changes or "jumps" in pH, A. Jagendorf and his colleagues were able to show that isolated chloroplasts can be induced to phosphorylate ADP in response to an experimentally imposed pH gradient. Specifically, if chloroplasts are placed in a low-pH bath (pH 4.0) to decrease their internal pH, and the pH of the external solution is then raised rapidly (to pH 8.0), a burst of phosphorylation can be obtained. These findings are in striking agreement with the predictions of the chemiosmotic hypothesis. It is also true, however, that the inner chloroplast membrane undergoes dramatic conformational changes upon illumination. Advocates of the conformational coupling model point out the similarities with the changes that occur in the inner mitochondrial membrane and suggest that both represent a means of converting the energy of electron flow to the synthesis of ATP. The eventual elucidation of the mechanism that couples electron flow to ATP generation will be a major advance in cellular bioenergetics. And when it comes, the answer will almost certainly apply to both mitochondria and chloroplasts.

With the conclusion of our discussion on photosynthesis, we come to the end of Part 2. Throughout these chapters our emphasis has been on the sources of energy for cells—on the means whereby cells generate the ATP so vital to every cellular function. Having explored in detail the chemistry and energetics of ATP production from substrates as   **Onward to Part 3** diverse as sugars, fats, amino acids, and sunlight, we are ready now to turn to the complementary question of the uses to which cells put that ATP and the energy it represents. And for that, we fill our buckets with ATP and proceed to Part 3.

Practice Problems

7-1. Rationale for Photosynthesis: From the higher plant's point of view, the "purpose" of photosynthesis is to use the energy of solar radiation incident upon the green leaves to produce sugars that can then be translocated to other portions of the plant (roots, stems, flowers, developing fruits) for nourishment. Viewed as a problem in bioenergetics, this means that the plant is using solar energy to make ATP in the leaves (by photophosphorylation), then converting the energy of ATP into bond energies of sugar molecules (by the Calvin cycle), moving those sugar molecules through the phloem to other parts of the plants (translocation), and using the sugar to regenerate ATP (respiratory metabolism) needed to support the growth of the plant. Thus ATP is used to make sugar and the sugar is used in turn to make ATP. It seems as if it would be simpler for the plant just to have the leaves make ATP and then translocate the ATP itself directly to other parts of the plants, thereby completely eliminating the need for a Calvin cycle, a glycolytic pathway, and a TCA cycle (and making life considerably simpler for cell biology students in the process). Why do you suppose this is in fact *not* the way a plant operates its energy economy? You should be able to think of at least two major reasons.

7-2. Practical Efficiency of Photosynthesis: From the chemotroph's point of view, the "purpose" of photosynthesis is to provide a source of carbon and energy for the nonphotosynthetic components of the biosphere. Thus we grow crops such as sugar beets that have been highly selected for their ability to store carbon in vast excess of the immediate needs of the plant, and we use that stored food as our link to the energy of the sun. Let us see how efficient a crop like sugar beets is at converting solar energy into sugar. The growing season for sugar beets in Wisconsin is from about May 15 to September 1, the average length of daylight during that period is 15 hr/day, and a respectable yield is 15 tons of beets per acre, of which about 20 percent is sucrose. Further useful information is as follows:

$$1 \text{ acre} = 43,560 \text{ ft}^2$$
$$1 \text{ inch} = 2.54 \text{ cm}$$
$$1 \text{ pound} = 454 \text{ g}$$
$$1 \text{ einstein} = 6.02 \times 10^{23} \text{ photons}$$

The quantum requirement of photosynthesis is about 10 quanta (photons) of light absorbed per molecule of CO_2 fixed and reduced.

Assume that of the total solar radiation impinging upon the upper atmosphere (1.94 cal/cm^2-sec, the solar energy constant), the amount actually reaching the earth's surface is 1.4 cal/cm^2-sec.

(a) What is the minimum amount of light energy in einsteins that would have to be absorbed during the growing season by an acre of sugar beets to produce the 3 tons of sucrose actually obtained?

(b) What is the maximum theoretical amount of light energy in einsteins available to the sugar beet crop during the growing season? (Assume an average energy content of 55 kcal/einstein for visible light.)

(c) What is the efficiency of conversion of available light energy to sugar?

(d) Why is the efficiency calculated in part (c) so low? You should be able to think of three or four major reasons.

7-3. Photosynthetic Carbon Metabolism: If a green leaf (of a C_3 plant) is illuminated in the presence of radioactive CO_2 ($^{14}CO_2$) for a few seconds and then extracted to isolate various compounds, glycerate-3-phosphate will be found to contain label predominantly in carbon atom 1. In which positions of glucose-6-phosphate would you expect to find the most ^{14}C? Why?

7-4. Carbon Fixation: The cells of your own body possess catabolic pathways that can give rise to $NADPH_2$ and ATP, and they also have enzymes that can carboxylate organic substrates. (Your cells can, for example, "fix" CO_2 onto phosphoenol pyruvate to form oxaloacetate just as the C_4 plants do.) Why, then, is it not possible for people to effect net synthesis of sugars and other organic compounds by reductive fixation of CO_2 just as the autotrophs do?

7-5. Energetics of Photosynthesis: Phosphorylation is coupled to electron transport in both respiratory metabolism and photosynthesis, but the ATP yields per electron pair are different. For respiratory transport, three ATPs are produced per electron pair (assuming $NADH_2$ as the electron donor), whereas the transport chain linking photosystems I and II yields only one ATP per electron pair. Although cyclic electron transport around photosystem I is shown in Figure 7-6 to couple to a single phosphorylation event per electron pair, there is still uncertainty among investigators as to whether one or two phosphorylation events are coupled to this transport chain.

(a) Explain the difference in ATP yield per electron pair between respiratory transport and the noncyclic transport chain linking photosystems I and II.

(b) Based on your answer to part (a), would you be more likely to expect cyclic electron flow to couple to one or to two phosphorylation events per electron pair?

(c) If cyclic electron transport coupled to two phosphorylation events per electron pair rather than to the single event shown in Figure 7-6, what effect would that have on the quantum requirement (number of photons needed per CO_2 fixed) of photosynthesis, which is assumed in Problem 7-2 to be 10?

7-6. Emerson Enhancement Effect: Photosystems I and II absorb light at slightly different wavelengths (and were, in fact, originally differentiated from each other because of this fact). At 720 nm, photosystem I absorbs but photosystem II does not; at 650 nm, the converse is true. Under experimental conditions it can be demonstrated that plastoquinone, cytochrome f, and plastocyanin all exist in the oxidized state when cells of the green alga *Chlamydomonas* are illuminated with monochromatic light of one of these two wavelengths and that all three carriers are converted to the reduced form when light of the other wavelength is used.

(a) At which wavelength (650 or 720 nm) will the carriers be oxidized?

(b) Why will they be converted to the reduced form at the other wavelength?

(c) Why does a single *Chlamydomonas* culture illuminated with both 650-nm and 720-nm light have a higher photosynthetic rate than the sum of two separate cultures, one illuminated with 650-nm light only and the other with 720-nm light only?

7-7. Bioenergetics of Photophosphorylation: Under optimal conditions, isolated spinach chloroplasts are able to phosphorylate ADP to ATP in the light at pH 7.6 and 25°C, to the extent that the following steady-state concentrations are achieved in the chloroplasts:

$$\text{ATP: } 138 \, \mu M \qquad \text{ADP: } 5 \, \mu M \qquad \text{P}_i \text{: } 670 \, \mu M$$

The energy for the photophorylation comes from light-driven photosynthetic electron transport through cytochromes and other electron carriers.

(a) What is the change in free energy for the formation of ATP under these conditions? (Use a value of 9.25 kcal/mole for the standard free-energy change at pH 7.6 and recall Equation 2-16.)

(b) Calculate the minimum oxidation-reduction potential difference ΔE between adjacent electron transfer chain components that would be required to yield the amount of free energy calculated in part (a). Assume that two electrons are transferred per phosphorylation event.

(c) Would it be sufficient to know the standard oxidation-reduction potential E_0 (at pH 7.6) of the various components in the cytochrome chain and to compare their differences with the value obtained in part (b) to determine at what step phosphorylation occurs? Explain.

‡7-8. Carbon Metabolism Again: In Problem 7-3, which two carbon atoms of the regenerated ribulose-1,5-diphosphate would remain unlabeled after one turn of the cycle?

‡7-9. Photosynthetic Electron Transport: Several mutants of the green alga *Chlamydomonas reinhardi* are known that can grow on acetate but cannot photosynthesize, apparently because of a block in the electron transport chain that links photosystem II to photosystem I. Here are several of these acetate-requiring (*ac*) mutants, with the components of the electron transport chain they are missing and the condition of two cytochromes in their chloroplasts:

		Oxidation State of Two Cytochromes	
Mutant	*Missing Factor*	*Cytochrome b-559*	*Cytochrome c-553*
ac-21	"M-component"	Reduced	Oxidized
ac-80a	Chlorophyll P-700	Reduced	Reduced
ac-115	Cytochrome *b*-559	Absent	Oxidized
ac-206	Cytochrome *c*-553	Reduced	Absent
ac-208	Plastocyanin	Reduced	Reduced
ac-216	Plastoquinone	Oxidized	Oxidized

The sequence of carriers can be determined from these data if you remember that photosystem I *oxidizes* carriers (because it uses their electrons to reduce NADP) and photosystem II *reduces* carriers (using electrons derived from the photooxidation of water). Thus if a given component is absent, everything between it and photosystem II will be reduced and all the carriers between it and photosystem I will be oxidized.

(a) How do you suppose it was determined that mutant *ac-206* lacked cytochrome *c-553*?

(b) What is the sequence of electron carriers?

(c) Why are these *Chlamydomonas* mutants able to grow on acetate?

(d) Do you think these mutants would be able to grow on glucose?

References

Bassham, J. A., "The Control of Photosynthetic Carbon Metabolism," *Science* (1971), *172*: 526.

Björkman, O., and J. Berry, "High-Efficiency Photosynthesis," *Sci. Amer.* (Oct. 1973), *229*: 80.

Calvin, M., and J. A. Bassham, *The Photosynthesis of Carbon Compounds* (New York: Benjamin, 1962).

Devlin, R. M., and A. V. Barker, *Photosynthesis* (New York: Van Nostrand Reinhold, 1971).

Emerson, R., R. V. Chalmers, and C. Cedenstrand, "Some Factors Influencing the Long-Wave Limit of Photosynthesis," *Proc. Nat. Acad. Sci. U.S.* (1957), *43*: 133.

Govindjee (ed.), *Bioenergetics of Photosynthesis* (New York: Academic Press, 1975).

Govindjee, and R. Govindjee, "The Absorption of Light in Photosynthesis," *Sci. Amer.* (Dec. 1974), *231*: 68.

Hatch, M. D., and C. R. Slack, "Photosynthesis by Sugar-Cane Leaves: A New Carboxylation Reaction and the Pathway of Sugar Formation," *Biochem. J.* (1966), *101*: 103.

Jagendorf, A. T., "Acid-Base Transitions and Phosphorylation by Chloroplasts," *Fed. Proc.* (1967), *26*: 1361.

Levine, R. P., "The Mechanism of Photosynthesis," *Sci. Amer.* (Dec. 1969), *221*: 58.

Zelitch, I., *Photosynthesis, Photorespiration, and Plant Productivity* (New York: Academic Press, 1971).

Zelitch, I., "Pathways of Carbon Fixation in Green Plants," *Ann. Rev. Biochem.* (1975), *44*: 123.

PART 3

Uses of Energy

Biosynthetic Work

8

In the preceding section the emphasis was on the sources of energy available to cells and the ways in which cells transduce, conserve, and store energy. Whether the nutritional mode is chemotrophic (Chapters 4 through 6) or phototrophic (Chapter 7), the immediate purpose of cellular energy metabolism is the conversion of the energy available from the environment (as chemical bond energy or solar radiation) into the high-energy phosphate bonds of ATP. All the foregoing is therefore essential to an understanding of the cellular energy economy because ATP is the immediate source of energy for almost all cellular needs. But, in a sense, the foregoing could also be regarded as prefatory, since the cell biologist is primarily interested in the cellular activities for which that ATP (or, more specifically, the hydrolysis thereof) provides the energy.

Having seen, then, how cells obtain energy and store it as ATP, we are ready to turn in this part of the book to the uses to which that energy is put—in other words, to the various kinds of biological work mentioned briefly in Chapter 1. Interest will be concentrated on what the energy is used for (the kinds of work the cell does) and also on how the energy of ATP is actually tapped and harnessed (the molecular mechanisms involved in the conversion of the energy of ATP hydrolysis into cellular work). In a sense, however, this discussion can do no more than introduce the topic of biological energy utilization, since energy-requiring processes are at the heart of virtually every topic in cellular biology and are almost certain to figure prominently in any future encounters you may have with biology or biochemistry. Thus whether the discussion turns to protein synthesis, nerve impulse transmission, membrane transport, chromosomal movement, or any of a host of other cellular activities, the utilization of ATP in the performance of cellular work will turn out to be a common theme, recurring fuguelike in almost every movement of the symphony of cellular function.

Biological Uses of Energy

Energy is required by living systems to do work or to cause change—changes in location or orientation (mechanical work), changes in concentration (osmotic and electrical work), or changes in chemical bonds (biosynthetic work). An underlying feature of each of these kinds of change is that they are nonrandom changes that generally move the system farther from thermodynamic equilibrium. Thus the changes are, by very definition, endergonic. The ordered movement of chromosomes during nuclear division, the uptake of a sugar molecule against a concentration gradient, the synthesis of a protein molecule from its constituent amino acids—all are thermodynamically improbable events requiring the input of free energy. Yet such processes occur routinely in cells and are in fact requisite to the continuation of life. From a thermodynamic point of view, life could be viewed as the continual expenditure of energy to maintain a system in an improbable state of order far from thermodynamic equilibrium.

All cells carry out all three major forms of cellular work (mechanical, concentration, and biosynthetic) to one extent or another, though each kind of work is obviously best exemplified in specific cell types specially adapted for a function that emphasizes a particular kind of activity. Our consideration of energy utilization begins in this chapter with the work of biosynthesis and is followed by discussions of concentration work in Chapter 9 and mechanical work in Chapter 10.

The Work of Biosynthesis

The work of biosynthesis is undoubtedly the most fundamental cellular activity. It involves, as we shall see, both the formation of all the chemical constituents of the cell from simple precursors available from the environment and also the eventual assembly of these constituents into the larger structures characteristic of cellular organization.

Biosynthesis, though a property of every living cell at virtually all times, is especially obvious in actively growing or differentiating cells or tissues, since the accumulation or rearrangement of cellular mass is quite conspicuous in such systems. But biosynthesis also occurs in the cells of nongrowing, mature organisms as well, for repair and maintenance purposes. It is, in fact, a good generalization that a thermodynamically improbable state, be it of concentration, structure, or configuration, requires energy not only to be attained initially but also to be maintained thereafter.

The term "biosynthesis" actually encompasses a hierarchy of synthetic activities, leading all the way from the simplest abiotic molecules to the living cell itself. It is possible, however, to distinguish several major categories of biosynthetic work, occurring at different levels, as shown in Figure 8-1. The cell and its organelles (level E) are made up of supramolecular structures (level D), and these in turn are ordered arrays of macromolecules (level C). The macromolecules of cells are themselves polymers of a relatively small group of organic molecules (level B), and these are synthesized from simple abiotic molecules available from the environment (level A). This scheme allows us to divide the overall work of cellular biosynthesis into four major processes: (1) synthesis of small organic molecules from simple inorganic precursors; (2) polymerization of such small organic monomers into large polymeric macromolecules; (3)

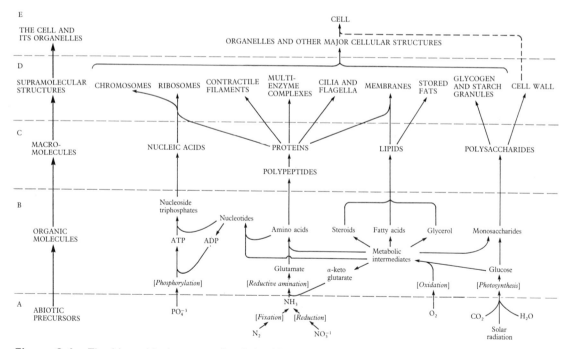

Figure 8-1   The hierarchical nature of cellular biosynthesis. The cell and its organelles (level E) are made up of supramolecular structures (level D), and these in turn are ordered arrays of macromolecules (level C). The cellular macromolecules are themselves polymers of small organic molecules (level B), and these are synthesized from simple abiotic precursors available from the environment (level A).

assembly of the macromolecules into higher-order structures; and (4) organization of such supramolecular structures into the complex functional units we recognize as organelles and cells. Thus the synthesis of an amino acid from simple starting materials is an example of the first category, the linking of that amino acid with others into a protein is representative of the second class, the insertion or deposition of that protein into a membrane characterizes the third kind of synthetic activity, and the presence of that membrane in a functional mitochondrion or other organelle illustrates the fourth category.

Each category of biosynthetic activity is needed to achieve the structure we ultimately recognize as a cell. Yet by considering chemical work in these subdivisions, it will be possible to look at prototype examples without getting into more biochemistry than is either needful or appropriate for present purposes. Furthermore the discussion will deal mainly with the synthesis of small organic molecules and macromolecules, for relatively little is known about the bioenergetics of the subsequent assembly of macromolecules into higher-order structures. What information *is* available suggests that such structures tend to be self-assembling, indicating that the macromolecules involved already possess the information and energy required to ensure spontaneous assembly into appropriate supramolecular structures.

Biopolymers as a Basic Cellular Strategy

Most of the form and order characteristic of living systems is a result of highly organized assemblies of large macromolecules synthesized by the polymerization of monomeric units. Indeed the construction of large molecules by joining smaller units in a repetitive manner is a basic "strategy" of cellular structure. Thus the enzymes responsible for catalysis of cellular reactions, the nucleic acids involved in the storage and readout of genetic information, the glycogen stored by your liver, and the cellulose that provides rigidity to a plant cell wall are all variations on the same design theme—each is a macromolecule made by linking together small repeating units. Other advantages of this strategy will become apparent later, but the fundamental advantage of the assembly of macromolecules by polymerization of small subunits is one of sheer economy in terms of the genetic information required. Bear in mind that every chemical modification required in a biosynthetic pathway must be carried out by a separate enzyme, that each enzyme is itself a molecule which must be synthesized, and that an enzyme is a costly entity in terms of both the genetic information necessary to specify its amino acid sequence and the energy needed to drive its synthesis.

Obviously it is to the advantage of the cell to use as few enzymes as possible in the synthesis of a molecule of a given size. Consider, for example, a molecule of starch as it occurs in a plant cell. Typically starch has a molecular weight somewhere in the range of 10^5 to 10^6 daltons and consists, therefore, of between 600 and 6000 glucose molecules linked together covalently. From Chapter 7 we already know that about 17 enzymes are required to synthesize glucose from CO_2 in the Calvin cycle. It takes a few more enzymes to accomplish the subsequent addition of a new glucose unit to a growing starch chain, but once the cell possesses the enzymatic competence to make a single glucose unit and link it to a growing starch molecule, that cell can synthesize starch molecules of almost unlimited size. In other words, the same 20 enzymes necessary to synthesize two glucose molecules and link them together to form a disaccharide with a molecular weight of about 350 daltons constitute the total complement of enzymes needed to make a starch molecule with a molecular weight of 10^6 daltons! The synthesis of a compound of this size by any mechanism other than the polymerization of repeating subunits would require a battery of enzymes that would rapidly exceed the information-storage capabilities of the cell.

The Biosynthesis of Small Organic Molecules

The repeating monomeric units used in the synthesis of macromolecules are always small organic molecules, usually with molecular weights of 350 daltons or less. They must themselves be synthesized from simpler molecules and, ultimately, from inorganic precursors such as CO_2, H_2O, NO_3^-, and PO_4^{---}, as shown in Figure 8-1, level A. Of course, not all cells can synthesize all the needed organic molecules from inorganic precursors: most chemotrophs need at least a few kinds of organic molecules, not only for energy but also as starting points for biosynthetic pathways. Nonetheless almost all cells make many if not most of their own organic molecules from simpler compounds and possess a variety of anabolic pathways for the biosynthesis of such molecules. Every sugar, every amino acid, every nucleotide, and every other monomeric molecule (as well as a number of specialized small molecules such as coenzymes and hormones, which are not actually used as building blocks for polymers) must be synthesized in the cell by a specific pathway, or series of enzyme-catalyzed reactions. A number of such synthetic pathways are shown schematically at level B of Figure 8-1. A detailed diagram of such *intermediary metabolism*, as it is called, would be very complicated. But since our emphasis is on the energetics rather than the chemistry of such synthetic pathways, we can legitimately leave many details of the chemistry to other texts and look instead at a few prototype examples chosen to illustrate points about the energy requirements of biosynthesis at this level.

In considering the synthesis of small organic molecules from simpler, often abiotic precursors, we can identify four distinctive features of the process: (1) it is thermodynamically unfavorable and inevitably requires the input of energy, usually as ATP; (2) it is usually a reductive process, since cellular components are generally more highly reduced than the precursors from which they are made; (3) it is accomplished by enzyme-catalyzed metabolic pathways that are often similar, but never identical, to the degradative pathways available for the same molecules; and (4) it is a regulated process, allowing the cell to adjust production to need.

Synthesis of Glucose

The four features listed above can be illustrated equally well in the pathways for the synthesis of sugars, fatty acids, amino acids, or nucleotides, but only the synthesis of glucose will be treated in detail here. Actually, glucose synthesis from CO_2 and H_2O, as it occurs in phototrophs, has already been considered in Chapter 7. Chemotrophs also carry out the synthesis of glucose from simpler starting compounds (*gluconeogenesis*), but the process is more restrictive in that the starting materials are not CO_2 and H_2O. Chemotrophs can in general synthesize glucose only from pyruvate, lactate, or similar molecules with at least three carbon atoms.

As shown in Figure 8-2, the synthesis of glucose from pyruvate proceeds by a route that is essentially the reverse of glycolysis, *consuming* ATP and $NADH_2$ instead of producing them. There are, however, several distinctions between the glycolytic and the gluconeogenic directions that are of supreme importance from both the thermodynamic and the regulatory points of view and that illustrate general principles of biosynthetic pathways well. As you may recall from Chapter 4, the glycolytic sequence is designed to accomplish the oxidative degradation of glucose and includes several steps that are so highly exergonic in the direction of degradation that they render the whole pathway essentially irreversible. That is, of course, a sound and, as it turns out, a very general design feature for a metabolic pathway: virtually every pathway in the cell is so contrived that at one or more steps it includes a reaction characterized by such a large release of free energy that there is essentially no chance for reversibility. This guarantees the unidirectionality of the entire pathway,

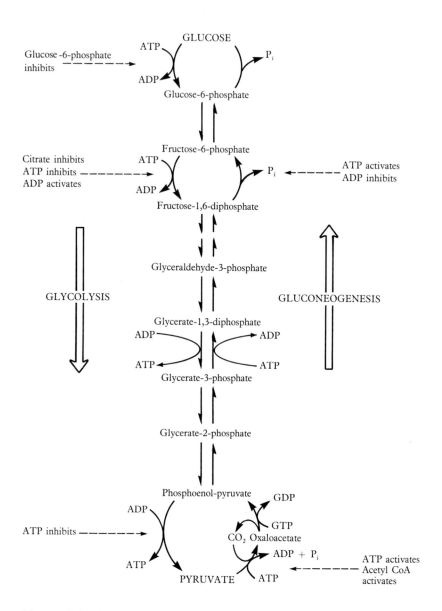

Figure 8-2 The pathways and regulation of glycolysis and gluconeogenesis. The glycolytic degradation of glucose to pyruvate is shown on the left (reading down), while the complementary process of glucose synthesis from pyruvate is illustrated on the right (reading up). For both processes, major sites of allosteric regulation by ATP, ADP, citrate, acetyl CoA, and glucose-6-phosphate are indicated.

since it is a dictate of thermodynamics that no pathway can proceed in a given direction unless every component reaction in that direction is exergonic under prevailing conditions.

To reverse most metabolic pathways, therefore, requires that such irreversible steps be bypassed. In the case of glycolysis, these steps can be readily pinpointed: the ATP-driven phosphorylation of both glucose and fructose-6-phosphate and the conversion of phosphoenol pyruvate to pyruvate (steps Gly-1, Gly-3, and Gly-10 of Figure 4-4) are highly exergonic and for all practical purposes irreversible. The two phosphorylation reactions (Gly-1 and Gly-3, catalyzed in the glycolytic direction by enzymes called *kinases*) are irreversible because they couple to ATP hydrolysis and would therefore have to drive the synthesis of ATP in the reverse direction. They can be rendered thermodynamically favorable in the gluconeogenic direction by using not kinases but *phosphatases*, enzymes that remove the phosphate group as inorganic phosphate. (This, you may recall, is also the way that the same reactions are rendered thermodynamically possible in the Calvin cycle; see Figure 7-1.) Thus the sequence between glucose and fructose-1,6-diphosphate can be exergonic in either direction, depending upon whether the phosphorylating (kinase) or phosphate-cleaving (phosphatase) enzymes are active:

In the glycolytic direction:

$$\text{Glucose} \xrightarrow[\text{kinase}]{\text{ATP} \quad \text{ADP}} \text{glucose-6-P} \xrightarrow[\text{isomerase}]{} \text{fructose-6-P} \xrightarrow[\text{kinase}]{\text{ATP} \quad \text{ADP}}$$

$$\text{fructose-1,6-diP} \qquad (8\text{-}1)$$

Net equation: glucose + 2ATP $\longrightarrow$
fructose-1,6-diP + 2ADP $\Delta G^0 = -7.3 \text{ kcal/mole}$ (8-2)

In the gluconeogenic direction:

$$\text{Fructose-1,6-diP} \xrightarrow[\text{phosphatase}]{\text{P}_i} \text{fructose-6-P} \xrightarrow[\text{isomerase}]{}$$

$$\text{glucose-6-P} \xrightarrow[\text{phosphatase}]{\text{P}_i} \text{glucose} \qquad (8\text{-}3)$$

Net equation: fructose-1,6-diP $\longrightarrow$
glucose + 2P$_i$ $\Delta G^0 = -7.3 \text{ kcal/mole}$ (8-4)

The other essentially irreversible reaction in the glycolytic pathway is the step from phosphoenol pyruvate to pyruvate; even though already "harnessed" to the generation of ATP, it is still sufficiently exergonic to be virtually unidirectional. To bypass this step, the gluconeogenic pathway uses a two-step sequence, with both steps coupled to the hydrolysis of a high-energy phosphate bond (one from ATP, the other from GTP). Involved is a carboxylation/decarboxylation sequence, with the four-carbon compound oxaloacetate (of TCA-cycle fame) as the intermediate. Thus a reaction that generates one high-energy phosphate bond in the catabolic direction is driven in the anabolic direction by a "double whammy" sequence that uses *two* high-energy phosphate bonds:

In the glycolytic direction:

$$ \text{Phosphoenol pyruvate} \xrightarrow[\text{pyruvate kinase}]{\overset{\displaystyle \text{ADP} \qquad \text{ATP}}{\curvearrowright}} \text{pyruvate} \qquad (8\text{-}5) $$

Net equation: phosphoenol pyruvate + ADP $\longrightarrow$

pyruvate + ATP $\qquad \Delta G^0 = -7.5 \text{ kcal/mole} \qquad (8\text{-}6)$

In the gluconeogenic direction:

$$ \text{Pyruvate} \xrightarrow[\text{carboxylase}]{\overset{\displaystyle \text{CO}_2 \ \text{ATP} \ \text{ADP} + \text{P}_i}{\curvearrowright}} \text{oxaloacetate} \xrightarrow[\text{carboxykinase}]{\overset{\displaystyle \text{GTP} \ \text{GDP} \ \text{CO}_2}{\curvearrowright}} $$

phosphoenol pyruvate $\qquad (8\text{-}7)$

Net equation: pyruvate + ATP + GTP $\longrightarrow$

phosphoenol pyruvate + ADP + GDP + P_i $\qquad \Delta G^0 = +0.2 \text{ kcal/mole}$

$(8\text{-}8)$

Note, then, that the synthesis of glucose from pyruvate occurs by a route that is essentially the reverse of glycolysis (and can as a result use seven out of ten already-existing enzymes), but those reactions which would otherwise ensure that the sequence is highly exergonic in the catabolic direction are bypassed by alternative reactions made exergonic in the anabolic direction by the input of energy from ATP and GTP.

By using similar, but not identical, pathways for the opposite processes of glycolysis and gluconeogenesis, the cell enjoys several advantages: (1) since most of the pathway is common to both directions, reversal of direction requires a minimum of new enzymes and is therefore genetically conservative; (2) yet at the same time the reaction sequence can be rendered essentially irreversible in a given direction simply by specifying which set of enzymes is available or active for the key bypass reactions; such that (3) the direction of the sequence can be controlled within a cell or subcellular location by regulating the presence or activities of these key enzymes. This, it turns out, is another general feature of the cellular metabolism of small organic molecules: biosynthetic pathways are often similar enough to the corresponding degradative pathways to make maximum use of common enzymes but differ at one or a few key steps to guarantee that the sequence can be rendered thermodynamically favorable in either direction. The actual direction in which the sequence proceeds is then dependent upon the set of enzymes present or active under specific circumstances.

Not surprisingly, the steps at which biosynthetic and degradative pathways differ for thermodynamic reasons are also the points at which the pathways are usually regulated allosterically. This makes good sense, of course, since it allows control to be exerted in a specific direction. In fact it is really only in the context of Figure 8-2 that phosphofructokinase, the enzyme that phosphorylates fructose-6-phosphate to fructose-1,6-diphosphate (Reaction Gly-3), is seen to be a particularly logical choice for allosteric regulation of glycolysis. The phosphofructokinase reaction is one of the three steps unique to the glycolytic direction, such that regulation of its activity affects the flow along the pathway in the catabolic direction only. (As shown in Figure 8-2, the other enzymes unique to the glycolytic direction are also allosteric control points, though of lesser overall significance than phosphofructokinase.)

By the same reasoning, the most logical candidates for allosteric regulation in the gluconeogenic direction ought to be the enzymes unique to that direction. And Figure 8-2 indicates that the prediction is a good one: regulation of glucose synthesis is in fact exerted at the phosphate-cleaving step that bypasses the phosphofructokinase reaction and at the pyruvate-carboxylating step used at the other bypass. The actual regulation is also inordinately logical, since an increase in ATP both inhibits the degradative (ATP-generating) pathway and activates the synthetic (ATP-consuming) direction.

Synthesis of Other Small Molecules

Almost all the other small organic molecules in the cell are synthesized by anabolic routes that begin with metabolic intermediates derived from the partial oxidation of glucose or other substrates. Thus in addition to their central role in catabolic metabolism, the glycolytic and TCA pathways also serve as points of departure for many synthetic pathways, leading, for example, to the lipids, amino acids, and nucleotides, as well as to monosaccharides other than glucose. Detailed consideration of specific anabolic pathways is clearly beyond the scope of the present discussion; synthesis of the amino acids alone, for example, requires 20 at least partially separate metabolic sequences, and nucleotide biosynthesis involves notably complicated pathways. A brief summary of biosynthetic pathways and their energy requirements appears in Table 8-1. As the details of these pathways are

TABLE 8-1   Summary of the Biosynthesis of the Major Classes of Monomeric Molecules Found in Cellular Macromolecules

Small Organic Molecule	Synthesized From	Synthetic Route	Typical Compound	Required for Synthesis	
				ATP	Reduced Coenzyme
Monosaccharides	1. CO_2 and H_2O (autotrophs)	Calvin cycle (Figure 7-1)	Glucose (from CO_2 and H_2O)	18	$12NADPH_2$
	2. Lactate or pyruvate (heterotrophs)	Gluconeogenesis (Figure 8-2)	Glucose (from pyruvate)	6	$2NADH_2$
Fatty acids	Acetyl CoA	Essentially by reversal of β-oxidation	Palmitate (from acetyl CoA)	7	$14NADPH_2$
Amino acids	Variety of intermediates, mostly from glycolytic and TCA pathways; nitrogen from amino groups, NH_3, NO_3^-, or N_2	Twenty distinct pathways, some with initial sequence in common	Glutamate (from nitrate and CO_2 in an autotroph)	17	$15\frac{1}{2}NADPH_2$ $2NADH_2$
			Glutamate (from α-ketoglutarate and NH_3 in a heterotroph)	0	$1NADPH_2$
Nucleotides	Amino acids, single-carbon fragments, CO_2, ammonia, and ribose-5-phosphate	Two complex pathways: one for purines, the other for pyrimidines	AMP (from ribose-5-phosphate, CO_2, glycine, glutamine, aspartate, and one-carbon fragments)	6	0

encountered in other contexts, you are likely to be impressed with how invariably they require energy in the form of ATP or related nucleotide triphosphates and reducing power in the form of $NADPH_2$ or related coenzymes. Also noteworthy is the common pattern of allosteric regulation at points of thermodynamic irreversibility.

Thus far we have seen how carbon, oxygen, and hydrogen (in the form of CO_2 and H_2O) become fixed into organic form by photosynthetic autotrophs, followed by extensive metabolic rearrangements in both

Metabolism of Phosphorus and Nitrogen

chemotrophs and phototrophs to give rise to the small organic molecules (especially monosaccharides, amino acids, and nucleotides)
required for formation of cellular polymers. But nothing has yet been
said about the way in which elements other than C, H, and O become
incorporated into organic form. Nitrogen and phosphorus are particularly important—nitrogen as an essential element in proteins, nucleic
acids, and coenzymes, and phosphorus in nucleic acids, phospholipids,
coenzymes, and the great variety of phosphorylated intermediates
involved in cellular metabolism.

Phosphorus (as inorganic phosphate, the ionized form of phosphoric acid) gains entry into organic molecules either by direct uptake
into substrates, as occurs in Reaction Gly-6, or more commonly by
phosphorylation of ADP to form ATP, as in both photo- and oxidative
phosphorylation. Accordingly the incorporation of inorganic phosphate
into organic form can be quite adequately represented by the phosphorylation of ADP to ATP, as shown in Figure 8-1.

Bringing nitrogen into organic form is a bit more complicated,
however, because of the greater variety of metabolic possibilities
involved. Nitrogen can exist in the three principal inorganic forms
shown at level A in Figure 8-1: gaseous nitrogen (N_2), nitrate anion
(NO_3^-), and ammonia (NH_3). Of these, only ammonia is generally
useful to most cells. Thus an important aspect of biological nitrogen
metabolism involves the initial conversion of the other two inorganic
forms into ammonia. The most abundant form of nitrogen—and the
least useful to most cells—is the nitrogen gas (N_2) that makes up
79 percent of the earth's atmosphere. Only certain critical microorganisms—a few bacteria, the blue-green algae, and a few fungi—
can use N_2 directly as a nitrogen source. The process is termed
nitrogen fixation and is responsible for the initial entrapment of most
of the usable nitrogen present in the earth's ecosystems. Some of the
nitrogen-fixing microorganisms live free in soil or water, while others
exist in a close symbiotic relationship with the roots of higher plants,
where they occur in prominent lumplike growths called nodules. In
either case they effect the reduction of N_2 to NH_3, probably via
hydroxylamine (NH_2OH):

$$N\equiv N \xrightarrow{+2H^+, +2e^-} HN=NH \xrightarrow{+2H_2O}$$

$$2NH_2OH \xrightarrow{+4H^+, +4e^-, -2H_2O} 2NH_3 \qquad (8\text{-}9)$$

The electron donor for this process can be any of a variety of compounds. Certain photosynthetic microorganisms can use $NADPH_2$ as
the reducing agent, but most other nitrogen fixers use electrons from
organic substrates. Although some N_2 is also converted into usable form

by electrical discharges such as lightning and by the manufacture of commercial fertilizers as well, biological nitrogen fixation is the most important source. This inordinately important process probably ranks second only to photosynthesis among activities essential for the continuation of life on this planet.

The nitrogen-fixing microorganisms make ammonia available to plants either directly, through a symbiotic relationship, or indirectly, through release into the soil or water. Ammonia is also produced by the microbial decomposition of the organic nitrogenous compounds in the bodies of dead plants and animals and in the urine and feces excreted by animals. Some of this free ammonia is picked up and used directly by plants, but the main source of inorganic nitrogen for higher plants is nitrate, NO_3^-, which must first be reduced to ammonia. The process, called *nitrate reduction*, occurs in photosynthetic tissue and is an expensive one for plants, requiring both reducing power and ATP. Nitrate reduction occurs via nitrite and involves the enzymes nitrate reductase and nitrite reductase:

$$NO_3^- \xrightarrow[\text{nitrate reductase}]{+2H^+, +2e^-, -H_2O} NO_2^- \xrightarrow[\text{nitrite reductase}]{+7H^+, +6e^-, -2H_2O} NH_3$$

$$(8\text{-}10)$$

Nitrate reductase is located in the cytoplasm and uses $NADH_2$ as the electron donor (probably by coupling to glycolysis, since the glycolytic pathway is the source of most cytoplasmic $NADH_2$). Nitrite reductase, on the other hand, is a chloroplast enzyme and depends for its source of reducing power upon ferredoxin, the same substance responsible for the reduction of NADP to $NADPH_2$ in the light reactions of photosynthesis (see Figure 7-6).

Whether supplied by nitrogen fixation or nitrate reduction, ammonia is the form in which most cells incorporate nitrogen into organic form. As indicated in Figure 8-1, the process of incorporation is called *reductive amination* and involves the reaction of ammonia with an α-keto acid and the subsequent reduction to yield the corresponding α-amino acid:

$$R\underset{\underset{O}{\|}}{-}C-COOH + NH_3 \xrightarrow[\text{amination}]{\quad H_2O \quad}$$

$$R\underset{\underset{NH}{\|}}{-}C-COOH \xrightarrow[\text{reduction}]{\quad NADPH_2 \quad NADP \quad} R\underset{\underset{H}{|}}{\overset{\overset{NH_2}{|}}{-}}C-COOH \qquad (8\text{-}11)$$

All three of the α-keto acids in the glycolytic and TCA pathways (pyruvate, oxaloacetate, and α-ketoglutarate) can be reductively aminated to form the corresponding amino acids, but α-ketoglutarate is by far the most common acceptor for NH_3, making this a key compound in amino acid metabolism, as illustrated in Figure 8-1.

Once incorporated into organic form by amination, an amino group can be transferred from one carbon backbone to another by the transamination process encountered in Chapter 6. As a result, all the amino acids can be generated from the glutamate that results when α-ketoglutarate is aminated, provided only that the organism in question can synthesize all the needed carbon backbones. The reduction, fixation, and amination processes shown in Figure 8-1 are vital not only to amino acid synthesis but to nucleotide production as well, since the nitrogen atoms present in the purine and pyrimidine bases of nucleic acids are derived from amino acids and, therefore, indirectly from the nitrogen that enters organic form upon amination of α-ketoglutarate.

The Biosynthesis of Macromolecules

At the first level, then, biosynthetic work in cells involves the formation of small organic molecules from simpler precursors—ultimately, from the inorganic compounds available to the biological world from its environment. With the necessary small organic molecules—notably the sugars, amino acids, and nucleotides—synthesized and in hand, we are now ready to move to the second category of biosynthetic work, the polymerization of these monomers into the large macromolecules so characteristic of living systems. Table 8-2 indicates the major bio-

TABLE 8-2   Principal Biological Polymers

Biopolymer	Repeating Unit of Polymer			
	Chemical Nature	Number of Different Kinds in a Given Macromolecule	Activated Form	Chemical Bond of Polymer
Polysaccharides	Monosaccharides	Usually 1	Sugar linked to nucleoside diphosphate	Glycosidic
Proteins	Amino acids	20	Aminoacyl tRNA	Peptide (amide)
Nucleic acids	Nucleotides	4	Nucleoside triphosphate	Phospho-diester

polymers synthesized in the cell, along with the kind, number, and activated form of repeating subunits required for polymer formation and the chemical nature of the bond that joins the repeating units together.

An important point to note at the outset is the basic distinction between storage and informational macromolecules. Nucleic acids (both DNA and RNA) and proteins are called *informational macromolecules* because the order of the nonidentical subunits of which they are composed is nonrandom and highly significant. The order of nucleotides in nucleic acids or that of amino acids in proteins is genetically determined and carries information that directs the function or utilization of these macromolecules. The protein insulin, for example, is a *protein* because it is a string of amino ácids linked together by peptide bonds, but it is *insulin* because it is a particular string of amino acids in a unique, specified order, and any variation in that sequence is very likely to have a deleterious effect on the ability of the protein to carry out the functions expected of an insulin molecule. For the informational macromolecules, polymerization of smaller subunits into biopolymers is not just an economical way of building large molecules but an essential part of the role they play in the cell. For the nucleic acids, synthesis by linkage of small repeating units (nucleotides) is their very *raison d'être*, since it is in the actual order of the monomeric units that these molecules store and transmit the genetic information of the cell.

Polysaccharides, on the other hand, are not informational macromolecules; for them the order of the monomeric units is enzymatically determined, carries no information, and is not essential to the function or utilization of the molecule in the sense that it is for proteins or nucleic acids. Instead polysaccharides are either *storage* or *structural macromolecules* and often consist of a single kind of repeating monomer, most commonly glucose. The most familiar storage polysaccharides are the starch of plant cells and the glycogen of animal cells. Both starch and glycogen consist exclusively of α-glucose units linked together by *α-glycosidic bonds* between carbon 1 of one glucose molecule and carbon 4 of the next glucose (or, at branch points, between carbon 1 of one glucose and carbon 6 of the next), as shown in Figure 8-3. The resulting straight-chain or branched-chain polymers serve as depots for glucose, providing for its storage in a compact yet accessible form. For these and other storage macromolecules, polymerization of small subunits into biopolymers has additional advantages beyond that of genetic conservation mentioned earlier. Specifically, polymer formation serves the double purpose of decreasing the osmotic pressure and increasing the energy yields of stored food. By storing 100 glucose

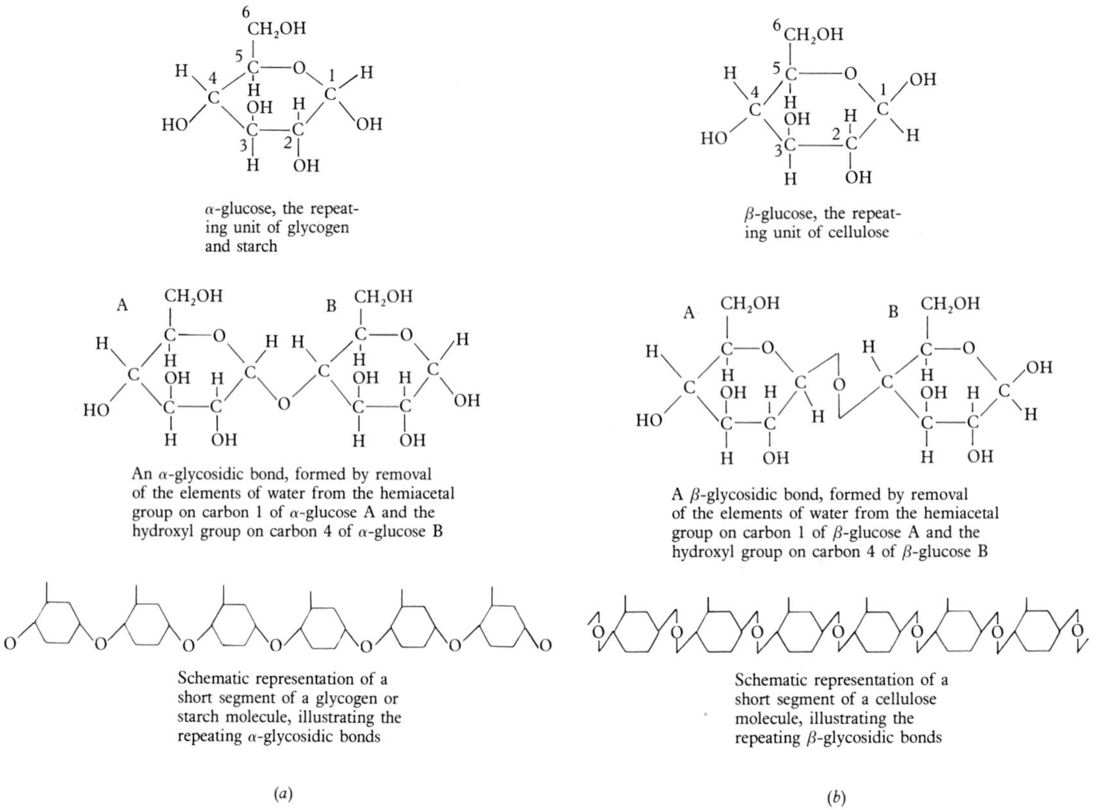

α-glucose, the repeat-
ing unit of glycogen
and starch

β-glucose, the repeat-
ing unit of cellulose

An α-glycosidic bond, formed by removal
of the elements of water from the hemiacetal
group on carbon 1 of α-glucose A and the
hydroxyl group on carbon 4 of α-glucose B

A β-glycosidic bond, formed by removal
of the elements of water from the hemiacetal
group on carbon 1 of β-glucose A and the
hydroxyl group on carbon 4 of β-glucose B

Schematic representation of a
short segment of a glycogen or
starch molecule, illustrating the
repeating α-glycosidic bonds

Schematic representation of a
short segment of a cellulose
molecule, illustrating the
repeating β-glycosidic bonds

(a)

(b)

Figure 8-3   The structural units, glycosidic bonds, and polymeric nature of
(a) the storage polysaccharides glycogen and starch and (b) the structural
polysaccharide cellulose. The schematic representations shown underscore
the polymeric structure of the compounds but do not indicate the configurational difference between the flexible helix of a starch or glycogen
molecule and the rigid, rodlike shape of cellulose.

units as a single glycogen molecule instead of as 100 glucose molecules,
the osmotic contribution from stored glucose is decreased by a factor
of 100 since osmotic pressure is strictly a function of the number of
molecules per unit volume. At the same time, the energy yield of those
glucose molecules upon subsequent glycolysis is also enhanced since,
as it turns out, the mobilization of glucose from glycogen requires
the input of one less ATP in the initial "pump-priming" steps of
glycolysis than does the activation of free glucose.

The single best-known example of a structural polysaccharide is the cellulose found in plant cell walls. Cellulose is an important polymer quantitatively, since more than half of all the carbon in higher plants is in cellulose. Like starch and glycogen, cellulose is also a linear polymer of glucose units. But, as shown in Figure 8-3, the glucose of cellulose is in the β form, with the —H and —OH groups on carbon 1 in a configuration opposite to that of α-glucose. The immediate consequence of the resulting β-glycosidic bond is that the cellulose molecule assumes the configuration of a rigid, rodlike chain, whereas starch and glycogen molecules have a rather flexible helical shape. Another aspect of the β-glycosidic bond is that mammalian cells do not possess an enzyme that can cleave it, although they can digest α-glycosidic bonds readily. This explains why humans can eat potatoes (starch) but not grass (cellulose). Animals such as cows and sheep that do eat grass cannot cleave β-glycosidic bonds either, but they harbor in their rumen (part of a compound stomach) a population of bacteria and protozoa that can digest such bonds. The microorganisms digest the cellulose and the host animal obtains, in turn, the end products of microbial fermentation.

Several important principles emerge from a consideration of the chemistry and energetics of polymer formation: (1) biopolymers are always synthesized by the stepwise polymerization of similar or identical small molecules onto a growing chain; (2) the addition of each monomeric unit occurs with the removal of a molecule of water and is therefore a condensation reaction; (3) the monomeric units that are joined together must be present in an activated, energized form before condensation can occur; and (4) activation of the monomers is driven by the hydrolysis of ATP—not to ADP and inorganic phosphate (P_i) but to AMP and inorganic pyrophosphate (PP_i), the advantage of which we shall see shortly.

Since the elimination of water is common to all biological polymerization reactions, it follows that the small molecules that are condensed together in such reactions must always have a reactive hydrogen (—H) on at least one functional group and a reactive hydroxyl group (—OH) elsewhere on the molecule. Schematically, then, the formation of a macromolecule by polymerization of a number of small molecules (SM) could be represented as follows:

$$H—\boxed{SM_1}—OH + H—\boxed{SM_2}—OH + H—\boxed{SM_3}—OH +$$

$$(n-1)H_2O$$

$$\cdots + H—\boxed{SM_n}—OH \nearrow \longrightarrow$$

$$H—\boxed{SM_1}—\boxed{SM_2}—\boxed{SM_3}—\cdots—\boxed{SM_n}—OH \qquad (8\text{-}12)$$

Long-chain polymer with n units

Actually, however, the polymerization process always occurs stepwise with a single monomeric unit added at a time. Thus a single such step, corresponding to the addition of the $(n + 1)$st monomeric unit to a macromolecule already consisting of n units, could be represented as follows:

$$\text{H}-\boxed{\text{SM}_1}-\boxed{\text{SM}_2}-\boxed{\text{SM}_3}-\cdots-\boxed{\text{SM}_n}(-\text{OH}+\text{H}-)\boxed{\text{SM}_{n+1}}-\text{OH}\xrightarrow{\text{H}_2\text{O}}$$

$$\text{H}-\boxed{\text{SM}_1}-\boxed{\text{SM}_2}-\boxed{\text{SM}_3}-\cdots-\boxed{\text{SM}_n}-\boxed{\text{SM}_{n+1}}-\text{OH} \qquad (8\text{-}13)$$

Even this formulation is not really adequate, though. The addition of the $(n + 1)$st unit decreases the entropy of that unit and is therefore an endergonic process that will not occur unless the incoming unit is first energized in an activation step driven by hydrolysis of ATP (or, in some cases, related nucleotide triphosphates). Inevitably this involves the coupling of the monomeric unit to some sort of high-energy *carrier*. The chemical nature of the energized monomer, the carrier, and the actual activation process differ for each of the biopolymers but can be generalized as follows:

1. *Activation of the incoming monomer:*

$$\text{X}-\text{OH} + \text{H}-\boxed{\text{SM}_{n+1}}-\text{OH}\xrightarrow[\quad\quad]{\text{ATP}\quad\text{AMP} + \text{PP}_i}\text{X} \sim \boxed{\text{SM}_{n+1}}-\text{OH} + \text{H}_2\text{O}$$

Carrier   Monomer   Activated monomer

$$(8\text{-}14)$$

2. *Addition of the activated monomer to the growing polymer chain:*

$$\text{H}-\boxed{\text{SM}_1}-\boxed{\text{SM}_2}-\boxed{\text{SM}_3}-\cdots-\boxed{\text{SM}_n}-\text{OH} + \text{X} \sim \boxed{\text{SM}_{n+1}}-\text{OH} \longrightarrow$$

$$\text{H}-\boxed{\text{SM}_1}-\boxed{\text{SM}_2}-\boxed{\text{SM}_3}-\cdots-\boxed{\text{SM}_n}-\boxed{\text{SM}_{n+1}}-\text{OH} + \text{X}-\text{OH}$$

$$(8\text{-}15)$$

As suggested by this formulation, the actual condensation step (Reaction 8-15) is an exergonic reaction proceeding spontaneously in the direction of chain elongation and driven by the energy of the bond between the incoming monomer and its carrier. The energy input clearly comes in the activation step (Reaction 8-14) when the high-energy bond between the monomer and its carrier is formed. Although the source of the energy, predictably, is the hydrolysis of ATP (or a related nucleotide triphosphate), a characteristic feature of activation reactions leading to biopolymer formation is that the ATP is cleaved not to ADP and P_i, as we have encountered so frequently in past chapters, but to AMP and pyrophosphate (PP_i) instead:

$$\text{Adenine-ribose}\!-\!O\!-\!\overset{\overset{\displaystyle O}{\|}}{\underset{\underset{\displaystyle OH}{|}}{P}}\!-\!O\!-\!\overset{\overset{\displaystyle O}{\|}}{\underset{\underset{\displaystyle OH}{|}}{P}}\!-\!O\!-\!\overset{\overset{\displaystyle O}{\|}}{\underset{\underset{\displaystyle OH}{|}}{P}}\!-\!OH \xrightarrow{\ +H_2O\ }$$

ATP

$$\text{adenine-ribose}\!-\!O\!-\!\overset{\overset{\displaystyle O}{\|}}{\underset{\underset{\displaystyle OH}{|}}{P}}\!-\!OH + HO\!-\!\overset{\overset{\displaystyle O}{\|}}{\underset{\underset{\displaystyle OH}{|}}{P}}\!-\!O\!-\!\overset{\overset{\displaystyle O}{\|}}{\underset{\underset{\displaystyle OH}{|}}{P}}\!-\!OH \qquad (8\text{-}16)$$

AMP $\qquad\qquad\qquad$ PP_i

The thermodynamic utility of this innovation is not immediately apparent since the two acid anhydride bonds of ATP are about equal in energy yield, such that the hydrolysis of ATP has essentially the same energy yield whether cleaved to ADP + P_i or AMP + PP_i. However, the acid anhydride bond of pyrophosphate is still a high-energy bond (due, again, to resonance stabilization of the products of hydrolysis), and the subsequent hydrolysis of the pyrophosphate is highly exergonic. Thus by the simple expedient of providing an enzyme (pyrophosphatase) capable of cleaving the pyrophosphate released in this reaction, the cell gains access to the energy of *both* the anhydride bonds of ATP, making the overall process of pyrophosphate cleavage doubly energetic:

(1)	$ATP + H_2O \rightarrow AMP + PP_i$	$\Delta G^0 = -7.3$ kcal/mole	(8-17)
(2)	$PP_i + H_2O \rightarrow 2P_i$	$\Delta G^0 = -7.3$ kcal/mole	(8-18)

$ATP + 2H_2O \rightarrow AMP + 2P_i$	$\Delta G^0 = -14.6$ kcal/mole	(8-19)

Biosynthetic reactions that result in the initial cleavage of pyrophosphate rather than phosphate from ATP lead ultimately to the hydrolysis of two high-energy phosphate bonds, resulting thereby in a much greater thermodynamic "pull" for the reaction to which the ATP hydrolysis is coupled. It is a valid generalization that essentially all biosynthetic reactions leading to polymer formation are characterized by one or more steps that release pyrophosphate, thereby providing the added free energy necessary to ensure that such reaction sequences are driven essentially to completion.

Polysaccharide Synthesis

To illustrate the principle of condensation polymerization, consider the biosynthesis of glycogen, the form in which glucose is stored in animal cells. In essence, glycogen synthesis (starch synthesis, too, for that matter) consists simply of adding successive α-glucose units to the growing polymer chain by means of α-glycosidic bonds. First, however, the incoming glucose unit must be activated by linkage to an energized carrier. For glycogen synthesis, the carrier is the nucleoside diphosphate UDP. UDP is similar in structure to ADP, but with the adenine replaced by uracil, a pyrimidine ring compound; the structure of the UDP-glucose complex is shown in Figure 8-4.

The actual starting point for glycogen synthesis is glucose-1-phosphate, which is formed from glucose by initial phosphorylation to glucose-6-phosphate (as in glycolysis) but followed by transfer of the phosphate group from carbon 6 to carbon 1:

$$\text{Glucose} + \text{ATP} \xrightarrow[\text{hexokinase}]{} \text{glucose-6-phosphate} + \text{ADP} \quad (8\text{-}20)$$

$$\text{Glucose-6-phosphate} \xrightarrow[\text{phosphoglucomutase}]{} \text{glucose-1-phosphate} \quad (8\text{-}21)$$

The resulting glucose-1-phosphate is then linked to its carrier in a somewhat complicated reaction, with UTP serving as the source of both the carrier and the energy required to drive the activation process:

$$\text{Glucose-1-phosphate} + \text{UTP} \xrightarrow[\substack{\text{UDP-glucose} \\ \text{pyrophosphorylase}}]{} \text{UDP-glucose} + \text{PP}_i \quad (8\text{-}22)$$

This reaction is accompanied by only a small standard free-energy change ($\Delta G^0 = +0.3\,\text{kcal/mole}$) and would be readily reversible except that, as indicated, it results in the liberation of pyrophosphate, which can then be further split to two phosphates to release additional energy:

Figure 8-4 The structure of UDP-glucose.

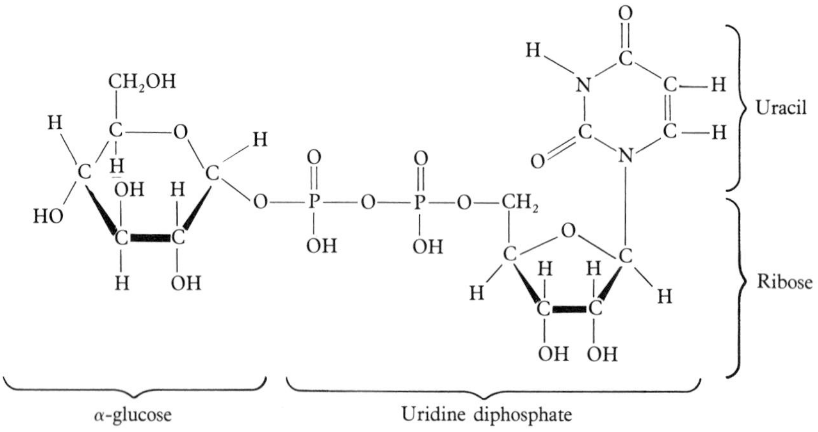

α-glucose   Uridine diphosphate

$$PP_i + H_2O \xrightarrow[\text{pyrophosphatase}]{} 2P_i \qquad \Delta G^0 = -7.3 \text{ kcal/mole} \quad (8\text{-}23)$$

Thus the formation of UDP-glucose occurs with an overall ΔG^0 of -7.0 kcal/mole, making it an essentially irreversible process.

Once linked to its UDP carrier, the glucose molecule has sufficient energy to add exergonically to the growing glycogen chain, the actual polymerization event:

$$(\text{Glucose})_n \quad + \quad \text{UDP-glucose} \xrightarrow[\text{synthetase}]{\text{glycogen}} (\text{glucose})_{n+1} \quad + \quad \text{UDP}$$

Glycogen with n Glycogen with $n + 1$
  glucose units   glucose units

$$(8\text{-}24)$$

And in the final step the UDP liberated in this reaction is converted back to UTP by phosphorylation with ATP:

$$\text{UDP} + \text{ATP} \xrightarrow[\text{diphosphokinase}]{\text{nucleoside}} \text{UTP} + \text{ADP} \qquad (8\text{-}25)$$

By summing these six reactions (8-20 through 8-25), we arrive at an overall equation for the addition of one glucose unit to a growing glycogen chain:

Glycogen with n glucose units + glucose + 2ATP →
$$\text{glycogen with } n + 1 \text{ glucose units} + 2\text{ADP} + 2P_i \qquad (8\text{-}26)$$

The use of two ATPs may seem extravagant since the actual energy required to add one glucose to a growing glycogen chain is only about $+4.4$ kcal/mole and the overall process therefore has a standard free energy of about -10 kcal/mole ($-14.6 + 4.4 = -10.2$). This, however, is the price the cell must pay to guarantee the irreversibility of polymer formation. Polysaccharide synthesis in plant cells is very similar to the process described for glycogen synthesis in animals, except that the carrier is ADP for starch synthesis and GDP for cellulose synthesis.

Nucleic Acid Synthesis

Nucleic acids are *polynucleotides*—polymers consisting of nucleotides linked together through phosphodiester bonds as shown in Figure 8-5. A nucleotide consists of a nitrogen-containing base (either a *pyrimidine* with a single aromatic ring or a *purine* with two joined rings), a five-carbon sugar, and a phosphate group, joined together as illustrated in the figure. For example, adenosine monophosphate (AMP) is a

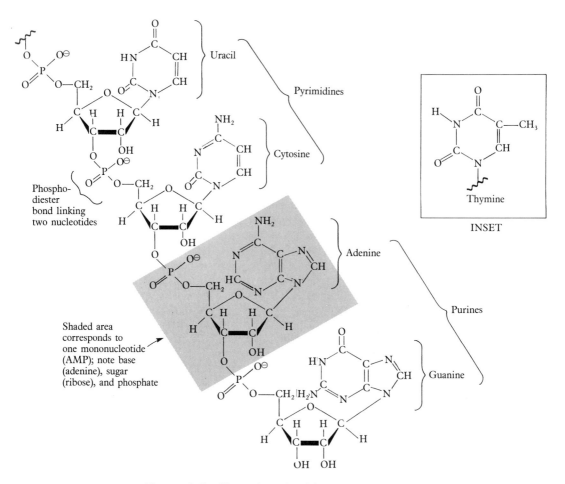

Figure 8-5 The polynucleotide structure of nucleic acids. The structure shown is the tetranucleotide sequence —U—C—A—G from RNA. For DNA the free hydroxyl group on carbon 2 of each ribose would be replaced by a hydrogen atom and the base uracil would be replaced by thymine (see inset).

nucleotide with which we are already familiar. Nucleic acids are of two types, depending upon the kind of sugar present. *Ribonucleic acid* (RNA) contains the sugar ribose in each of its nucleotides; *deoxyribo-nucleic acid* (DNA) contains 2-deoxyribose instead. (The sugar 2-deoxyribose has the same structure as ribose but with the hydroxyl group on carbon 2 replaced by a hydrogen atom.) DNA is the genetic material of almost all organisms (a few viruses use RNA instead), and RNA is involved at several points in the process of translating the genetic information encoded in the nucleotide sequence of the DNA into the eventual amino acid sequence of the protein. The two kinds of nucleic acids also differ somewhat in the nitrogenous bases that are present in their nucleotides: RNA uses the two purines *adenine* and *guanine* and the two pyrimidines *cytosine* and *uracil*, whereas *thymine* is substituted for uracil in DNA.

From a chemical point of view, nucleic acid synthesis is quite straightforward since the activated form of a nucleotide is simply the triply phosphorylated derivative. Thus the necessary precursors for RNA synthesis are ATP, GTP, CTP, and UTP, while for DNA the required precursors are dATP, dGTP, dCTP, and dTTP (the "d" before each indicates that the sugar is deoxyribose rather than ribose). If we let XMP stand for a generalized nucleotide, then the activated form is the triphosphate XTP and the activation process consists of two successive phosphorylations, each using ATP as the phosphate donor:

First phosphorylation:	$XMP + ATP \rightleftharpoons XDP + ADP$	(8-27)
Second phosphorylation:	$XDP + ATP \rightleftharpoons XTP + ADP$	(8-28)

Sum:	$XMP + 2ATP \rightleftharpoons XTP + 2ADP$	(8-29)

The addition of each successive nucleotide unit to a growing polynucleotide chain occurs as shown schematically in Figure 8-6. In abbreviated form, the polymerization process can be represented as follows:

Polynucleotide with n nucleotides $+ XTP \xrightarrow{\text{polymerase}}$

$$\text{polynucleotide with } n + 1 \text{ nucleotides} + PP_i \qquad (8\text{-}30)$$

Again the net free-energy change of the actual polymerization event is not great, but the process is rendered irreversible in the synthetic direction by the subsequent cleavage of pyrophosphate:

$$PP_i + H_2O \xrightarrow{\text{pyrophosphatase}} 2P_i \qquad (8\text{-}31)$$

By summing the equations for activation, polymerization, and pyrophosphate cleavage (8-29 through 8-31) we can write an overall equation for nucleic acid synthesis as follows:

Polynucleotide with n nucleotides $+ XMP + 2ATP \rightarrow$

$$\text{polynucleotide with } n + 1 \text{ nucleotides} + 2ADP + 2P_i \qquad (8\text{-}32)$$

Again the energy input of two ATPs per monomer is far in excess of the actual thermodynamic requirements of the polymerization process and therefore ensures the unidirectionality of the process.

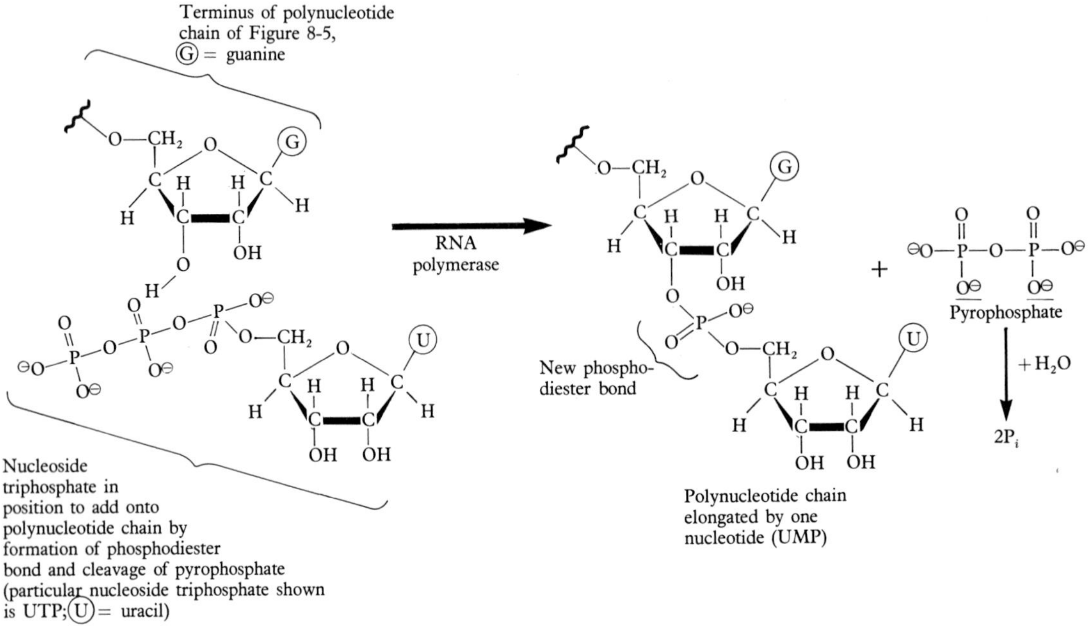

Figure 8-6 Phosphodiester bond formation as it occurs during polynucleo-tide synthesis. The particular step illustrated is the addition of the nucleotide UMP to the terminus of the polypeptide shown in Figure 8-5, such that the sequence —U—C—A—G is elongated to —U—C—A—G—U.

Thus far the synthesis of nucleic acids bears at least a formal resemblance to glycogen synthesis, since both processes entail the addition of a monomer unit in a process involving initial activation and subsequent polymerization, with the net consumption in both cases of two ATPs per monomer. There is a vital distinction to be made, though. Glycogen is a storage macromolecule with a single kind of repeating unit and requires for its synthesis only the several enzymes involved in the activation and polymerization events. Nucleic acids, on the other hand, are informational macromolecules with four different kinds of subunits. And since the *order* of the nucleotides is all-important to the function of both RNA and DNA, it follows that, in addition to the activated precursors and the necessary enzymes, nucleic acid bio-synthesis requires a *template* capable of specifying uniquely which of the four possible nucleotides is to be added in a given position.

This is a topic that belongs more appropriately to a discussion of molecular genetics than of bioenergetics, and you will, in fact, find a detailed account of the genetic aspects of DNA and RNA synthesis in virtually any biochemistry, genetics, or cellular biology text. For our purposes it will suffice simply to note that DNA serves as the template for both its own synthesis and that of RNA, with the nucleotide sequence of the DNA template serving to order the sequence of the incoming precursors as synthesis progresses. In other words, the specific nucleotide to be added at a given point in the synthesis of an RNA or DNA chain depends uniquely and exclusively upon the nucleotide already present at the corresponding position in the template. This means, of course, that there must be a specificity that allows the nucleotide of the template to recognize the incoming nucleotides, selecting one and rejecting the other three. This specificity lies in the hydrogen-bonding possibilities of the four different bases, A, G, C, and T (or U). The chemistry of the bases is such that A pairs always and only with T (or, for RNA synthesis, with U), and C pairs always and only with G.

Thus the nucleotide sequence of the template specifies the nucleotide sequence of the product in a unique and complementary manner, as illustrated in Figure 8-7. This specificity of base pairing is essential both to the *replication* (DNA → DNA) and to the *transcription* or readout (DNA → RNA, or in a few viruses RNA → DNA) of genetic information in all organisms. And, as we shall see shortly, it also plays a role in the *translation* of the encoded genetic message into the amino acid sequence of proteins.

Clearly, then, for informational macromolecules two considerations are important: as a biopolymer, the chemistry, mechanism, and energetics of monomer addition are of interest in the context of cellular work and uses of ATP; but as an informational biopolymer, the order of monomer addition and the means whereby specific ordering is achieved are topics with special genetic significance as well.

Protein Synthesis

Proteins are *polypeptides*—polymers of α-amino acids joined to one another through peptide (or amide) bonds as shown in Figure 8-8. The "R" group indicated in the generalized amino acid structure can be any one of the 20 different side groups shown, ranging from a simple hydrogen atom to relatively complex aromatic groups (as in phenylalanine and tryptophan). These side groups give the various amino acids distinctively different properties (especially of charge, polarity, and affinity for an aqueous environment) and confer characteristic properties upon the proteins of which they are the monomeric components. Proteins, like nucleic acids, are informational molecules; the order of amino acids in the polypeptide is all-important in deter-

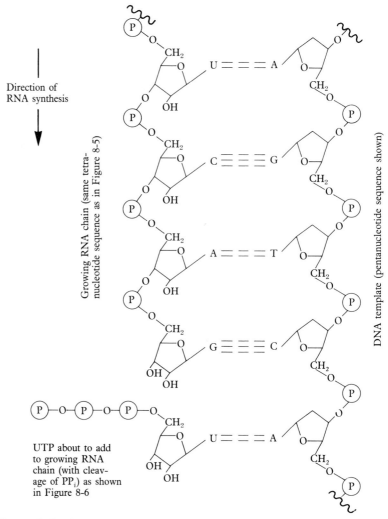

Direction of
RNA synthesis

Growing RNA chain (same tetranucleotide sequence as in Figure 8-5)

DNA template (pentanucleotide sequence shown)

UTP about to add
to growing RNA
chain (with cleav-
age of PP$_i$) as shown
in Figure 8-6

Figure 8-7 The role of DNA as a template in RNA synthesis. The base sequence A—G—T—C—A serves as a template for the synthesis of the RNA sequence U—C—A—G—U. Illustrated here is the same step (addition of U) shown in Figure 8-6. The dashed lines connecting the base pairs U—A, C—G, and A—T represent hydrogen bonds.

Figure 8-8 The polypeptide structure of proteins and the structures of the individual amino acids. The structure shown at the top is a tetrapeptide sequence from a protein chain. The individual amino acids may be any of the 20 structures shown in the lower portion of the figure. In each case the carboxyl and amino carbons are shown in black and the variable "R" side groups are shown shaded. Note that eight of the amino acids are nonpolar (hydrophobic), five have side groups that can carry a net positive or negative charge, and the remainder are uncharged but nonetheless polar (hydrophilic) in nature.

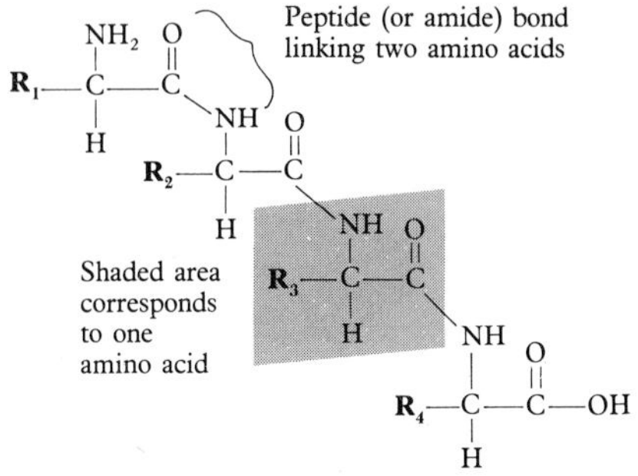

Peptide (or amide) bond linking two amino acids

Shaded area corresponds to one amino acid

Nonpolar Amino Acids (Hydrophobic)

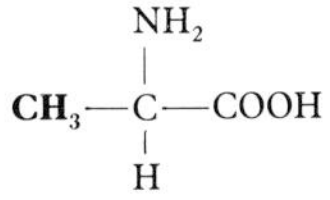

Alanine 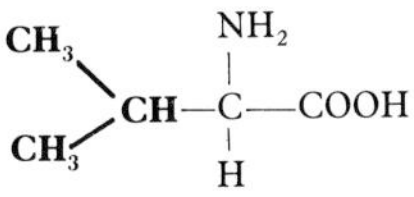 Valine 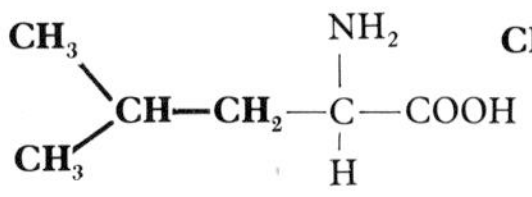 Leucine 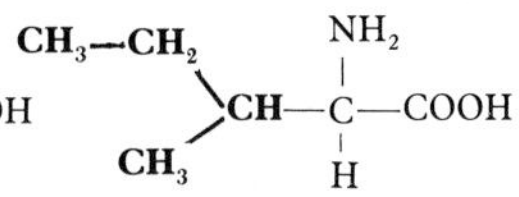 Isoleucine

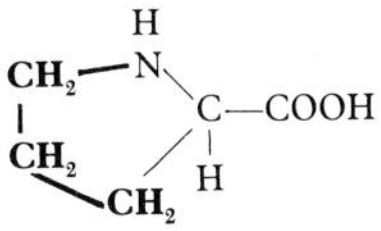

Proline 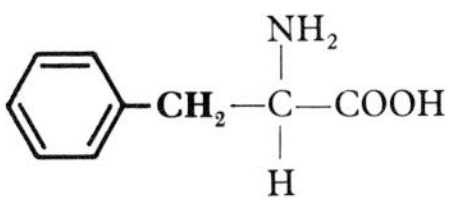 Phenylalanine 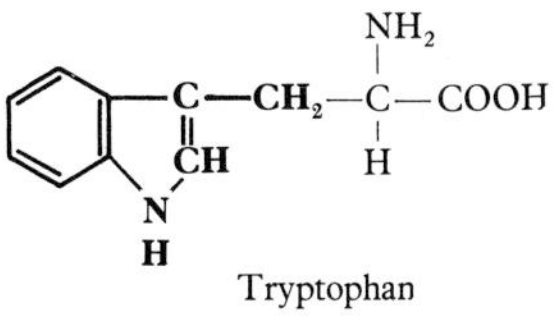 Tryptophan 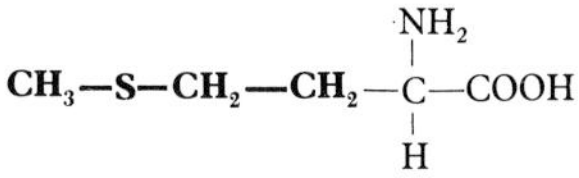 Methionine

Polar, Uncharged Amino Acids (Hydrophilic)

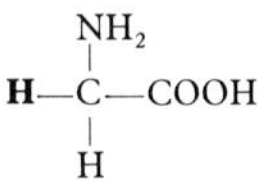

Glycine 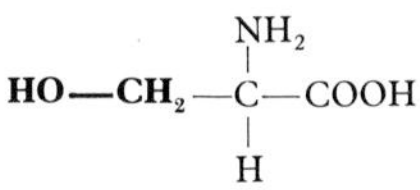 Serine 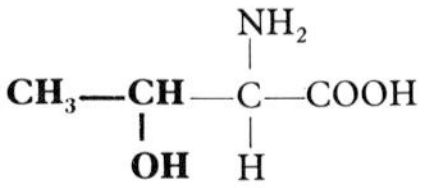 Threonine 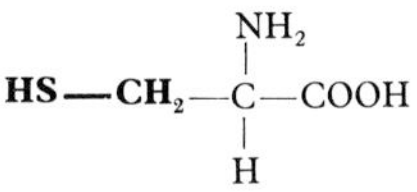 Cysteine

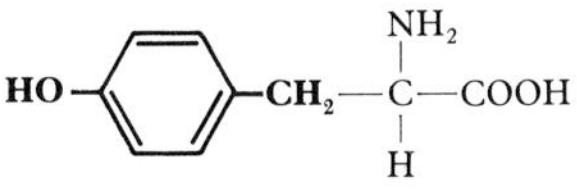

Tyrosine 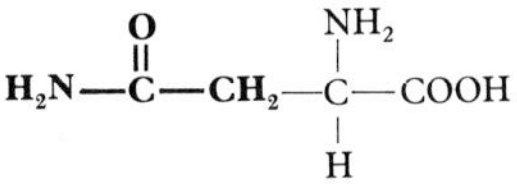 Asparagine 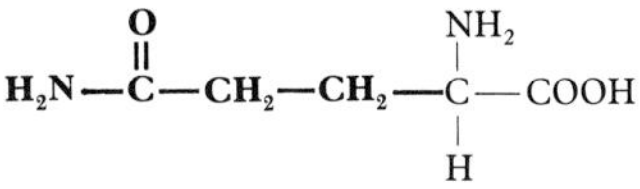 Glutamine

Polar, Charged Amino Acids (Hydrophilic)

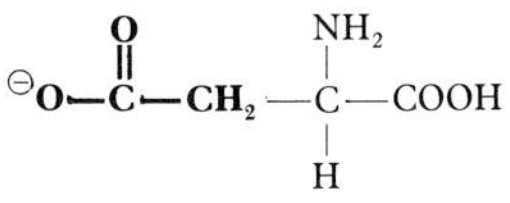

Aspartate 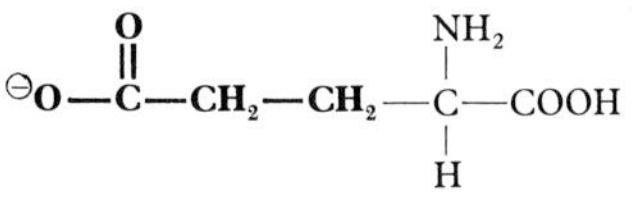 Glutamate

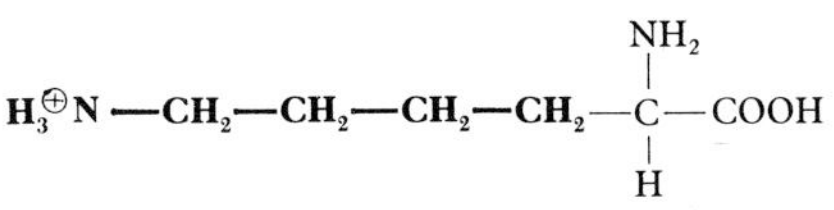

Lysine 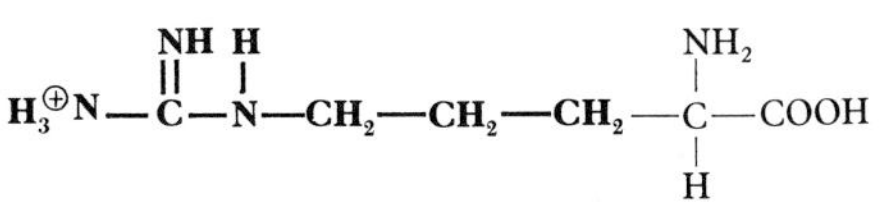 Arginine

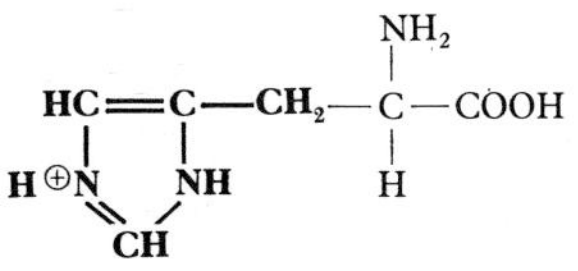

Histidine

mining the structure and function of the resulting protein, and this order must be specified by a template at the time of polypeptide synthesis.

From a purely chemical point of view, however, polypeptide synthesis can be summarized quite adequately in terms of the now-familiar pattern of activation followed by polymerization. The carrier for amino acid is a special kind of RNA molecule called a *tRNA*. (The "t" stands for *transfer* and indicates the informational role of the carrier in transferring the correct amino acid into position for polymerization.) The structure of a tRNA molecule is indicated schematically in Figure 8-9. An amino acid is linked to its tRNA carrier by an ester bond, and the complex is called an *aminoacyl tRNA*.

The formation of the ester bond between the amino acid and the tRNA is the actual activation event, driven, as expected, by the cleavage of ATP into AMP and pyrophosphate:

$$
\begin{array}{c}
\overset{H_2N}{\underset{|}{}}\quad \overset{O}{\underset{\parallel}{}} \\
R\!-\!CH\!-\!C\!-\!OH + HO\!-\!tRNA + ATP \xrightarrow[\text{synthetase}]{\text{aminoacyl tRNA}} \\
\text{Amino acid}\qquad\quad \text{tRNA}
\end{array}
$$

$$
\begin{array}{c}
\overset{H_2N}{\underset{|}{}}\quad \overset{O}{\underset{\parallel}{}} \\
R\!-\!CH\!-\!C\!-\!O\!-\!tRNA + AMP + PP_i \qquad (8\text{-}33) \\
\text{Aminoacyl tRNA}
\end{array}
$$

The pyrophosphate is hydrolyzed to render the activation sequence irreversible, and the AMP is returned to the ADP level by phos-

Figure 8-9 Schematic illustration of the structure of a tRNA molecule and its linkage via an ester bond to the amino acid for which it serves as both activator and adaptor. Each of the small circles in part (*a*) represents a nucleotide unit of the tRNA molecule; each tRNA contains about 80 nucleotides and has a molecular weight of about 25,000 daltons. (Note the discrepancy in scale; a tRNA molecule is in fact about 250 times larger than an amino acid.) Cross-links between nucleotides in paired regions of the structure represent hydrogen bonds between complementary bases (A with U, C with G). As indicated very schematically in part (*a*) and in greater detail in part (*b*), the tRNA molecule contains both paired regions and unpaired segments, the latter protruding as loops. The triplet anticodon that recognizes the codon of the messenger RNA is located on one of these unpaired loops. The amino acid for which a given tRNA serves as the specific carrier is linked to the free —OH group on carbon 3 of the ribose group of the last nucleotide (always AMP) at one end of the molecule. The specificity required to ensure that a particular tRNA becomes esterified only with the proper amino acid resides in the tRNA synthetase enzyme that catalyzes the formation of the ester bond between amino acid and tRNA carrier.

phorylation at the expense of ATP:

$$PP_i + H_2O \xrightarrow[\text{pyrophosphatase}]{} 2P_i \qquad (8\text{-}34)$$

$$AMP + ATP \xrightarrow[\text{adenylate kinase}]{} 2ADP \qquad (8\text{-}35)$$

The actual polymerization reaction then involves transfer of the incoming amino acid from its tRNA carrier to the growing polypeptide chain. The transfer itself is exergonic, but the process of polypeptide synthesis, to be discussed shortly, is such that two molecules of GTP are split (to GDP + P_i) during the addition of each amino acid group:

Polypeptide with n amino acids + aminoacyl tRNA + 2GTP $\xrightarrow[\text{polypeptide synthesis}]{}$

polypeptide with $n + 1$ amino acids + tRNA + 2GDP + 2P_i $\qquad$ (8-36)

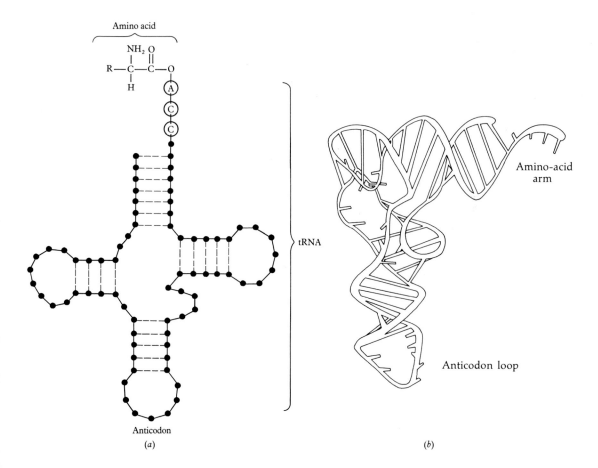

(a) (b)

And the GTP can, of course, be regenerated from GDP by receiving a phosphate group from ATP:

$$2GDP + 2ATP \xrightarrow[\substack{\text{nucleoside} \\ \text{diphosphokinase}}]{} 2GTP + 2ADP \qquad (8\text{-}37)$$

If we now sum these five reactions (8-33 through 8-37), we arrive at the following overall expression for protein synthesis:

Polypeptide with n amino acids + amino acid + 4ATP $\longrightarrow$

$$\text{polypeptide with } n + 1 \text{ amino acids} + 4ADP + 4P_i \qquad (8\text{-}38)$$

Again, as for both polysaccharide and nucleic acid synthesis, the actual energy required to add each successive monomeric unit ($\Delta G^0 = +5.5$ kcal/mole for the peptide bonds of protein synthesis) is much less than that expended in the process ($\Delta G^0 = -29.2$ kcal/mole for four ATPs), ensuring that the synthetic process goes virtually to completion.

Because proteins are informational macromolecules, their synthesis, like that of the nucleic acids, is a fairly complex, template-mediated process. A detailed description of the process exceeds the intended scope of the present discussion, but we shall look briefly at the essential features as summarized in Figure 8-10. The template for protein synthesis is a molecule of messenger RNA, previously transcribed from DNA by the complementary base-pairing process

Figure 8-10 Addition of one amino acid to a growing polypeptide chain. Illustrated here is the cyclic process required to select and add the next amino acid in the process of polypeptide synthesis as carried out on the ribosome. The particular step shown is the addition of the fourth amino acid to the polypeptide chain for a specific protein, the coat protein, of the bacteriophage R17 (a protein for which both the amino acid sequence of the polypeptide and the nucleotide sequence of the mRNA are known). The A (aminoacyl) site is shaded in each case. Amino acids, their abbreviations, and codon assignments are as follows: methionine (met): AUG; alanine (ala): GCU; serine (ser): UCU; and asparagine (asn): AAC. The incoming amino acid (asparagine in the step shown) arrives in activated form as the aminoacyl tRNA, having been previously esterified to its proper tRNA carrier by the tRNA synthetase reaction of Equation 8-33, driven by the splitting of ATP to AMP and PP_i. The cyclic process illustrated here is repeated for each successive amino acid until one of the chain-terminating codons (UAG, UAA, or UGA; see Table 8-3) is encountered, at which point the final peptidyl-tRNA bond is hydrolyzed to release the completed polypeptide.

described earlier. Viewed as a problem in information transfer, the question becomes one of using the linear sequence of nucleotides in the RNA template to dictate a unique linear sequence of amino acids in the protein product. Since there are 20 different kinds of amino acids and only 4 different kinds of nucleotides, a simple one-to-one pairing mechanism is clearly not adequate. Instead the nucleotides are "read" in groups of three called *codons*, each codon specifying a particular amino acid. Thus the codon sequence GGA specifies the

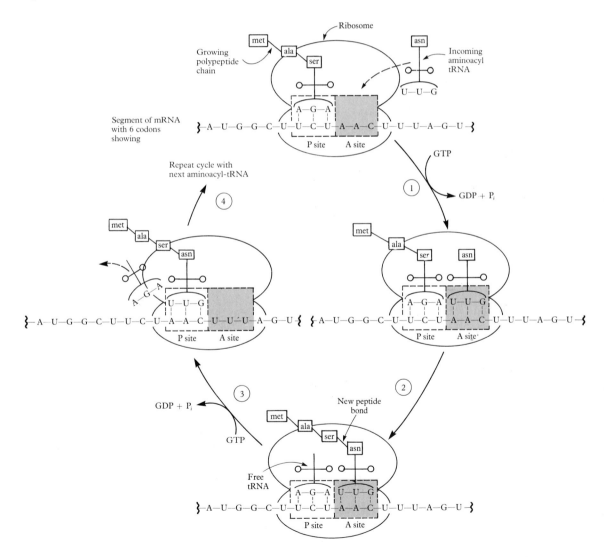

insertion of glycine in the polypeptide chain, UUU codes for phenylalanine, etc.

Actually, four different nucleotides arranged in groups of three account for $4 \times 4 \times 4 = 64$ different codons, more than enough to code for 20 different amino acids. Most amino acids have, in fact, two or more codons, and several codons are reserved for "punctuation" purposes. Table 8-3 presents a complete list of triplet codons

TABLE 8-3 The Genetic Code

First Base in Codon	Second Base in Codon			
	U	C	A	G
U	UUU—phenylalanine UUC—phenylalanine UUA—leucine UUG—leucine	UCU—serine UCC—serine UCA—serine UCG—serine	UAU—tyrosine UAC—tyrosine UAA—c.t. UAG—c.t.	UGU—cysteine UGC—cysteine UGA—c.t. UGG—tryptophan
C	CUU—leucine CUC—leucine CUA—leucine CUG—leucine	CCU—proline CCC—proline CCA—proline CCG—proline	CAU—histidine CAC—histidine CAA—glutamine CAG—glutamine	CGU—arginine CGC—arginine CGA—arginine CGG—arginine
A	AUU—isoleucine AUC—isoleucine AUA—isoleucine *AUG—methionine	ACU—threonine ACC—threonine ACA—threonine ACG—threonine	AAU—asparagine AAC—asparagine AAA—lysine AAG—lysine	AGU—serine AGC—serine AGA—arginine AGG—arginine
G	GUU—valine GUC—valine GUA—valine *GUG—valine	GCU—alanine GCC—alanine GCA—alanine GCG—alanine	GAU—aspartate GAC—aspartate GAA—glutamate GAG—glutamate	GGU—glycine GGC—glycine GGA—glycine GGG—glycine

Note: This arrangement shows the correspondence between the base sequence in the triplet codons of the messenger RNA and the specific amino acids inserted into the growing polypeptide chain during protein synthesis. Of the 64 possible codons, 61 specify amino acids and 3 serve as signals for chain termination, abbreviated c.t. in the table. The two codons marked with an asterisk can, under appropriate conditions, serve as chain-initiating signals.

and the corresponding amino acids or punctuation signals. This relationship between triplet codons of nucleotides and the specific amino acids they code for is called the *genetic code*; the "cracking" of this code, mainly in the laboratories of M. Nirenberg and H. G. Khorana, is one of the most exciting chapters in all of modern biology and earned the principals a Nobel prize.

As you may already be aware, the tRNA carrier plays a key role in ensuring that the triplet codon is read correctly and the proper amino acid is added to the growing polypeptide chain. This is possible because each amino acid is linked to a specific tRNA carrier, and each tRNA molecule has a specific three-nucleotide sequence called an *anticodon* by which it recognizes the appropriate codon in the mRNA, using the same kind of complementary base-pairing that is involved in nucleic acid synthesis. Thus to recognize the phenylalanine codon UUU, the phenylalanine-bearing tRNA has the sequence AAA in its anticodon—and that tRNA is never used to carry any other amino acid except phenylalanine. This means, of course, that there must be at least as many different kinds of tRNAs as there are different amino acids. In fact most amino acids have several tRNAs because of the need for several different anticodons to recognize several different codons. The crucial process of linking the correct amino acids to the correct tRNAs obviously occurs during the activation or "charging" step and requires specific activating enzymes, one for each of the 20 amino acids. Thus in protein synthesis the activation step serves both to energize the amino acid for polymer formation and to link it to the right tRNA, thereby guaranteeing that it will be inserted at the correct position along the growing polypeptide chain.

The final element needed to understand the process of protein synthesis is the *ribosome*. Ribosomes are cellular particles with particle weights of about 2.8×10^6 daltons in prokaryotes and 4.2×10^6 daltons in eukaryotes. Each ribosome consists of two subunits, and the subunits in turn are complexes of *ribosomal RNA* (rRNA) and proteins. Functionally, ribosomes are the "workbenches" on which protein synthesis is carried out, though the involvement of the ribosome itself in the synthetic process is probably not so passive as the term "workbench" might suggest. Each ribosome has two sites involved in protein synthesis: an A (for aminoacyl) site at which the incoming aminoacyl tRNA binds, and a P (for peptidyl) site at which the growing polypeptide chain is attached. In addition the ribosome binds a messenger RNA molecule, which serves as the template to direct the selection of incoming aminoacyl tRNAs at the A site.

The actual process of protein synthesis as diagrammed in Figure 8-10 therefore involves, at a minimum, a messenger RNA (mRNA) template, a ribosome as the site of synthesis, 20 different amino acids, specific tRNA carriers for the amino acids, and GTP as an energy source. Once the messenger RNA has become attached to the ribosome and the peptide chain is initiated, the following cycle of events is required in order to add one amino acid to the growing polypeptide chain:

Step 1 : The tRNA bearing the correct amino acid (asparagine in the case shown in Figure 8-10) binds to the A site of the ribosome by complementary base-pairing between the codon of the messenger RNA at the A site (AAC) and the anticodon of the tRNA

(UUG). This binding step requires the hydrolysis of one GTP to GDP and P_i.

Step 2: The growing polypeptide chain attached to the tRNA that brought in the previous amino acid is now transferred from that tRNA at the P site to the amino group of the newly arrived amino acid at the A site, thus adding the new amino acid to the growing chain and effectively transferring the chain to the A site.

Step 3: The tRNA bearing the polypeptide chain is then transferred from the A site to the P site, displacing the free tRNA at the P site. Concomitantly the mRNA is moved one codon further along, driven by the cleavage of a second GTP to GDP and P_i.

Step 4: This leaves the A site vacant, ready for the next incoming aminoacyl tRNA, as specified by the new codon (UUU) now at the A site.

This cycle of events is repeated for the addition of each new amino acid, of which there are usually 100 to 500 per polypeptide. Each cycle results in the movement of the ribosome and its attached polypeptide one codon further along the messenger RNA, a physical translocation driven by GTP hydrolysis. Special events are involved in the initiation and termination of the chain, but the cyclic sequence of Figure 8-10 represents at least the essence of the synthetic process.

The complete process whereby the genetic information encoded in the nucleotide sequence of DNA in a cell is expressed eventually as the amino acid sequence of a protein involves both *transcription* (DNA → RNA) and *translation* (RNA → protein). It is, as this discussion illustrates, a complex process indeed. It is also very expensive energetically: a polypeptide of molecular weight 20,000 daltons consists of 200 amino acids and requires 800 ATPs (200 amino acids × 4 ATPs/amino acid) for its synthesis. In addition, synthesis of the polypeptide presumes the presence of an RNA message with at least 600 nucleotides, the synthesis of which requires a further 1200 ATPs (600 nucleotides × 2 ATPs/nucleotide), though once synthesized a given mRNA can give rise to multiple copies of the protein as it is reutilized by successive ribosomes. The combined synthesis of mRNA and polypeptide therefore requires 2000 ATPs, which in turn demands that about 53 molecules of glucose with an aggregate molecular weight of about 9500 daltons be combusted completely to CO_2 and H_2O. And when you realize that even a small bacterial cell may be synthesizing more than 1000 protein molecules per second, you begin to get an idea of the magnitude of the work of macromolecular synthesis and the energy demands it can place upon a cell.

So far we have seen how the small organic molecules of cells are synthesized from simpler precursors (ultimately from inorganic compounds present in the environment) and then linked into the polymers that are the characteristic units of cellular structure. We have noted that the synthesis of monomers, such as the sugars, amino acids, and nucleotides, requires both reducing power (usually $NADPH_2$) and energy (almost always ATP), while the subsequent polymerization events require only energy, usually provided by pyrophosphate cleavage to guarantee the irreversibility of the polymerization process. For the most important cellular polymers, the proteins and nucleic acids, synthesis also requires information, in the form of a preexisting template that dictates the subunit order so crucial to the function of these macromolecules. Appreciative of the investment of $NADPH_2$, ATP, and information that the cellular polymers represent, we are now ready to consider the steps that lie beyond—the processes whereby these biopolymers become organized into the supramolecular assemblies and organelles characteristic of cellular structure. Or, in terms of Figure 8-1, we have reached level C and are ready to move on to levels D and E en route to the intact cell.

We can begin our consideration by noting two general features of supramolecular assembly processes: they are *hierarchical* and they occur *spontaneously*. The hierarchical nature of such processes should come as no surprise, since it is just an extension of the overall plan of cellular organization and hierarchical synthesis illustrated in Figure 8-1. But that such processes are spontaneous may seem surprising indeed, and for two distinct but related reasons: entropy and information. On the one hand, it is clear that the assembly of highly organized cellular structures from a random collection of biopolymers results in a striking decrease in entropy, and this we have learned to associate with nonspontaneous processes. On the other hand, these assembly processes have a specificity and reproducibility that suggest substantial input of information, yet they occur in most cases without a template or other external source of information. The apparent paradox posed by such spontaneous assembly processes is underscored by comparing the assembly of several polypeptides into a multi-molecular complex with the synthesis of any one of those polypeptides by sequential polymerization of amino acids. Both processes clearly result in greater order (lower entropy) and both require information. Yet while polypeptide synthesis requires both energy (four high-energy phosphate bonds per amino acid, in fact) and a template, subsequent supramolecular assembly occurs spontaneously, requiring neither energy nor information input.

In brief, the answer to these apparent paradoxes is that both the thermodynamic driving force and the information required for such assembly processes is already present in the polymers, and the spontaneousness is possible because the decrease in entropy of the *system* (the polypeptide) is more than offset by increases in the entropy of the *surroundings*. This means, of course, that in terms of biosynthetic work the synthesis of a cell is complete when all its macromolecular components have been synthesized, since all subsequent assembly processes appear to be thermodynamically spontaneous, occurring without further input of energy. The topic can still be of legitimate interest to us, however, both to illustrate the importance of hydrophobic interactions as a driving force in higher-order assembly processes and also to emphasize the extent to which a detailed understanding of such processes remains a challenge for the future.

A useful prototype for assembly processes can be found in the folding of a one-dimensional polypeptide chain into a three-dimensional protein. Although the distinction is not always clearly made, the immediate product of amino acid polymerization is a polypeptide and not a protein. To become a functional protein, one or more such linear polypeptide chains must coil and fold in a predetermined manner to assume the unique three-dimensional configuration necessary for biological activity, as illustrated in Figure 8-11. The striking feature of such coiling and folding is that it occurs spontaneously as the polypeptide chain reels off the ribosome during synthesis. By the time the fully elongated chain is released from the ribosome, it has already assumed a stable, predictable, three-dimensional structure, without any input of energy or information beyond the polymerization process. Furthermore the folding is unique in that each polypeptide with the same amino acid sequence will fold in an identical manner under identical conditions. If a protein is *denatured* (treated with a chemical agent such as urea or high salt that disrupts its three-dimensional structure) and the resulting polypeptide is then returned to favorable conditions, it will refold spontaneously to the original (*native*) configuration.

The unique three-dimensional structure of a protein is a direct and inevitable consequence of its particular amino acid sequence. This is because of the different functional groups that the various amino acids possess (the notorious "R-groups" of Figure 8-8) and the variety of interactions possible between them. Every possible interaction between a given amino acid and any other amino acid (or between an amino acid and the water molecules that surround it in the medium) is characterized by a specific free energy, and the final configuration is simply the one with the lowest possible free energy and hence the greatest stability. No work is done during the folding process—the polypeptide simply folds and coils as its constituent side groups inter-

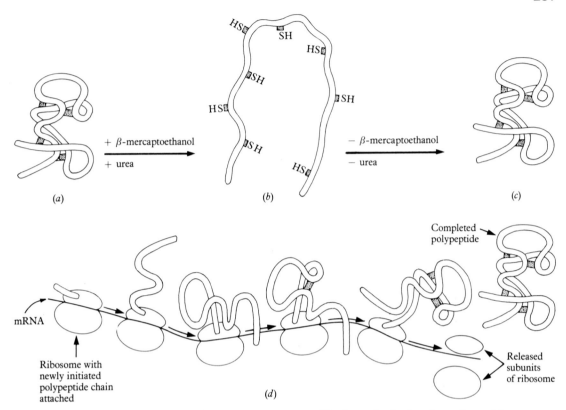

Figure 8-11   The three-dimensional structure of proteins. The hypothesis that the amino acid sequence of a polypeptide contains sufficient information to determine the unique three-dimensional structure of the protein can be tested by treating the intact (*native*) protein (*a*) with a reducing agent such as β-mercaptoethanol (to reduce disulfide bonds) and a denaturing agent such as 8M urea (to disrupt hydrogen bonds). This results in a completely unfolded or *denatured* molecule (*b*), with no fixed shape and no biological activity. Upon gradual removal of the mercaptoethanol and urea, the denatured molecule refolds spontaneously (*renatures*) to the original native configuration (*c*), which is once more stabilized by reformation of the correct disulfide cross-links. The specific protein shown is ribonuclease, the first protein to be denatured and then renatured in this way. The native configuration of a protein is actually achieved progressively during synthesis of its polypeptide chain, as shown in (*d*). Each of the ribosomes moving along the mRNA chain (such a multi-ribosome structure is called a *poly-ribosome*) has attached to it a partially complete polypeptide chain, which grows in length as the ribosome moves from the initiating end of the mRNA to the terminating end. The polypeptide chain is not released from the ribosome until all of its amino acids have been added (124 for ribo-nuclease), but spontaneous folding of the chain begins shortly after synthesis and continues as the polypeptide elongates. By the time the completed polypeptide is released from the ribosome, it has already attained its appropriate three-dimensional shape.

act with each other and seek out the configuration with the lowest free energy.

The key to understanding the driving forces behind this folding process lies in the relative affinities of the various amino-acid side groups for water. As indicated in Figure 8-8, some of the side groups are *hydrophilic* (from the Greek, meaning "water-loving"); they are polar in nature, have a strong affinity for water, and tend to seek out positions near the surface of the protein, where they can interact maximally with the water molecules of the medium. Others, however, are *hydrophobic* (Greek again, "water-fearing"); they are essentially nonpolar, have little or no affinity for water, and tend therefore to gravitate toward the center of the protein molecule, where they can interact with each other in an essentially nonaqueous milieu and avoid exposure to water. Valine, leucine, and isoleucine are examples of amino acids with hydrophobic side groups; as you can see from Figure 8-8, their side groups are hydrocarbon in nature and would not be expected to mix with water if present as separate molecules.

To a good first approximation, protein structure is the result of a balance between the tendency of the hydrophilic groups to seek an aqueous environment near the surface and that of the hydrophobic groups to minimize contact with water by associating with each other in the center of the molecule. Clearly, if all or even most of the amino acids in a protein were hydrophobic, the protein would be virtually insoluble in water and would seek out a nonpolar environment. (Indeed, membrane proteins are probably localized in membranes for this very reason.) Similarly, if all or most of the amino acids were hydrophilic, the polypeptide would most likely be quite content to remain in a fairly distended, random configuration, allowing maximum access of each amino acid to its aqueous environment. But precisely because the polypeptide chain contains both hydrophilic and hydrophobic groups (in a specified order, remember), some will tend to seek out an aqueous, and others a nonaqueous, environment. Thus some portions of the molecule are drawn toward the surface while other portions are driven toward the interior, and the final structure of the protein is the inevitable outcome of these tendencies.*

This kind of balance between opposing hydrophilic and hydro-

---

*Although the actual folding process whereby a protein attains its three-dimensional state is driven by the thermodynamic tendency to minimize interaction between hydrophobic regions and the aqueous environment, other intramolecular interactions are often involved in further stabilizing the protein once the correct configuration has been attained. Especially important in this regard are intramolecular *disulfide bonds* between cysteine side groups. Such side groups may be located quite a distance from one another along the polypeptide chain but are brought in close proximity by the folding process. Proteins stabilized by disulfide bonds frequently lose function if these important intramolecular links are broken (by reduction to the free sulfhydryl groups). This is why the denaturation of a protein known to contain disulfide bonds requires both a denaturing agent such as urea (to disrupt hydrogen bonds) and a reducing agent such as β-mercaptoethanol (to reduce disulfide bonds). This feature is shown in Figure 8-11 for the protein ribonuclease, which is known to be stabilized by specific disulfide bonds.

phobic tendencies is probably very important in other assembly processes as well; its importance in the organization of cellular membranes, for example, is emphasized in the next chapter. Since the chemistry of the amino-acid side groups and their interactions is well known, it ought to be possible, in theory at least, to predict the single most stable thermodynamic configuration for a protein once the amino acid sequence of the component polypeptide(s) is known. In fact, however, that is not yet possible, even with the use of computer technology.

The same properties and interactions of amino-acid side groups responsible for the folding pattern of an individual polypeptide are apparently also important in higher-order assembly processes involving proteins. These include the association of multiple subunits in the many proteins containing more than one polypeptide chain (lactate dehydrogenase, the enzyme that interconverts pyruvate and lactate, for example, has four subunits), the organization of individual proteins into multimolecular complexes, and the formation of higher-order complexes between proteins and other cellular components (such as membranes, ribosomes, or virus particles). In each case it is presumed that the characteristic molecular associations involved in such complexes are dictated by (1) the specific amino-acid side groups present at or near the surface of the individual folded polypeptide chains and (2) the possibilities for interaction represented by hydrophobic "patches" on the surface, which would rather associate with similar such regions on the surface of another polypeptide than with an aqueous environment. Ultimately, then, all the interactions of which proteins are capable are programmed into its amino acid sequence, such that the gene coding for a specific protein determines far more than the linear sequence of the amino acids in the component polypeptide. Inherent in the amino acid sequence of each polypeptide must be all the information necessary to specify all the interactions of which that polypeptide is capable.

As Figure 8-1 indicates, some of the major supramolecular structures of cells are composed mainly, if not exclusively, of proteins. These include not only the multienzyme complexes like that involved in pyruvate oxidation but also the various filament- and tubule-containing structures involved in the contractile and motile processes to be discussed in Chapter 10. The assembly of such structures will almost certainly turn out to be understandable in terms of the hydrophobic/hydrophilic forces involved in the folding and association of polypeptides.

The ubiquity and importance of proteins should not obscure the fact that many of the characteristic structures of the cell are complexes between two or more different kinds of biopolymers and clearly involve interactions that are chemically distinct from those of polypeptide folding and association. Ribosomes, for example, contain both RNA and proteins (the larger of the two subunits, for instance, consists of two RNA molecules and about 30 different proteins); membranes are made up of both lipid and protein; and even a plant cell wall, though composed mainly of cellulose fibrils, also contains a small but apparently crucial protein component. The chemical details of such associations are not yet well understood, but they are coming increasingly under experimental scrutiny as techniques become available for isolating, dissociating, and reconstituting such structures. Few definitive statements can yet be made, but the data from recent studies, particularly on the assembly of ribosomes, viral particles, and artificial membranes, support the validity of the general principles established earlier. Specifically, even these larger structures seem to assemble in a hierarchical, spontaneous manner. And despite the obvious differences in chemistry involved with polymers as different as proteins, nucleic acids, and polysaccharides, the interactions that drive these supramolecular assembly processes seem to be essentially the same as those that dictate the folding of individual protein molecules.

In most cases the information required to specify the exact configuration of the end product seems to lie entirely within the biopolymers that contribute to the structure. Certainly enzymes with more than one polypeptide chain, multienzyme complexes, ribosomes, and at least the simpler viruses are in this category. As in the case of protein folding, the final structure is simply the inevitable consequence of the thermodynamic drive toward minimum free energy. Self-assembly systems therefore achieve thermodynamically stable structures without additional information input because the information content of the component biopolymers is adequate to specify the complete assembly process.

There is, however, also evidence that some assembly systems require that some information be supplied by preexisting structure. In such cases the ultimate structure arises not by *de novo* assembly of the components but by ordering these components into the matrix of an existing structure. Examples of cellular structures that are routinely built up by the addition of new material to existing structures include membranes, cell walls, and chromosomes. On the other hand, such structures are in general not yet sufficiently well characterized to ascertain whether the presence of preexisting structure is actually obligatory or whether under the right conditions the components might actually be capable of self-assembly. Evidence from studies with artificial membranes and with chromatin (isolated chromosomal components), for example, suggests that preexisting structure, though routinely present *in vivo*, may not be an actual requirement for the assembly process. Much additional insight will be necessary before it can be determined definitively whether and to what extent external information is required or exploited in cellular assembly processes.

Practice Problems

8-1. Gluconeogenesis: The synthetic route to glucose from pyruvate is essentially the glycolytic pathway functioning in reverse but rendered thermodynamically feasible in the direction of synthesis by the use of alternative reactions at three points. The following standard free-energy changes are relevant to the energetics of glycolysis and gluconeogenesis:

$$\text{Glucose} + 6O_2 \rightarrow 6CO_2 + 6H_2O \qquad \Delta G^0 = -686 \text{ kcal/mole}$$

$$\text{Pyruvate} + 2\tfrac{1}{2}O_2 \rightarrow 3CO_2 + 2H_2O \qquad \Delta G^0 = -273.5 \text{ kcal/mole}$$

$$\text{NADH}_2 + \tfrac{1}{2}O_2 \rightarrow \text{NAD} + H_2O \qquad \Delta G^0 = -52 \text{ kcal/mole}$$

$$\text{ATP} + H_2O \rightarrow \text{ADP} + P_i \qquad \Delta G^0 = -7.3 \text{ kcal/mole}$$

(a) Write an overall equation for the catabolic degradation of glucose ($C_6H_{12}O_6$) to pyruvate ($C_3H_4O_3$). Be sure to indicate the yield of $NADH_2$ and ATP.

(b) Write an overall equation for the anabolic synthesis of glucose from pyruvate. Be sure to indicate the requirement for $NADH_2$ and ATP.

(c) Calculate the ΔG^0 for glycolysis (glucose to 2 pyruvate).

(d) Calculate the ΔG^0 for gluconeogenesis (2 pyruvate to glucose).

(e) Why is it essential that the anabolic route be different from the catabolic route?

(f) The activity of the enzyme that converts pyruvate to oxaloacetate in the gluconeogenic pathway is known to be regulated allosterically by the intracellular concentrations of acetyl CoA. Would you expect acetyl CoA to activate or to inhibit this enzyme? Why?

8-2. Serine Biosynthesis: The amino acid serine can be synthesized from glycerate-3-phosphate in a three-step sequence involving (1) oxidation of the hydroxyl group of glycerate-3-phosphate, (2) transamination of the resulting α-keto acid, and (3) cleavage of the phosphate group as P_i.

(a) Draw the structures of the two intermediates in serine biosynthesis and write a summary equation. Your equation should indicate a need for both NAD and glutamate in serine biosynthesis.

(b) Recalling that glycerate-3-phosphate is an intermediate in the gluconeogenic pathway (Figure 8-2), devise a reaction sequence to account for the net synthesis of serine from pyruvate in a chemotrophic cell and write a balanced equation for the process.

(c) Recalling that glutamate can be regenerated by reductive amination (Figure 8-1), devise a scheme to account for the complete synthesis of one molecule of serine from pyruvate and NH_3. Write a balanced equation for the process

O
‖
C—OH
|
H—C—OH
|
CH_2—O—℗

Glycerate-
3-phosphate

O
‖
C—OH
|
H—C—NH_2
|
CH_2—OH

Serine

and determine the number of molecules of ATP required per serine. Explain why there is no net need for reducing power in the summary equation.

8-3. Nitrogen Metabolism: The main source of inorganic nitrogen for higher plants is the nitrate anion, NO_3^-, which the plant takes up from the soil through its roots and reduces to ammonia (NH_3) in its leaves, a process that requires both reducing power and ATP. The supply of nitrate in the soil is continuously replenished by the oxidation of soil ammonia. This process, called *nitrification*, is usually accomplished by two different groups of soil bacteria, one of which oxidizes ammonia to nitrite (NO_2^-) and the other of which oxidizes the nitrite to nitrate, which is then released into the soil and becomes available to plant roots:

$$NH_3 \xrightarrow[\text{related bacteria}]{\text{\textit{Nitrosomonas} and}} NO_2^- \xrightarrow[\text{related bacteria}]{\text{\textit{Nitrobacter} and}} NO_3^-$$

Both *Nitrosomonas* and *Nitrobacter* obtain their carbon by CO_2 fixation.

(a) Since cellular metabolism operates strictly on a capitalistic basis, with the motive behind each process being to benefit the individual cell maximally rather than to contribute to the good of the ecosystem, there must clearly be something in it for nitrifying bacteria when they convert NH_3 to NO_3^- other than the social gesture of providing nitrate for plants. Can you think what it might be?

(b) Why are the nitrifying bacteria described as "chemotrophic autotrophs"?

8-4. Polysaccharide Synthesis: The activated form of glucose used in glycogen synthesis is UDP-glucose, which derives one of its two phosphates from glucose-1-phosphate and the other from UTP. Glycogen synthesis can be carried out *in vitro*, provided the following components are included in the incubation mixture: glucose, a partial glycogen chain, UDP, ATP, and the six enzymes required for all the steps in the synthetic process (Reactions 8-20 through 8-25). Assume that you have set up such a system using ATP labeled with ^{32}P in the terminal (γ) phosphate group only. After a suitable period of incubation, the following compounds are isolated and examined for the presence of ^{32}P:

Glucose	UDP-glucose	UTP
Glucose-1-phosphate	pyrophosphate (PP_i)	UDP
Glucose-6-phosphate	phosphate (P_i)	ATP
Glycogen		ADP

(a) Which of these compounds would you expect to contain ^{32}P? Explain how the label gets into each.

(b) Repeat part (a) but with the ATP labeled exclusively in the innermost (α) phosphate group.

(c) If you set up a similar system, but with ADP and UTP instead of ATP and UDP, would you expect it to be able to synthesize glycogen? Why or why not?

(d) If in the system of part (c) the UTP were labeled with ^{32}P in the middle

(β) phosphate group only, which of the compounds listed above would be labeled with ^{32}P after a reasonable incubation time?

8-5. DNA Synthesis: An *Escherichia coli* cell contains about 4.5×10^{-15} g of DNA, which consists of nucleotides with an average molecular weight of 320 daltons. Under favorable growth conditions, cultured *E. coli* cells can divide every 20 min which requires, of course, that each cell must replicate (make a copy of) its DNA within the 20-min period.

(a) How many nucleotides are present in the DNA of an *E. coli* cell?

(b) How many molecules of ATP are consumed by a single cell during the replication of all its DNA in preparation for cell division? (Assume the presence of dAMP, dCMP, dGMP, and dTMP in the culture medium.)

(c) At what average rate (in molecules per second) must the cell oxidize glucose to provide the energy necessary for DNA synthesis under aerobic conditions?

(d) What would be the consequences for the cell if the nucleotide precursors of DNA were *not* present in the culture medium?

‡8-6. Label Chasing: The anabolic route from pyruvate to phosphoenol pyruvate (Reaction 8-7) involves successive carboxylation and decarboxylation reactions, with oxaloacetate as an intermediate. However, the carbon atom added in the carboxylation step is the same one that is removed again in the decarboxylation that follows. This suggests, of course, that radioactively labeled CO_2 should never be incorporated into phosphoenol pyruvate or other intermediates in gluconeogenesis. Surprisingly, however, if a chemotrophic gluconeogenic system (liver, for example) is allowed to synthesize glucose from pyruvate in the presence of $^{14}CO_2$ (as bicarbonate, $H^{14}CO_3^-$), label does in fact appear in phosphoenol pyruvate, in all successive intermediates, and ultimately in carbons 3 and 4 of glucose. Curiously, several of the four-carbon intermediates of the TCA cycle also become labeled in the process. Explain!

References

Anfinsen, C. B., "The Formation of the Tertiary Structure of Proteins," *Harvey Lectures* (1967), *61*: 95.

Arnott, S., "The Structure of Transfer RNA," *Progress in Biophysics and Mol. Biol.* (1971), *22*: 179.

Casjens, S., and J. King, "Virus Assembly," *Ann. Rev. Biochem.* (1975), *44*: 555.

Dickerson, R. E., and I. Geis, *The Structure and Action of Proteins* (New York: Harper & Row, 1969).

Ingram, V. M., *Biosynthesis of Macromolecules*, 2nd ed. (Menlo Park, Ca.: Benjamin, 1971).

Khorana, H. G., "Polynucleotide Sequence and the Genetic Code," *The Harvey Lectures, 1966–1967* (New York: Academic Press, 1968).

Kushner, D. J., "Self-Assembly of Biological Structures," *Bacteriol. Rev.* (1969), *33*: 302.

Lewin, B. M., *The Molecular Basis of Gene Expression* (London: Wiley-Interscience, 1970).

Nirenberg, M., "Genetic Memory," *J. Am. Med. Assoc.* (1968), *206*: 1973.

Petermann, M. L., "How Does a Ribosome Translate Linear Genetic Information?" *Subcellular Biochemistry* (1971), *1*: 67.

Safrany, D. R., "Nitrogen Fixation," *Sci. Amer.* (Oct. 1974), *231*: 64.

Sussman, J. L., and S. H. Kim, "Three-Dimensional Structure of a Transfer RNA in Two Crystal Forms," *Science* (1976), *192*: 853.

Watson, J. D., *Molecular Biology of the Gene*, 3rd ed. (Menlo Park, Ca.: Benjamin, 1976).

Concentration Work

9

Every cell has a discrete, well-defined boundary and maintains within its confines a great variety of cellular components at concentrations that are both strikingly constant and very often markedly higher than those in the surrounding milieu. In fact, life could be defined not only as a state of improbable order but of improbable concentration as well (both are actually just different ways of expressing decreased entropy). Few indeed are the cellular reactions or processes that could occur at reasonable rates if they depended upon the concentrations of needed substrates present in the cell's surroundings. Essential to life as we know it, then, is the concentration both within cells and also at specific intracellular locations of the molecules and ions required as substrates and regulators for the enzyme-catalyzed reactions of cellular metabolism and function.

This concentration of needed components within a well-defined volume is possible only because each cell (and most eukaryotic organelles as well) is surrounded by a differentially permeable membrane. This membrane serves as the boundary between the cell or organelle and its environment and is equipped with means for transporting materials across it in a selective, controlled, and often directional way. Thus the membrane is essential to cellular existence both because it confines the contents of the cell or organelle and because it provides for restricted ingress and egress of molecules, thereby allowing controlled, selective accumulation of desired molecules as well as elimination of wastes and secretory products.

Knowing that the concentrations of cellular components are often higher inside than outside the cell, we can anticipate a need for energy input to drive the continued uptake of molecules and ions, since such uptake clearly results in a decrease in entropy (i.e., increase in order) and is therefore inherently endergonic. In fact the transport of solute molecules or ions across membranes against a concentration gradient is an important and universal use of the chemical energy of ATP in biological systems. Such *concentration work* is often less conspicuous than the mechanical work of the next chapter or even the biosynthetic work of the preceding chapter, but it is an important and fundamental activity of all cells and often makes a large claim upon the cellular energy economy. To quantitate this claim, consider the epithelial cells of your own kidney: about two-thirds of the ATP generated by these cells is used to transport sodium from the blood and concentrate it in the urine. Even considering your body as a whole, about one-third of the ATP you expend when at rest is used to maintain the Na^+ and K^+ gradients across cell membranes, a vital example of concentration work that we shall encounter later.

The Chemistry of Membranes

It is the surrounding membrane, then, that makes selective transport of components into the cell or organelle both possible and necessary. Membranes, however, must not be thought of simply as sieves that exclude big molecules and admit small ones. Whether or not a molecule or ion is admitted depends not only upon its size but also upon its charge, shape, chemical properties, and solubility. For many substances, movement across a membrane depends critically upon whether the membrane possesses a transport system specific for that particular molecule or ion. In addition the rate of entrance is greatly influenced by the concentration at which the molecule or ion in question is already present, both inside and outside the cell or organelle.

Most membranes are similar in gross composition. They are composed of 50 to 60 percent protein and 40 to 50 percent lipid by weight, with carbohydrates and other minor components sometimes also present. Because of the prominence of lipids and their significance in determining membrane properties, it is both necessary and worthwhile to consider lipid chemistry briefly. Unlike carbohydrates, proteins, and nucleic acids, lipids are not necessarily related to each other structurally, since they are defined simply by their common physical property of being soluble in organic (nonpolar) solvents. We shall be concerned primarily with *phospholipids* here because they are one of the principal structural components of all cell membranes. (*Steroids* are also distinctive membrane lipids, but we shall concentrate primarily on the phospholipids because far more is known about their involvement in membrane structure and chemistry.)

Phospholipids can be most readily understood as a variation of the structure of the neutral fats already encountered in Chapter 6. You may want to turn back to Figure 6-1 and refresh your memory of the chemistry of a *triglyceride*, which is really just a molecule of glycerol esterified with three fatty acids. As shown in Figure 9-1, a phospholipid is structurally similar except that it is a derivative of *glycerol phosphate* instead of glycerol and therefore has fatty acids esterified on two carbons but a phosphate ester on the third. In addition the phosphate is also linked to another group (designated "X" in Figure 9-1), which is usually an alcohol with a positively charged amino group on it. The two most common such alcohols are ethanolamine and choline, shown in Figure 9-1. In membranes the most common phospholipid is *phosphatidyl choline* (also called *lecithin*), which has the phospholipid structure shown in Figure 9-1 but with palmitate (a 16-carbon saturated fatty acid) and oleate (18 carbons, single unsaturated site) as R_1 and R_2, respectively, and choline as the nitrogenous alcohol (X) attached to the phosphate.

Inasmuch as the long-chain fatty acids have long, nonpolar hydrocarbon "tails," whereas the phosphate group and nitrogenous alcohol component are charged groups, phospholipids can be thought of as having both a highly polar end and a nonpolar end. This feature is illustrated schematically in Figure 9-2 (which also depicts the more uniformly nonpolar nature of a neutral fat). Molecules with both a polar

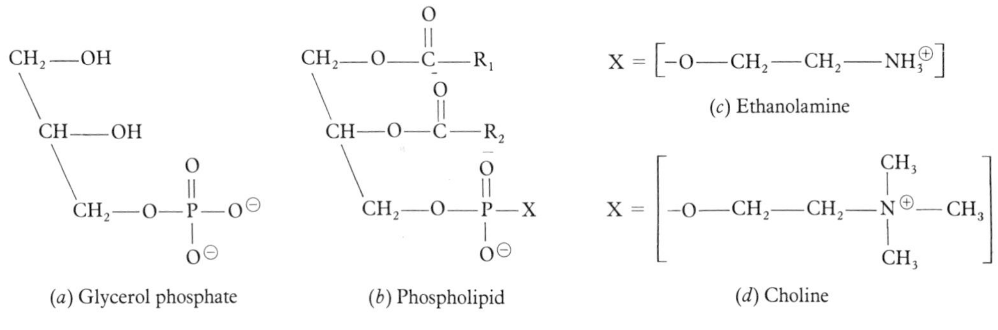

Figure 9-1 Phospholipid chemistry. The phospholipid molecule (*b*) is a derivative of glycerol phosphate (*a*), with two long-chain fatty acids esterified to the free hydroxyl groups of the glycerol. The R_1 and R_2 groups of the phospholipid structure are therefore the long-chain hydrocarbon "tails" of fatty acids, with or without one or a few double bonds. The X of the phospholipid structure is usually an alcohol bearing a positively charged amino group. The two most common nitrogenous alcohols esterified to the phosphate of a phospholipid are ethanolamine (*c*) and choline (*d*).

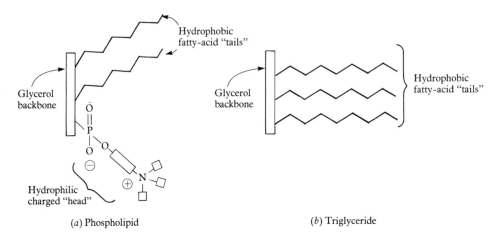

(a) Phospholipid   (b) Triglyceride

Figure 9-2   The amphipathic nature of the phospholipids. The phospholipid molecule (a) has both a nonpolar end (the two long hydrocarbon "tails" of the fatty-acid groups) and a polar end (the charged phosphate and amino groups at the "head" of the molecule). This amphipathic feature of the phospholipids is of special significance in membrane structure because it makes possible the phospholipid bilayer shown in Figure 9-3. By comparison the triglyceride molecule (b) is more uniformly nonpolar and has no special role in membrane organization. For chemical details of phospholipid and triglyceride structure, see Figures 9-1 and 6-1, respectively.

and a nonpolar end are referred to as *amphipathic* (the Greek prefix *amphi* means "of both kinds"). The amphipathic nature of the phospholipids gives these molecules distinctive properties that turn out to be very important in membrane chemistry. In particular the long, nonpolar fatty-acid tails are very insoluble in water and render that end of the phospholipid molecule hydrophobic ("water-fearing," remember?). The phosphate-X end of the molecule, on the other hand, is highly polar. (Both the phosphate group and the amino group of choline or ethanolamine exist in charged form at the neutral pH of most cells, so that this end of the molecule has both a positive and a negative charge.) Moreover it is readily soluble in water, making it hydrophilic ("water-loving").

When exposed to an aqueous milieu, phospholipids spontaneously arrange themselves such that their polar heads are facing outward toward the aqueous phase but their hydrophobic tails are hidden from the water. The resulting structure is a double layer of phospholipid molecules, with the heads of both layers facing outward and the fatty-acid tails extending inward toward those of the other layer, forming a continuous hydrophobic phase as shown schematically in Figure 9-3. The *lipid bilayer*, as this structure is frequently called, represents the minimum-energy configuration for phospholipids in water, since exposure of the hydrophobic regions to the aqueous environment is thereby minimized. This means that the formation of bilayers from random dispersions of phospholipid in water is accompanied by the release of free energy and is therefore thermodynamically favorable. Such structures, which form most readily in apertures separating two aqueous compartments, are sometimes regarded as *artificial membranes* inasmuch as they display some of the same structural and functional features of natural membranes.

Views on membrane structure have changed so rapidly in recent years that research papers, review articles, and text discussions dealing with this topic can often be dated quite accurately just by noting their concept of membrane organization. Early research on membranes stressed the importance of the lipid bilayer, a concept that has proved a useful framework for much of our thinking about membrane structure

The Structure of Membranes

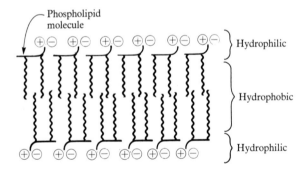

Figure 9-3 The phospholipid bilayer thought to provide the basis of membrane structure. Because of their amphipathic nature, phospholipids in an aqueous milieu orient themselves into a double layer, with the hydrophilic heads pointing outward toward the aqueous phase and the hydrophobic hydrocarbon chains buried on the inside. The driving force for bilayer formation is the same as that which underlies protein folding: the tendency to minimize exposure of hydrophobic regions of the structure to the aqueous environment. Originally proposed as a model for membrane structure more than fifty years ago, the phospholipid bilayer came into prominence in the Danielli-Davson model (Figure 9-4) and continues to play a central role in such contemporary views of membrane structure as the fluid mosaic model (Figure 9-5).

right to the present day. A landmark en route to current views of membranes came in 1935 with the model of J. F. Danielli and H. Davson, in which an inner lipid bilayer was considered to be coated on both sides by thin layers of protein. Their model, illustrated in Figure 9-4, accounted for the known presence of both lipid and protein in membranes and also explained discrepancies in surface tension that arose when simple lipid bilayers were compared with actual cellular membranes. When technical advances of the early 1950s permitted the direct observation of membranes in the electron microscope, a characteristic three-layered structure was seen, with a low-density region in the middle and an electron-dense layer on either side. This was taken as striking visual confirmation of the Danielli-Davson model of the membrane as a molecular "sandwich" with a bilayer of amphipathic lipids inside (their nonpolar tails accounting for the light inner zone of the membrane) and layers of densely staining protein and polar lipid heads on both surfaces.

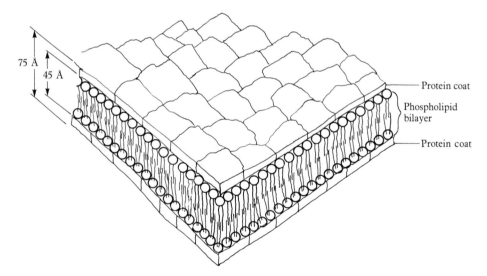

Figure 9-4 The unit membrane concept, which played an important historical role in the development of current views on membrane structure. The Danielli-Davson model, first proposed in 1935, viewed the membrane as a phospholipid bilayer covered on both sides with relatively thin layers of protein. On the strength of electron microscopic evidence, this concept was generalized and formalized by J. D. Robertson into the unit membrane hypothesis. The Danielli-Davson-Robertson model dominated thinking about membrane structure until the mid- or even late 1960s, eventually giving way to contemporary views of the membrane as a fluid mosaic (Figure 9-5) or a composite of lipoprotein subunits.

An additional feature of membrane structure that emerged from electron microscopic studies in the 1950s was the similarity in appearance of membranes from different sources, including both prokaryotes and eukaryotes as well as a variety of membrane-bounded eukaryotic organelles such as the nucleus, the mitochondrion, and the chloroplast. Based on these similarities, J. D. Robertson postulated his famous *unit membrane* concept, suggesting a common basic structure (essentially that of the Danielli-Davson model) for all membranes. The evidence supporting the unit membrane theory seemed so convincing that for a period of time extending into the mid- or even late 1960s a general confidence prevailed that membrane structure was already well understood. Increasingly, however, data began to accumulate that were incompatible with various aspects of the model. This development led to a variety of objections and criticisms of the unit membrane concept that, by the late 1960s, had thrown the whole field of membrane structure into turmoil.

Especially awkward were the difficulties encountered in trying to reconcile the emerging properties of membrane proteins with the surface location assigned to proteins in the Danielli-Davson-Robertson model. For example, biophysical evidence suggested strongly that membrane proteins were globular in shape, just like most enzymes. This agreed well with the diverse functional roles recognized for membrane proteins and with the need for a proper three-dimensional folding of polypeptide chains that such functions imply. But it did not seem at all consistent with the unit membrane view of a thin protein surface stretched over a lipid bilayer. Moreover membrane proteins were increasingly recognized to be quite hydrophobic in nature, a property that clearly dictates an internal location for such proteins, where they can be protected from the aqueous environment. Further doubts were cast on the validity of the unit membrane concept by the freeze-etching technique described in Chapter 7. Recall that this technique essentially splits a membrane open down the middle, allowing direct visualization of its interior structure. Freeze-etching studies revealed the presence of numerous particles of protein in the interior of many (though not all) kinds of membranes, a finding clearly also at variance with the unit membrane model.

Disenchantment with the Danielli-Davson-Robertson model reached its peak in the late 1960s and led eventually to the formulation of a variety of alternative models. In general these models agree on the need to regard proteins as more integral constituents of membrane structure, but they differ in the manner in which this is accomplished.

Some of the alternative models reject the unit membrane concept entirely and suggest that membranes consist not of continuous, discrete layers of lipid and protein but of repeating subunits of lipoprotein. Such *lipoprotein subunit models* obviously presume a much more intimate association between lipid and protein than did the unit membrane theory, thus explaining the generally lipophilic nature of many membrane proteins. A special attraction of these models is the ease with which membranes that consist of repeating subunits can be imagined to assemble upon synthesis. In essence such models simply extend to membranes a principle already well established for other cellular structures—that of self-assembly from subunits. A subunit organization for membranes is supported by observations made with the electron microscope, since particulate substructure can be readily visualized in many membranes under appropriate conditions. It is also true, however, that such repeating patterns are seen most readily in the inner membranes of mitochondria and chloroplasts. We already know (from Chapters 5 and 7) that these membranes carry out highly specialized functions of electron transport and phosphorylation and that the proteins and pigments involved in these functions are organized into repeating structural units. It seems possible, therefore, that these membranes may have a unique structure dictated by their specialized function and may not be good prototypes for general membrane design.

The single model that seems at present to explain membrane organization most successfully is the *fluid mosaic model* proposed by S. J. Singer and G. L. Nicolson in 1972. A measure of the merits of their model can be seen in the widespread acceptance it has already been accorded. As shown in Figure 9-5, this model also presumes a more intimate contact between lipid and protein than is expected of a unit membrane. But the proteins, rather than being organized into discrete repeating subunits, float somewhat more freely in and on the phospholipid bilayer, creating a mosaic design rather than a predictable pattern. Thus the fluid mosaic model retains the lipid bilayer of the historical unit membrane but allows the membrane proteins to penetrate into the lipid, thereby interrupting the continuity of the lipid bilayer.

The freedom accorded the protein molecules in this model agrees well with the generally fluid nature of membranes, as seen both from the way membranes appear to flow together when cells or organelles fuse and also from the way new material seems to be added evenly throughout the membrane during cell growth and expansion. The model also accommodates proteins with differing surface charge properties. The more hydrophilic proteins will float near the surface of the lipid, adding to the membrane thickness and approximating the protein coat of a unit membrane, while those with an essentially hydrophobic surface can submerge themselves in the nonaqueous milieu of the membrane interior. The model distinguishes, in fact, between two

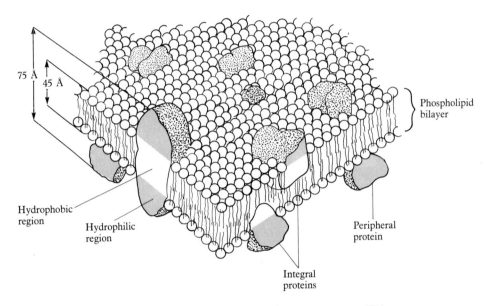

75 Å

45 Å

Phospholipid
bilayer

Hydrophobic
region

Hydrophilic
region

Peripheral
protein

Integral
proteins

Figure 9-5 The fluid mosaic model of membrane structure. This contemporary view of the membrane retains the phospholipid bilayer of the unit membrane model (Figure 9-4), but membrane proteins, rather than being assigned a surface location only, are allowed to penetrate into the interior of the individual protein molecule. Integral proteins are amphipathic and orient themselves such that hydrophobic regions are sequestered in the interior of the membrane, while hydrophilic regions assume a surface location. Peripheral proteins are more uniformly hydrophilic and are as a consequence localized at or on the membrane surface.

broad classes of membrane proteins. *Integral proteins* are embedded within the lipid matrix and are regarded as an intrinsic part of the membrane continuum. *Peripheral proteins*, on the other hand, are only weakly associated with the membrane surface. Furthermore the integral proteins are considered to be amphipathic, with their nonpolar regions buried in the membrane interior and their polar regions at the aqueous surface.

Thus by modifying the original concept of a lipid bilayer coated by protein (Figure 9-4) into that of a lipid matrix interrupted by amphipathic proteins varying in their hydrophobicity (Figure 9-5), we emerge with a model of membrane structure that overcomes many of the problems encountered by earlier concepts. Moreover the fluid mosaic model seems to provide a framework that can accommodate at least some of the features of other membrane models. If the fluid mosaic model is accepted as a generalized view of membrane structure, then the original unit membrane model could be interpreted as a specialized case in which most of the proteins are hydrophilic and therefore reside peripherally at or near the lipid surface. Similarly the subunit model could be considered to result from the special repeating clusters of proteins and pigments required for electron transport and phosphorylation, which might confer a repeating pattern upon what would otherwise be a more mosaic array of proteins.

Permeability Properties of Membranes

Whatever the details of membrane structure, all membranes have generally recognized properties of permeability that are of great relevance to a discussion of transport. As might be expected from the prominence of phospholipids with long hydrophobic tails, membranes are generally quite permeable to nonpolar hydrophobic compounds. There is, in fact, a fairly good relationship between the relative solubility of a compound in a nonpolar solvent like vegetable oil and the rate at which it will move across a membrane. (A glaring exception to this general rule is water itself, which is insoluble in such solvents yet moves freely across membranes, a point to which we shall return later.) Such nonpolar materials can often cross membranes by simple diffusion, since they essentially dissolve in the membrane on one side and move to the other.

For most compounds, however, the membrane represents an impermeable barrier to movement, and specific provision must be made for transport. Included in this category are inorganic ions and most of the molecules involved in cellular metabolism, such as organic acids, amino acids, and phosphorylated compounds. It is this very impermeability to all the common cellular components that makes the membrane an effective means of defining the boundary of a cell or organelle and of keeping the contents inside. If you look, for example, at the glycolytic pathway, the TCA cycle, or the Calvin cycle, you will find that all the intermediates in each of these pathways contain at least one carboxyl or phosphate group. All these molecules are therefore highly polar and are very effectively contained by the membrane of the cell, mitochondrion, or chloroplast. Thus the hydrophobic lipid phase is vital to the role of the membrane as a barrier between the essentially aqueous contents of the cell and the aqueous environment in which most cells live.

Transport Across Membranes

The same properties of chemistry and structure that make membranes an effective means of defining and containing cells also dictate the need for specific transport mechanisms if polar compounds are to be moved into and out of cells and organelles. As a result, transport of materials across membranes is an important activity of all cells and is often an energy-requiring process.

We can recognize three different kinds of transport—different not in the mechanisms likely to be involved but in what is accomplished in the process. *Cellular* transport involves the exchange of materials between a cell and its environment and includes both the uptake of nutrients and the removal of wastes and secretory products. Cellular transport clearly involves movement across the *cell membrane*, frequently also called the *plasma membrane*. *Intracellular* transport, on the other hand, involves movement across membranes of organelles inside the cell. It is therefore mainly a eukaryotic phenomenon and includes the molecular traffic into and out of such organelles as the nucleus, mitochondrion, chloroplast, lysosome, peroxisome, Golgi body, and endoplasmic reticulum. A third kind of transport is *transcellular*. Here the object is not simply to move a substance into the cell from the outside but to move it in on one side and out on the other, thereby accomplishing net transport across the cell. Transcellular movement is obviously found only in multicellular organisms and involves whole cell layers that serve as barriers to movement. The epithelial cells that line the gastrointestinal tract of animals and the cells of plant roots responsible for absorption of water and mineral salts are examples of cell layers specialized to move a substance across the cell by transporting it in on one side and out on the other.

Movement of molecules across membranes can occur by three different means: simple diffusion, passive transport, and active transport. The thermodynamic and kinetic properties of each of these processes are summarized in Table 9-1. *Diffusion* means spontaneous, unmediated

Diffusion

TABLE 9-1 Comparison of the Properties of Simple Diffusion, Passive Transport, and Active Transport

	Simple Diffusion	Passive Transport	Active Transport
Solutes Transported:			
Nonpolar solutes	Yes	Yes	Yes
Polar solutes	No	Yes	Yes
Thermodynamic Properties:			
Direction with respect to			
concentration gradient	Down	Down	Up
Effect on entropy	Increased	Increased	Decreased
Energy required	No	No	Yes
Directionality	No	No	Yes
Kinetic Properties:			
Carrier involved	No	Yes	Yes
Dependence of rate on			
solute concentration	Linear	Hyperbolic	Hyperbolic
Michaelis-Menten			
kinetics	No	Yes	Yes

movement in a direction and at a rate dictated by the difference in the concentration of a specific substance on the two sides of the membrane. As already mentioned, nonpolar compounds soluble in phospholipids usually traverse membranes by simple diffusion. These compounds essentially dissolve in the hydrophobic phase of the membrane and move across to the other side. Water is one of the exceptions, in that it is clearly not soluble in nonpolar solvents yet seems to diffuse through membranes. The reason for this anomalous permeability is not yet understood, but it has been suggested that water moves through small hydrophilic channels in the membrane. Such an explanation seems plausible, since water, with a diameter of less than 3 Å, is the smallest biologically important molecule. It might be able to pass freely through channels with strategically small diameters that exclude larger molecules.

A key distinguishing feature of diffusion is that the net rate of transport for a specific substance is directly proportional to the concentration difference for that substance across the membrane, as indicated by Equation 9-1:

$$v_D = k(C_{out} - C_{in}) \qquad (9\text{-}1)$$

where v_D = rate of diffusion of substance from outside to inside of cell
or organelle (or, if v_D is negative, from inside to outside)
C_{out} = concentration of substance on outside
C_{in} = concentration of substance on inside
k = constant

Simple diffusion is therefore characterized by a linear relationship between the concentration gradient and the rate of movement across the membrane. This, it turns out, is a distinctive property of diffusion and is useful in distinguishing diffusion from facilitated transport. To sum up, then, simple diffusion usually involves nonpolar substances (and water), requires no mediating agent, and results in a linear relationship between rate and concentration.

Facilitated Transport   Most cellular components do not diffuse across membranes rapidly enough to satisfy the metabolic needs of the cell. Most substances required by cells are polar in nature and move across membranes only because cells and organelles have specific means of mediating or *facilitating* that movement. *Facilitated transport* therefore differs from diffusion in that specific mechanisms are required to effect the passage across membranes of substances to which the membrane is otherwise impermeable. To accomplish this the cell or organelle has carrier molecules embedded in its membrane. These carriers are highly specific, often for a single compound though sometimes for a small group of closely related compounds. As might be expected from this specificity and from what you already know about membrane chemistry, these carriers are proteins; indeed no other common cellular component would be likely to have the molecular variety required to provide such striking specificity. Such carriers are called *transport proteins* or *permeases*. The latter term is especially apt because the suffix "ase" suggests a similarity between transport proteins and enzymes that turns out to be quite valid. Similar to an enzyme-catalyzed reaction, carrier-facilitated transport involves (1) an initial binding of carrier and "substrate," (2) a process mediated on the protein surface (a chemical reaction in one case, a physical translocation of solute in the other), (3) a subsequent release of "product," and (4) a reduction in the activation energy of the "reaction." As might be expected, carrier proteins can be saturated, just like enzymes. This means that carrier-mediated transport follows Michaelis-Menten kinetics, with an upper limiting velocity and a Michaelis constant corresponding to the concentration of transportable solute needed to achieve one-half of the maximum rate of transport. A plot of the rate of membrane passage versus concentration is therefore linear for simple diffusion but hyperbolic for facilitated diffusion. This difference, illustrated in Figure 9-6, is an important means of distinguishing between diffusion and facilitated transport.

Transport may be either passive or active. *Passive transport* refers to carrier-mediated movement of a solute through a membrane in the direction dictated by the existing gradient of solute concentration— i.e., *down* the concentration gradient. Passive transport really only provides a mechanism for a process that is already thermodynamically spontaneous but is prevented from occurring by the impermeability of the membrane. *Active transport*, on the other hand, involves the movement of compounds into or out of cells or organelles in the direction of increasing concentration—i.e., *up* a concentration gradient. This obviously requires energy. The transport proteins involved in active transport must therefore not only provide for the translocation of the solute across the membrane but must also couple that translocation to an energy-yielding reaction, almost invariably the hydrolysis of a high-energy phosphate bond. We shall look briefly at passive transport for some general features of transport systems and then turn to active transport for the actual work of concentration this chapter is intended to consider.

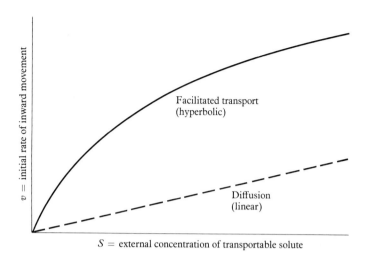

Figure 9-6 plot: y-axis labeled "v = initial rate of inward movement"; curves labeled "Facilitated transport (hyperbolic)" and "Diffusion (linear)"; x-axis labeled "S = external concentration of transportable solute"

Figure 9-6 A comparison of the kinetics of membrane transport by simple diffusion and by carrier-facilitated passage. Transport by simple diffusion shows a linear relationship between transport velocity (v) and solute concentration (S) over a broad concentration range, while carrier-mediated transport is characterized by the hyperbolic relationship predicted by Michaelis-Menten kinetics. (Note that the simple example illustrated here assumes an initial internal solute concentration of zero. For the more realistic case of nonzero internal concentration, correction must be made using Equation 9-1 for diffusion or Equation 9-2 for carrier-facilitated transport.)

Passive Transport   Passive transport is also called *carrier-facilitated diffusion*. This is an apt term in the thermodynamic sense, since the driving force for passive transport is the same as that for diffusion: the tendency toward maximum entropy and, therefore, the decrease in free energy that accompanies the movement of ions or molecules from a region of high concentration to one of lower concentration. The job of the carrier proteins of passive transport is to catalyze diffusion of polar solutes across an otherwise impermeable membrane. The carrier proteins probably do this by binding to the solute molecules in such a way as to shield the polar groups of the solutes from the nonpolar membrane interior.

Passive transport can be thought of in four steps: (1) the solute collides with and binds to the carrier protein on the outer surface of the membrane; (2) the carrier-solute complex diffuses across the membrane; (3) the solute dissociates from the carrier at the inner surface of the membrane; and (4) the carrier diffuses across the membrane again. The process might be likened in overall function to a revolving door, the compartments of which "bind" and transport people across an otherwise impermeable barrier (a department store wall, for example) and release them on the other side.

The analogy with the revolving door can in fact be extended further, since both passive transport mechanisms and revolving doors are intrinsically nondirectional and readily reversible. In both cases the net direction of transport depends solely on the relative concentrations of solute (people) inside and outside the cell (building). If the concentration is higher outside, net flow will be inward; if the higher concentration exists inside, net flow will be outward.

As an example of passive transport, consider the movement of glucose into your red blood cells. The level of glucose in blood is usually in the range 65–90 mg/100 ml, or about 3.6–5.0 mM. This means that the red blood cell (or any other cell in contact with the blood) can meet its need for glucose uptake simply by providing a glucose carrier in its membrane and keeping the intracellular glucose concentration below 3.6 mM. The low intracellular level of glucose needed for passive inward flow is possible because the glycolytic pathway begins with the phosphorylation of glucose (Reaction Gly-1), a reaction that is highly exergonic in the direction of glucose consumption ($\Delta G^0 = -4$ kcal/mole) and is catalyzed by an enzyme with a low Michaelis constant for glucose (hexokinase; $K_m = 0.15$ mM). That the glucose uptake by the red blood cell is indeed a passive process is confirmed by its insensitivity to inhibitors of glycolysis, the only energy-yielding process of which this cell type is capable. (Red blood cells have no mitochondria and are therefore not capable of aerobic metabolism.)

Active Transport

Though diffusion and passive transport are both useful means for ensuring movement of solutes across membranes, neither process is capable of maintaining the concentration of a specific solute within a cell or organelle at a level different from that of the same solute in the surrounding milieu. The vital ability to move solutes *up* a concentration gradient, whether into or out of the cell, requires active transport— the coupling of the endergonic transport process to an energy-yielding system. Active transport serves three major functions in cells or organelles: (1) it makes possible the uptake of fuel molecules and other essential nutrients from the environment or surrounding fluid, even when their concentrations are very low; (2) it allows various substances, such as metabolic waste products and sodium ions, to be removed from the cell or organelle, even when the concentration outside is greater than that inside; and (3) it permits the maintenance of constant, optimal internal concentrations of inorganic electrolytes, particularly potassium and hydrogen ions.

Like passive transport, active transport involves highly specific carrier proteins, occurs (at least conceptually) in four steps, and follows Michaelis-Menten kinetics. Active transport differs from passive transport in three important ways, however, as summarized in Table 9-1. The most obvious distinction is that active transport is energy-requiring and must therefore couple to an exergonic reaction. The usual energy source is ATP, supplied either by respiratory metabolism or, in some instances, by glycolysis alone. This requirement for concomitant functioning of an ATP-generating system can be readily demonstrated by the use of a respiratory or glycolytic poison. For example, the active accumulation of the disaccharide lactose by cells of *Escherichia coli* can be quickly stopped with azide, an agent known to uncouple oxidative phosphorylation.

The second important distinction between active and passive transport concerns the direction of transport. Passive transport occurs in the direction dictated by the prevailing concentration gradient and is readily reversible. Active transport, on the other hand, has an intrinsic directionality to it—a transport system designed to "pump" a solute across a membrane in a given direction will not be able to transport that solute actively in the other direction. This directionality is readily exemplified by the cellular mechanisms for active transport of potassium and sodium ions: K^+ can only be pumped in, and Na^+ can only be pumped out. Thus active transport is said to be a *unidirectional* or *vectorial* process. (In the absence of an energy supply, however, active transport systems can in some cases also be used to facilitate passive transport of the same solutes. Such passive transport can then occur in either direction, as dictated by the concentration gradient.)

The third—and most important—difference between active and passive transport lies in what they accomplish. Passive transport (and simple diffusion also) results in a greater and greater similarity between solute concentrations on both sides of the membrane, leading to an equilibrium state of equal concentrations inside and outside. Active transport, on the other hand, represents a means of creating deliberate concentration differences across membranes and leads to a non-equilibrium steady state. The active transport mechanisms of a cell are, in fact, finely tuned to keep the internal milieu of the cell remarkably constant despite fluctuations in the external environment.

The energy requirement of active transport arises because of the decrease in entropy that occurs when a solute is concentrated. Not surprisingly, the equation used to quantitate the energy requirement of concentration work relates that requirement to the magnitude of the concentration gradient against which transport must be effected. The free-energy change for the transport of a solute from a compartment at concentration C_1 to another compartment at concentration C_2 is calculated as follows (with R and T assuming their usual meanings as the gas constant and the absolute temperature, respectively):[*]

$$\Delta G = RT \ln \frac{C_2}{C_1} = (2.303)(1.987)(T) \log_{10} \frac{C_2}{C_1} \qquad (9\text{-}2)$$

Note that if C_2 is less than C_1, then ΔG is negative, indicating that the process is exergonic and may occur spontaneously, as would be expected for transport down a concentration gradient (passive transport). But if C_2 is greater than C_1, transport is up the gradient (active transport) and the amount of energy required to effect the movement is indicated by the positive value of ΔG. Suppose, for example, that the concentration of the sugar lactose within a bacterial cell is to be maintained at 10 mM, while the external lactose concentration is only 0.025 mM. The energy requirement at 25°C can be calculated as

$$\Delta G = +(2.303)(1.987)(273 + 25) \log_{10} \frac{10}{0.025}$$

$$= 1364 \log_{10} (400) = 3549 \text{ cal/mole} = +3.55 \text{ kcal/mole}$$

Clearly this is an energy requirement that can be met comfortably by coupling the transport to ATP hydrolysis ($\Delta G^0 = -7.3$ kcal/mole), which probably provides the driving force for lactose uptake by bacterial cells.

When we take up possible mechanisms of active transport, we are dealing with an area of biology that has elicited a great deal of current interest and excitement but has only recently begun to yield definitive answers. Much of the initial difficulty lay in the lipophilic nature of

Mechanisms of Active Transport

---

[*]Actually ΔG for the transport of a specific solute A can be regarded as the free-energy change accompanying the following reaction:

$$A_{\text{compartment 1}} \rightleftharpoons A_{\text{compartment 2}}$$

By complete analogy with Equation 2-11, the full expression for ΔG ought to be

$$\Delta G = \Delta G^0 + 2.303 RT \log_{10} \frac{[A]_{\text{compartment 2}}}{[A]_{\text{compartment 1}}}$$

But since the concentration of A is the same in both compartments at equilibrium, the equilibrium constant K is obviously 1.0 and ΔG^0 is therefore zero. Thus the expression for ΔG becomes that of Equation 9-2.

most transport proteins and the problems involved in extracting them from the membrane in an active form. Recently, however, it has proved possible to isolate and purify a number of transport proteins. Research in this area has also been hampered by the lack of an adequate assay system for transport proteins once they were solubilized. But for at least several systems, this problem has been solved by incorporating the isolated transport proteins into artificial phospholipid membranes for assay of transport function. Because of these and other recent technical advances, substantial progress has been made toward an understanding of the molecular basis of active transport.

One of the first principles to emerge from such studies is that all active transport clearly does not occur by the same mechanism. It is, in fact, possible to distinguish at least four different modes of active transport, differing not only in what is actually accomplished but also in the energy source to which the transport process is coupled. The usual source of energy to drive the active accumulation or depletion of a solute against its concentration gradient is the high-energy phosphate bond of ATP, but the link between ATP hydrolysis and the actual transport process need not always be direct. Moreover other high-energy phosphate compounds are also known to be involved in active transport, and some transport systems can be driven directly by electron transport.

At present it is possible to distinguish at least four different modes of active transport. *Simple active transport* involves the unidirectional pumping of a solute across a membrane against its concentration gradient without the simultaneous transport of any other molecule. The uptake of the disaccharide lactose by bacterial cells appears to be representative of this mode of transport. In *exchange transport* two solutes are moved simultaneously but in opposite directions, each being moved up its respective concentration gradient. The best-understood example of this type of process is the Na^+/K^+ exchange pump essential in maintenance of cellular electrolyte levels. *Cotransport* is also characterized by the simultaneous movement of two solutes, but in this case both solutes flow in the same direction, the "downhill" movement of one being used to drive the "uphill" transport of the other. Sodium-driven uptake of sugars, amino acids, and other metabolites by animal cells occurs by this means. *Phosphorylating transport*, the fourth mechanism of active transport, differs from the other three in that the solute is phosphorylated as an inherent part of the transport process. This is the mechanism used by bacteria for the uptake of glucose and other sugars.

We shall examine each of these mechanisms in turn, seeking in each case to learn how the energy of ATP hydrolysis (or alternative exergonic reactions) is used to drive an otherwise endergonic transport process. Throughout this discussion the term "pump" will be used to refer to the various mechanisms responsible for active transport of molecules and ions across membranes. This is in keeping with general literature and textbook usage, but no functional analogy with mechanical pumps is intended or implied. On the contrary, mechanical pumps invariably effect a mass flow from one location to another, whereas membrane pumps selectively transport specific components from one mass to another.

The active uptake of the disaccharide lactose by the bacterium *Escherichia coli* appears to fit the criteria of simple active transport well. Lactose transport is energy-dependent, can occur against a large concentration gradient, and is not linked to the transport of other solutes. Based on these properties, the existence of a specific membrane transport protein for lactose was predicted well in advance of the actual isolation of such a protein. The transport protein was given the name *galactoside permease* because lactose is, in chemical terms, a galactoside.* A further property of lactose uptake by *E. coli* is that it is an *inducible* process—cells grown in a medium without lactose cannot actively transport this sugar but acquire the ability to do so shortly after exposure to lactose. Galactoside permease is in fact one of the three proteins coded for by genes within the *lac* operon, a cluster of bacterial genes well known to every student of bacterial genetics.

Simple Active Transport: The Uptake of Lactose by Bacteria

This property of inducibility was exploited to good advantage in the isolation of the lactose transport protein by E. P. Kennedy and colleagues. Using a reagent called *N*-ethylmaleimide (NEM) that binds irreversibly to sulfhydryl (—SH) groups of proteins, they were able to identify and isolate a membrane-associated protein (*M protein*), the synthesis of which is specifically induced upon addition of lactose to the medium. To do so, they first treated uninduced bacterial cells with NEM to ensure that all proteins already present in the cells had reacted with the reagent. Lactose was then added to the medium to induce the synthesis of galactoside permease. At the same time, the culture was treated with radioactively labeled NEM. With all the preexisting proteins already saturated with nonradioactive NEM, the labeled NEM could bind only to newly synthesized proteins. The

---

*A *galactoside* is a derivative of the sugar galactose in which the hydrogen atom of the hydroxyl group on carbon 1 of the sugar has been replaced by some other group. Since lactose consists of a glucose molecule linked through an oxygen bridge to carbon 1 of galactose, lactose qualifies as a galactoside.

labeled NEM thus provided a radioactive "tag" that led to the detection of M protein in a plasma membrane fraction prepared from the cells and greatly facilitated the eventual isolation of the protein. Since the M protein isolated in this way was shown to be absent both in uninduced cells and in mutant cells that cannot transport galactoside, it is highly probable that the isolated M protein is galactoside permease.

How this transport protein actually functions to move lactose unidirectionally from the surrounding medium into the cell is not yet clear. A possible mechanism is illustrated in Figure 9-7. According

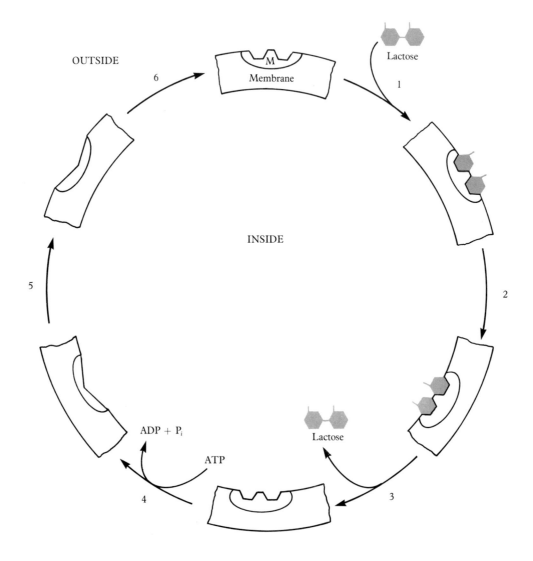

to this model the M protein can exist in two forms, only one of which can bind lactose effectively. The lactose-protein complex formed on the outer membrane surface moves across the membrane to the inner surface, where it dissociates to discharge lactose to the interior of the cell. The protein is then converted to the nonbinding form in an energy-requiring process and moves back across the membrane in this form. Once at the outer surface, the protein reverts back to the lactose-binding form. Presumably it is the rapid conversion of free M protein into the nonbinding form that renders the process unidirectional. This conversion may be driven by coupling to ATP hydrolysis as shown in Figure 9-7, but it has also been suggested that the energy may come directly from the oxidation of metabolic intermediates.

Not all active transport processes involve the unidirectional transport of a single solute. In some cases, two solutes are moved at once but in opposite directions. One such system is the pump involved in the maintenance of the high intracellular levels of potassium and the relatively low intracellular levels of sodium characteristic of many plant, animal, and microbial cells. High levels of potassium and low levels of sodium are requisite to proper cell function, since potassium is required for a number of vital cellular processes, notably ribosome function and activation of a variety of enzymes, and sodium often inhibits these same functions. Potassium levels are usually maintained

Exchange Transport: The Na$^+$/K$^+$ Pump

Figure 9-7 A model mechanism for the simple active transport of lactose in the bacterium *Escherichia coli*. The inward transport of lactose is mediated by a carrier called the M protein. Lactose transport can occur against a large concentration gradient, is directly dependent upon an energy source, and is independent of the transport of other molecules. Here ATP is shown as the energy source, but it has also been suggested that the transport process couples directly to the oxidation of a metabolic intermediate rather than to ATP hydrolysis. A hypothetical model for the transport process is shown in six conceptual steps arranged around the periphery of a "cell": (1) lactose binds to the M protein at the outer membrane surface; (2) the carrier protein translocates the lactose to the inner surface; (3) the lactose dissociates from the binding site, with the lactose-carrier interaction drawn in the direction of dissociation by the continuous depletion of binding sites, which occurs in the next step; (4) the carrier protein undergoes an energy-requiring (ATP-driven?) conformational change, altering the binding site and effectively abolishing its affinity for lactose; (5) the M protein returns to the outer membrane surface; where (6) it reverts to the lactose-binding form.

at about 100–150 mM inside cells, even though external levels of potassium are often much lower than that and may fluctuate widely. Conversely the intracellular concentration of sodium is usually maintained at low levels, often considerably less than in the surrounding medium. The pump mechanism responsible for maintaining these characteristic levels of Na^+ and K^+ was first discovered in the mid-1950s and represented the first documented case of active transport. The pump was shown to possess ATPase activity and to couple ATP hydrolysis to the inward transport of potassium ion and the outward transport of sodium ion. The ATPase, in fact, requires both K^+ and Na^+ for activity, differing in this respect from most other potassium-activated enzymes in the cell, which are almost invariably inhibited by sodium.

The *Na$^+$/K$^+$ exchange pump*, as this transport system is known, has been studied in the greatest detail in the membranes of red blood cells, but it occurs in the cell membranes of virtually all cells in vertebrate tissues. Like other active transport systems, the Na^+/K^+ exchange pump is a vectorial system; it has inherent directionality or "sidedness" to it in that potassium is pumped only inward and sodium only outward. In fact it has been demonstrated that sodium and potassium ions stimulate the ATPase associated with this transport system only on the side of the membrane from which they are transported, sodium on the inside and potassium on the outside.

Figure 9-8 A model mechanism for the Na^+/K^+ exchange pump. The outward transport of sodium ion is coupled to the inward transport of potassium ion, both against their respective concentration gradients. The driving force is provided by ATP hydrolysis, probably via phosphorylation of the carrier protein. Selective transport of Na^+ and K^+ in opposite directions is possible because Na^+ activates the ATPase of the carrier protein only on the inner membrane surface, whereas K^+ activation occurs only at the outer surface. The transport process is represented schematically in six conceptual steps arranged around the periphery of a "cell": (1) three Na^+ ions bind to the sodium sites of the carrier at the inner membrane surface, thereby inducing a conformational change in the protein that facilitates phosphorylation; (2) ATP hydrolysis drives the translocation of the binding sites to the outer membrane surface, probably by phosphorylation of the carrier; (3) the Na^+ ions dissociate at the outer surface; (4) two K^+ ions bind to the potassium sites of the carrier at the outer surface; (5) the binding sites are translocated to the inner membrane surface, probably with concomitant dephosphorylation of the carrier; and (6) the K^+ ions dissociate at the inner surface, leaving the carrier ready to accept Na^+ again, as in step 1.

Because the gradients against which sodium and potassium are pumped by animal cells seldom exceed 50:1, the energy requirement per ion moved is relatively low (usually less than 2 kcal/mole; $\Delta G = +RT \ln (50) = 2.22$ kcal/mole at 25°C). The possibility of coupling the transport of several ions to the hydrolysis of a single ATP that this suggests is actually realized by the cell, since the apparent stoichiometry of the transport process is $3Na^+$ moved out and $2K^+$ moved in per ATP hydrolyzed.

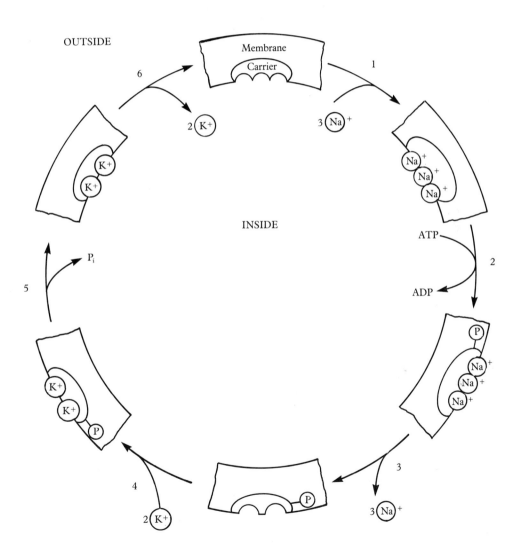

The model illustrated in Figure 9-8 is a mechanism consistent with the data available at present. The mechanism involves the initial binding of Na^+ ions on the inner side of the membrane, followed or accompanied by phosphorylation of the transport protein by ATP. Phosphorylation seems to cause the bound Na^+ to be translocated to the outer membrane surface and released to the outside. Then K^+ ions from the outside bind to the protein and are translocated to the inside as the carrier is dephosphorylated.

Cotransport: Sodium-driven Metabolite Transport in Animal Cells

Whereas the active uptake of potassium ions by cells seems always to be coupled to the active pumping of sodium in the *opposite* (outward) direction, other transport systems are known that also couple obligatorily to sodium transport, but with the flow of sodium in the *same* (inward) direction as that of the other solute. A unifying feature emerging from studies of the active uptake of sugars and amino acids by animal cells such as those of the intestinal epithelium is that the transport is obligatorily coupled to the concomitant inward *cotransport* of sodium. It is, in fact, only as we consider such cotransport systems that we appreciate the full significance of the continuous outward movement of sodium accomplished by the Na^+/K^+ pump. For it is the steep gradient of sodium concentration maintained by the exchange pump that serves as the driving force for the uptake of a variety of sugars and amino acids. Thus uptake of such compounds by animal cells is not coupled directly to ATP hydrolysis but occurs because these organic molecules are carried along by a transport system powered by the sodium gradient. Ultimately, of course, such uptake is still dependent upon ATP, since the maintenance of the sodium gradient depends upon the Na^+/K^+ pump—and that, as we have seen, is an ATP-driven system.

The carriers involved in sodium-driven metabolite transport have not yet been isolated in purified form, but several of their properties can be predicted from what is already known about the process in which they are involved. Clearly the carrier must have two sites, one specific for Na^+ and the other for the metabolite to be cotransported. This, of course, argues strongly for a protein carrier, since it is hard to imagine how such specificity could be achieved in any other way. Furthermore the binding of Na^+ to its site must somehow affect the affinity of the other site for the desired metabolite, since the metabolite should bind when Na^+ binds and be released when Na^+ dissociates if its transport is to follow the Na^+ gradient faithfully. Such a property is strongly suggestive of an allosteric protein, illustrating once again the almost invariable involvement of allostery whenever the topic turns to cellular control mechanisms.

To visualize the mechanism whereby sodium-driven uptake of a compound like glucose occurs (Figure 9-9), we need only conceive of a membrane protein with an Na^+ site and a glucose site. When Na^+ binds to its site on the outside of the membrane, the glucose site acquires a high affinity for glucose and binds a glucose molecule from the outside. The protein then undergoes a translocation or configurational change that brings the binding sites to the inside of the membrane. There the sodium dissociates in response to the low sodium concentration, and the glucose will also be ejected from its site regardless of the cellular glucose level, provided only that the affinity of that site for glucose is greatly reduced upon dissociation of the sodium ion.

The prominence of sodium-driven transport systems in animal cells raises the question of their possible universality in all cells. This seems unlikely, however, since sodium transport does not appear to be a prominent feature of plant cells, and bacterial transport systems generally couple directly to the hydrolysis of high-energy phosphate bonds, as we have seen already and are about to explore further.

Thus far we have considered several different mechanisms for the movement of organic substrates such as glucose into cells. Red blood cells, as we saw earlier, can rely upon passive transport for glucose uptake because of the high external concentration of glucose in the surrounding medium (the blood). Other animal cells depend upon a gradient of sodium for the cotransport of glucose, amino acids, and other compounds into the cell, such that the inward transport of such substances is coupled indirectly to a single kind of energy-requiring process, the outward transport of sodium by the Na^+/K^+ pump.

Phosphorylating Transport: Sugar Phosphorylation as an Uptake Mechanism in Bacteria

Bacterial cells, on the other hand, live in environments that can vary greatly and change rapidly and thus cannot rely upon gradients of glucose, sodium, or any other substance. This may well be at least part of the reason why the uptake of ions and other nutrients by bacterial cells usually couples directly to the hydrolysis of high-energy phosphate bonds.

We have already encountered one such directly coupled bacterial transport system—the inducible uptake of lactose, probably driven by ATP hydrolysis (Figure 9-7). This is not the only known mechanism for active transport of sugar into bacterial cells, however. In fact the best-characterized system uses as its energy source not ATP but phosphoenol pyruvate (PEP), the high-energy phosphate compound already familiar to us from glycolysis (Chapter 4), the Hatch-Slack pathway (Chapter 7), and gluconeogenesis (Chapter 8).

The PEP-driven uptake of sugars by bacterial cells differs from the other modes of active transport described so far in that a chemical modification of the sugar is an inherent part of the transport mechanism. The actual modification is a phosphorylation, and the process is accord-

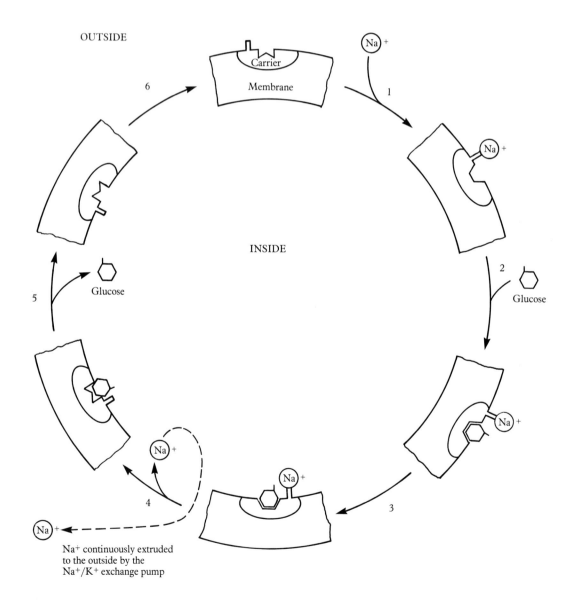

ingly termed *phosphorylating transport*. The result is that sugar molecules are both actively transported and simultaneously phosphorylated. In the case of glucose, for example, the sugar is taken up at the outer membrane surface as free glucose but released at the inner membrane surface as glucose-6-phosphate.

As shown in Figure 9-10, this mode of transport across a bacterial membrane involves a system with three components. The actual transport protein, designated E_{II}, is located in the membrane and is responsible for the specific recognition of particular solutes. There are probably as many kinds of E_{II} proteins in the membrane as there are different kinds of molecules to be transported by this mechanism. The molecule to be transported inward (glucose in the example of Figure 9-10) binds to the appropriate E_{II} carrier at the outer membrane surface, and the complex then moves across the membrane for release at the inner surface. Release of the sugar depends upon its phosphorylation. The phosphate group for this purpose is contributed by the phosphorylated form of a small cytoplasmic protein called HPr. Rephosphorylation of HPr is then accomplished by transfer of the high-energy phosphate group from phosphoenol pyruvate, catalyzed by the cytoplasmic enzyme E_I. Unlike E_{II}, both E_I and HPr appear to be nonspecific proteins that serve as the common phosphorylating system for a variety of substrates, each of which is transported inward by a specific E_{II} carrier.

Figure 9-9 A model mechanism for the cotransport of sodium and organic solutes in animal cells. The inward transport of organic solutes such as glucose against a concentration gradient is driven in animal cells by cotransport of sodium down its concentration gradient. According to the model shown here, the sodium-dependent carrier protein has two binding sites, one for Na^+ and the other for the organic solute to be transported. The process may occur in either direction, but preferential inward movement is driven by the high external concentration of sodium ion. Thus the immediate driving force necessary to move the organic solute into the cell is provided by the Na^+ gradient. But that gradient is in turn maintained only by the continued outward extrusion of Na^+ by the ATP-driven Na^+/K^+ exchange pump (Figure 9-8). The cotransport process (with glucose as a model solute) is shown in six conceptual steps arranged around periphery of a "cell": (1) Na^+ binds to one of the two sites on the carrier at the outer membrane surface, thereby enhancing the affinity of the other site for glucose; (2) glucose binds at the second site; (3) the carrier translocates to the inner membrane surface, driven by the steep Na^+ gradient; (4) Na^+ dissociates due to the low intracellular Na^+ concentration, thereby reducing the affinity of the other site for glucose; (5) glucose is ejected from its binding site; and (6) the carrier is free to return to the outer membrane surface.

The sequence of events diagrammed in Figure 9-10 for glucose transport by this mechanism can be summarized as follows:

Initial binding: $glucose_{(out)} + E_{II} \rightarrow E_{II}\text{-}glucose_{(out)}$ (9-3)

Translocation: $E_{II}\text{-}glucose_{(out)} \rightarrow E_{II}\text{-}glucose_{(in)}$ (9-4)

Release: $E_{II}\text{-}glucose_{(in)} + HPr \sim \textcircled{P} \rightarrow$
$E_{II} + HPr + glucose\text{-}6\text{-}phosphate_{(in)}$ (9-5)

Recharging HPr: $HPr + phosphoenol\ pyruvate \rightarrow$
$HPr \sim \textcircled{P} + pyruvate$ (9-6)

Net: $glucose_{(out)} + phosphoenol\ pyruvate \rightarrow$
$glucose\text{-}6\text{-}phosphate_{(in)} + pyruvate$ (9-7)

Electrical Work (Charge Concentration)

Electrical work is probably important enough as a cellular activity to warrant a separate discussion, but it is really just a special case of concentration work applied to ions. The cell membrane is never equally permeable to cations and anions, nor are they ever accumulated to the same extent. For this reason cells show a slight separation of charge across the plasma membrane, giving the cell a *transmembrane potential* of about -70 to -100 millivolts (i.e., the inner surface is more negative than the outer surface by about 70 to 100 millivolts). In animals this electrical difference is utilized for communication along nerve cells by the transmission of an electrical impulse that takes the form of an abrupt change in membrane potential. The establishment and maintenance of a deliberate charge separation across the membrane is obviously an energy-requiring process, since electrical charge is being "concentrated" against a gradient. The transmembrane potential is, in fact, dependent upon ATP-driven ion pumps, especially the Na^+/K^+ exchange pump already discussed. Electrical work is characteristic of every cell that maintains such a charge separation, but for an especially dramatic case it is hard to beat the electric eel. This creature expends considerable ATP to establish charge separations capable of delivering a very painful shock that serves both as a means of stunning prey and as a highly successful protective mechanism.

Active Transport in Summary

Various mechanisms are used for the active transport of needed nutrients into cells (and of undesirable metabolites out of cells). The mechanisms differ quite strikingly in details of design and chemistry, but it is nonetheless possible to identify some unifying features:

1. In each case a carrier molecule, most likely a protein, is involved in the actual movement of the solute across the hydrophobic

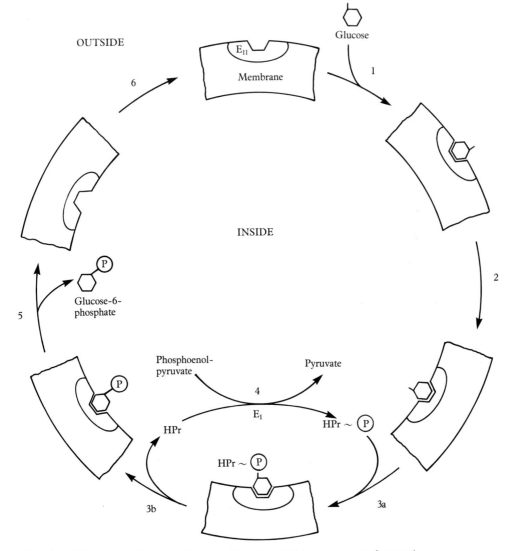

Figure 9-10 A model mechanism for phosphorylating transport of organic solutes in bacterial cells. The inward transport of organic solutes such as glucose is driven in bacterial cells by phosphorylation of the solute, with phosphoenol pyruvate as the donor. The driving force is provided by the high-energy bond of phosphoenol pyruvate, and the transported solute is phosphorylated as an inherent part of the uptake mechanism. The transport process is shown in six conceptual steps arranged around the periphery of a "cell": (1) glucose binds to the specific E_{II} carrier protein at the outer membrane surface; (2) the carrier translocates the glucose to the inner surface; (3) the phosphorylated form of the HPr protein, HPr $\sim$ ⓟ, is used to phosphorylate the glucose; and (4) is regenerated by accepting a high-energy phosphate group from phosphoenol pyruvate, in a reaction catalyzed by the enzyme E_I; (5) the phosphorylated glucose dissociates from the binding site on the inside; and (6) the carrier is free to return to the outer membrane surface.

membrane. Whether this requires an actual physical movement of the protein-solute complex across the phospholipid center of the membrane or is instead accomplished by a configurational change in the protein remains a topic for future research.

2. Because a saturable carrier is involved, active transport mechanisms follow Michaelis-Menten kinetics, with a hyperbolic relationship between reaction rate and concentration of the transportable solute.

3. The transport in each case is driven ultimately by the hydrolysis of a high-energy phosphate bond (usually, but not always, ATP), but the immediacy with which that hydrolysis is linked to the actual transport event varies greatly.

4. In all cases of transport against a concentration gradient, entropy is lowered, energy is required, and work is therefore done. Thus active transport depends ultimately upon the continued generation of high-energy phosphate bonds and upon the cellular energy metabolism that makes this possible.

Practice Problems

9-1. Transport Comparisons: For each of the statements below, circle D if the statement is true of simple diffusion, circle P if it is true of passive transport, and circle A if true of active transport. Any, all, or none of the choices may be appropriate for a given statement.

D P A   (a) Requires a carrier molecule localized within the membrane.
D P A   (b) Doubling the concentration gradient of the molecule to be transported will double the rate of transport over a broad range of concentrations.
D P A   (c) Work is done during the transport process.
D P A   (d) $\Delta G^0 = 0$.
D P A   (e) Transport can occur in either direction across the membrane, depending only upon the concentration gradient prevailing at the moment.

9-2. Transport Mechanisms: Some microorganisms can use either ethanol $(CH_3—CH_2—OH)$ or acetate $(CH_3—COO^-)$ as the sole carbon and energy source. Assume that the following data were obtained for the movement of ethanol and acetate across uniform-sized segments of the cell membrane for such an organism:

Concentration in Culture Medium (*mmoles/liter*)	Rate of Membrane Passage per Membrane Segment	
	Ethanol (*μmoles/min*)	*Acetate* (*μmoles/min*)
0.1	2.5	9
1.0	25	50
10.0	250	91
100.0	2500	99

(a) Plot rate of movement v as a function of concentration S. What can you conclude about the mode of movement of each of these compounds across the membrane?

(b) In what way are your conclusions from part (a) consistent with known properties of ethanol and acetate?

(c) Can you exclude active transport as a mechanism for ethanol movement under the conditions used to obtain these data? What about for acetate?

(d) Why might it be difficult to ascertain whether active transport is involved in the uptake of these compounds by experimentally poisoning the respiratory metabolism of the organism?

9-3. The Case of the Acid Stomach: One of the most impressive active transport systems known is the secretion of H^+ ions from the blood plasma (pH 7.0) into the gastric juice (pH 1.0) by the parietal cells in the epithelial lining of the stomach of mammals.

(a) Is this an example of cellular, intracellular, or transcellular transport?

(b) What is the concentration gradient of H^+ ions across the epithelium?

(c) Calculate the free-energy change associated with the secretion of 1 mole of H^+ into gastric juice at 37°C.

(d) Do you think that H^+ can be driven by ATP hydrolysis at the ratio of one ATP per H^+ transported?

9-4. Na^+/K^+ Transport: The red blood cells that contain the hemoglobin responsible for oxygen transport in your body are highly specialized cells lacking most of the typical eukaryotic organelles, including the nucleus and mitochondria. Given in the following table are the concentrations of K^+ and Na^+ ions in red blood cells and in the surrounding blood plasma:

Ion	In Red Blood Cells	In Blood Plasma
K^+	150 mM	5 mM
Na^+	25 mM	145 mM

Fluoride ion (F^-) is known to inhibit strongly the glycolytic enzyme enolase, which catalyzes the reaction

$$2\text{-Phosphoglycerate} \underset{\text{enolase}}{\rightleftharpoons} \text{phosphoenol pyruvate} + H_2O$$

(a) What do enolase and the glycolytic process have to do with the levels of K^+ and Na^+ in red blood cells?

(b) What do you predict will happen to the concentration of K^+ and Na^+ in red blood cells if fluoride is added to blood? Explain.

(c) Do you think a similar effect would be seen if fluoride were added to a culture of illuminated green algal cells that are known to be actively concentrating inorganic salts from their culture medium? Explain.

9-5. Ouabain Inhibition: *Ouabain* is a very specific inhibitor of the active transport of Na^+ out of the cell and is therefore a valuable tool in studies of membrane transport mechanisms. Which of the following processes occurring in your own body would you expect to be sensitive to inhibition by ouabain?

Explain.

(a) Passive transport of glucose into a muscle cell.

(b) Active transport of dietary phenylalanine across the intestinal mucosa.

(c) Uptake of K^+ ions by red blood cells.

(d) Active uptake of lactose by the bacteria living in your intestine.

(e) Respiratory metabolism in the kidney epithelium.

‡9-6. **Intracellular Transport:** As an example of intracellular transport, consider the problem facing the eukaryotic cell that generates some of its $NADH_2$ in the cytoplasm but oxidizes all its $NADH_2$ in the mitochondrion. The mitocondrial inner membrane is impermeable to $NADH_2$ and no $NADH_2$ permease is known. Both malate and aspartate can pass through the mitochondrial membrane, however, presumably by means of specific transport proteins. *Malate dehydrogenase* and *aspartate transaminase*, the enzymes responsible for the following reactions, are present both inside and outside the mitochondrion:

$$\text{Malate} + \text{NAD} \underset{\text{malate dehydrogenase}}{\rightleftarrows} \text{oxaloacetate} + NADH_2$$

$$\text{Aspartate} + \alpha\text{-ketoglutarate} \underset{\text{aspartate transaminase}}{\rightleftarrows} \text{oxaloacetate} + \text{glutamate}$$

(a) Why does the eukaryotic cell have to effect net transport of $NADH_2$ (or its equivalent in reducing power) from cytoplasm into mitochondrion? Does the prokaryotic cell face a similar transport problem?

(b) Why is the movement of malate and aspartate across the mitochondrial membrane "presumably by means of specific transport proteins"?

(c) Devise a scheme that allows net movement of "reducing equivalents" of $NADH_2$ across the membrane, consistent both with the inability of $NADH_2$ itself to pass across the membrane and with the flow of malate and aspartate, known to be equimolar, but in opposite directions. What other molecules does your scheme predict should also be transported across the membrane?

(d) Is your scheme likely to be reversible? Suppose the mitochondrial malate dehydrogenase coupled to $FAD/FADH_2$ instead of to $NAD/NADH_2$ (just a hypothetical case; an FAD-linked malate dehydrogenase has not been reported to exist). Would you expect your scheme to be reversible or not? Explain.

‡9-7. **Dual Pattern of Ion Uptake in Plants:** Throughout this chapter it has been assumed that a single mechanism of transport is operative for a given solute. That this need not always be the case is known from data on the uptake of certain ions in plants when studied over a broad range of concentrations. Figure 9-11 shows the rate of absorption of potassium ion (radioactive $^{42}K^+$) by excised barley roots as a function of the concentration of KCl in the solution. Both components of this apparently biphasic* uptake curve can be shown to

---

*The data of Figures 9-11 and 9-12 were published in 1963; the second component (operative at the higher concentrations) has since been shown to be resolvable into a number of inflections, suggesting several active sites with somewhat different affinities for potassium. This does not, however, detract from the pedagogic value of Problem 9-7, except as a caution that our treatment of the second component is undoubtedly oversimplified.

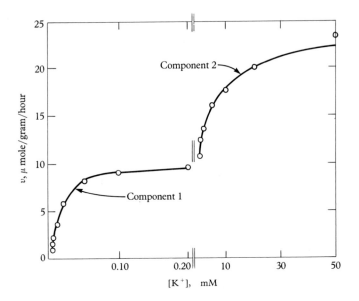

Figure 9-11 The biphasic nature of potassium uptake by excised barley roots. Roots were detached from young barley seedlings and suspended in KCl solutions at a variety of concentrations, each labeled with radioactive $^{42}K^+$. Roots were rinsed after exposure and assayed for radioactivity. The data were expressed as counts per minute of $^{42}K^+$ taken up per gram of root tissue and converted to micromoles per gram per hour. When plotted as a function of K^+ concentration, the rate of uptake displayed an apparently biphasic uptake profile.

follow Michaelis-Menten kinetics, and the overall velocity v_T can be closely approximated by summing the Michaelis-Menten equations for the two components. Shown in Figure 9-12 are separate double-reciprocal plots for the two components of the uptake curve. Also indicated on the double-reciprocal plots are the effects on the kinetics of K^+ uptake of Na^+ present at a fixed concentration.

(a) Describe how you think these experiments were done.

(b) From Figure 9-12 determine the V_{max} and K_m values for components 1 and 2 of the uptake curve (use the $-Na$ data).

(c) Calculate v_T for a series of K^+ concentrations bracketing the concentration range of Figure 9-11 (0.005 to 50 mM) and superimpose a plot of v_T versus K^+ concentration on Figure 9-11. How well does your calculated curve fit the expected data?

(d) What can you conclude about the relative effects of Na^+ on the K^+ transport mechanism operative at low K^+ concentrations (component 1)? On the high-K^+ mechanism (component 2)?

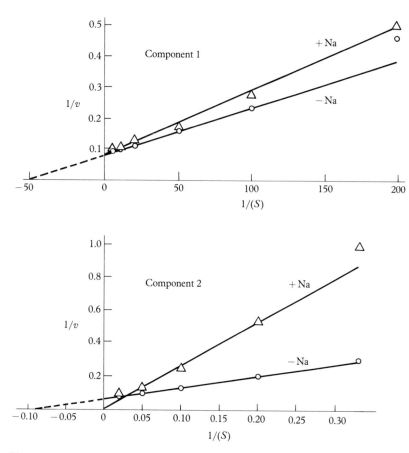

Figure 9-12 Double-reciprocal plots for component 1 (upper graph) and component 2 (lower graph) of the biphasic uptake data shown in Figure 9-11, in the presence and absence of Na$^+$ at a fixed concentration (0.5 mM for component 1 ; 2.0 mM for component 2).

References

Avery, J. (ed.), *Membrane Structure and Mechanism of Biological Energy Transduction* (New York: Plenum, 1973).

Branton, D., and D. W. Deamer, *Membrane Structure* (New York: Springer-Verlag, 1972).

Finean, J. B., "The Development of Ideas on Membrane Structure," *Subcellular Biochem.* (1972), *1* : 363.

Fox, C. F., "The Structure of Cell Membranes," *Sci. Amer.* (Feb. 1972), *226* : 30.

Green, D. E., and R. F. Brucker, "The Molecular Principles of Biological Membrane Construction and Function," *BioScience* (1972), *22* : 13.

Green, D. E., and J. H. Young, "Energy Transduction in Membrane Systems," *Amer. Sci.* (1971), *59* : 92.

Kimmich, G. A., "Coupling between Sodium and Sugar Transport in Small Intestine," *Biochim. Biophys. Acta* (1973), *300*:31.

Kornberg, H. L., "Carbohydrate Transport by Micro-organisms," *Proc. Royal Society* (1973), *183*: 105.

Nakao, M., "Active Transport," *Selected Papers in Biochemistry*, vol. 9 (Baltimore: University Park Press, 1972).

Neame, K. D., and T. G. Richards, *Elementary Kinetics of Membrane Carrier Transport* (New York: Wiley, 1972).

Robertson, J. D., "The Unit Membrane and the Danielli-Davson Model," *Intracellular Transport*, K. B. Warren, ed. (New York: Academic Press, 1966).

Rothfield, L. I., *Structure and Function of Biological Membranes* (New York: Academic Press, 1971).

Siekevitz, P., "Biological Membranes: The Dynamics of Their Organization," *Ann. Rev. Physiol.* (1972), *34*:117.

Simoni, R. D., and P. W. Postma, "The Energetics of Bacterial Active Transport," *Ann. Rev. Biochem.* (1975), *44*:523.

Singer, S. J., "The Molecular Organization of Membranes," *Ann. Rev. Biochem.* (1974), *43*:805.

Singer, S. J., and G. L. Nicolson, "The Fluid Mosaic Model of the Structure of Cell Membranes," *Science* (1972), *175*:720.

Mechanical Work

10

Having considered in the past two chapters the work of biosynthesis and concentration that is so characteristic and universal a feature of cells, we come now to the third major kind of cellular activity for which energy is required—the mechanical work of movement. This is by far the most obvious use of energy in the cell, since it involves a physical change in the location or orientation of an organism, a cell, or a part thereof, and such changes are often readily noticeable and easily measured. Moreover the direct conversion of chemical energy into mechanical energy is a special feature of living organisms; artificial machines that carry out such energy conversions (as, for example, the combustion of coal to run a locomotive) almost always require an intermediate form of energy, usually heat or electricity. Cells, in contrast, carry out these conversions directly, and the mechanisms involved are intriguing indeed.

Functionally, mechanical work can be thought of as occurring at the supracellular, cellular, and subcellular levels. It is at the supracellular level that we encounter the most conspicuous examples of mechanical work, for the muscle tissue common to most animals consists of cells specifically adapted for the task of moving other cells or tissues, and the results are often obvious, whether manifested as the movement of an arm or leg, the beating of a heart, or the closing of a clam shell. Since motility is such a prominent part of the animal mode of existence, it should not be surprising that about 40 percent of your body by weight is skeletal muscle and that mechanical work at this level represents a quantitatively significant component of your total energy needs.

At the cellular level, the emphasis is on the movement of the cell through its environment (or, in some cases, the movement of its environment past the cell). Cellular motility is a phenomenon observed most commonly in single cells or multicellular organisms consisting of but a few cells. It occurs among cell types as diverse as flagellated bacteria, ciliated protozoa, and motile sperm. Essential in each of these cases is some sort of cellular appendage adapted for cellular propulsion. In eukaryotic cells these may be sperm tails, cilia, or flagella, all of which have a similar structure and mechanism of movement. In bacteria the motile appendage is the bacterial flagellum. Other examples of motility at the cellular level include amoeboid motion, cell migration during animal embryogenesis, and the invasive movements of cancer cells in malignant tumors.

Less dramatic than the motility of a whole organism or cell but of even greater general significance is the movement of cellular components, which might be regarded as mechanical work at the subcellular level. For example, contractile elements in the cytoplasm play a key role in the separation of chromosomes during cell division. In addition many cells display a phenomenon called *protoplasmic streaming*, in which the cytoplasm undergoes rhythmic patterns of flow. This seems to serve a stirring and mixing function and is, of course, a further example of mechanical work at the subcellular level. Other instances of mechanical work at the subcellular level include the characteristic movements of molecular structures that take place during cell growth and differentiation. An example of such a process is the transport of cellulose fibrils to the growing wall of a dividing or differentiating plant cell.

Systems of Motility   In all these cases the change in position or orientation of the cell or organism requires energy, and that need is usually met by coupling the contractile or motile process to the hydrolysis of ATP. The major systems used to accomplish this can be classified at the biochemical level into three groups based, respectively, on filaments of actin and myosin, on microtubules, and on bacterial flagella. The best-known example of an *actin/myosin system* is muscle, since all muscles contract by means of the relative sliding of thick and thin filaments containing the proteins *myosin* and *actin*, respectively. In addition amoeboid movement, protoplasmic streaming, and related cytoplasmic movements in higher plant and animal cells are all based on actin and myosin, though the detailed mechanisms are not known in these latter cases.

Microtubule-based systems include the spindle fibers involved in chromosome movements during cell division, as well as the appendages (sperm tails, cilia, and flagella) responsible for motility of eukaryotic cells. *Microtubules* are slender, tubelike structures composed of many molecules of the protein tubulin and, like the actin/myosin system, they appear to function by a relative sliding between adjacent tubules.

The third system of cellular motility involves *bacterial flagella*, which are chemically and structurally unrelated to eukaryotic flagella despite the similarity of terminology.

Without a doubt, the best-known actin/myosin system—indeed the best-known example of mechanical work in general—is muscular contraction. Of the several different kinds of muscle common to mammals (skeletal, cardiac, and smooth), skeletal muscle is the best understood and will form the basis of our discussion. *Skeletal muscle* is a tissue consisting of bundles of long, parallel *fibers*. Each fiber is actually a large multinucleate cell, highly specialized for its contractile function. The multinucleate state arises from fusion of many cells during muscle differentiation; this cell fusion also accounts at least in part for the striking length of muscle cells, which may run into centimeters. At the subcellular level each fiber (cell) contains numerous *fibrils*, the basic functional units of contraction. Each fibril is 1–2 μ in diameter and, as shown in Figure 10-1, consists of bundles of *filaments* that are in lateral register with each other and give the fibrils a pattern of alternating dark and light bands under the polarizing light microscope.[*] It is, in fact, this characteristic pattern of dark and light bands, or striations, that gives *striated* muscle its name. The dark bands are called *A bands* (because of their *anisotropic* or birefringent nature) and the light bands are termed *I bands* (because of their *isotropic* or non-birefringent nature). In the middle of each I band there appears a dark *Z line* (Z for the German word *zwischen*, meaning "between"). The distance from one Z line to the next defines a *sarcomere*, the repeating unit along the fibril. To complete the nomenclature, we need two further terms, both of German derivation: the somewhat lighter zone in the center of the A band is the *H zone* (for *hell*, meaning "light"), and the line in the middle of the H zone (and hence in the middle of both the A band and the sarcomere as well) is the *M line* (for *mittel*, a German word with an obvious meaning).

Muscle: The Best-Understood Actin/Myosin System

---

[*] The nomenclature describing the structure and appearance of muscle fibrils was developed from observations made originally with the polarizing light microscope. *Anisotropy* means that the refractive index (extent to which a beam of plane-polarized light is deflected or "bent") varies depending upon the direction at which the measurement is made; *birefringence* or *double refraction* refers to the same property.

(a)

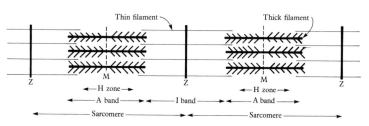

EXTENDED MUSCLE
(Sarcomere length = 3.2 μ)

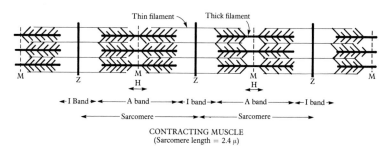

CONTRACTING MUSCLE
(Sarcomere length = 2.4 μ)

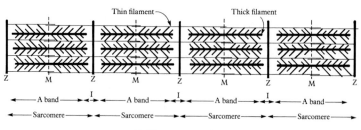

CONTRACTED MUSCLE
(Sarcomere length = 1.8 μ)

(b)

With that admittedly complex terminology in hand, we can now ask what happens to this characteristic structure during the contraction process. In the contracting muscle the A bands of the fibrils remain fixed in length at about 1.6 μ, while the I bands, which are almost 2.0 μ long in maximally extended muscle, shorten progressively and may virtually disappear in the fully contracted state. Thus the length of one sarcomere (one sarcomere $= \frac{1}{2}I + A + \frac{1}{2}I$) decreases from nearly 3.5 μ in the fully extended state to less than 2 μ in the fully contracted form, meaning that the individual fibrils (and hence the muscle cells and tissue) can shorten by almost 50 percent during contraction.

The striated pattern of skeletal muscle and the observed shortening of the I bands during contraction can be explained in molecular terms. Muscle fibrils actually consist of two different kinds of filaments. These are interspersed in regular, alternating pattern with each other and are interdigitated as shown in Figure 10-1b. The *thick filaments* are about 150 Å in diameter and about 1.6 μ long. As you might guess from the length, the thick filaments are responsible for the dark A bands of the fibril. Each thick filament consists of many molecules of *myosin*, one of the two quantitatively most significant muscle proteins. The thick filaments are symmetric in the sense that the myosin molecules are oriented in opposite directions in the two halves of the filament. Each myosin molecule of the thick filaments is a long, thin molecule with a molecular weight of about 470,000 daltons. It has a complex structure, with a double-headed "lollipop" on one end and a long "tail" consisting of two helical chains supercoiled around each other. Localized in the lollipop head is the ATP-splitting (ATPase) activity responsible for coupling the exergonic hydrolysis of ATP to

Figure 10-1   The structure and contraction of striated muscle fibrils. Each fibril consists of alternating dark and light bands, as seen in the electron micrograph of skeletal muscle in (*a*) and illustrated schematically in (*b*). The dark A bands correspond to the lengths of individual thick (myosin) filaments. The light I bands correspond to that portion of the thin (actin) filaments not interdigitated into the thick filaments. The dark Z lines which appear in the middle of each I band define the sarcomere (S), or repeating unit of the fibril. The M line marks the midpoint of the A band and is surrounded by a relatively light H zone that corresponds to the midregion of the thick filaments not overlapped by thin filaments. During contraction (*b*), the lengths of both thick and thin filaments remain unchanged, and shortening of the fibril is accomplished by an interdigitation of thick and thin filaments. The increasing overlap of thick and thin filaments leads to a progressive decrease in the lengths of both the I band and the H zone as the degree on interdigitation increases during contraction. (Photo courtesy of F. A. Pepe.)

the endergonic process of contraction. The myosin molecules are organized such that the head of each molecule projects out of the thick filament. This means that barblike projections with ATPase activity occur at regular intervals along the thick filament, as shown in Figure 10-2. The significance of this design feature will become evident shortly.

The thick filaments of the fibril interdigitate with the second kind of muscle filament, the *thin filaments* (Figure 10-1). These are only about 60 Å in diameter, have a length of about 1 μ, and, where not interdigitated with thick filaments, make up the light I bands of the muscle fibril. Each I band consists in fact of two sets of thin filaments, one set on either side of the Z line, with each filament attached to the Z line and extending toward and into the A band in the center

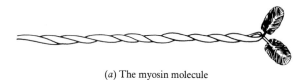

(*a*) The myosin molecule

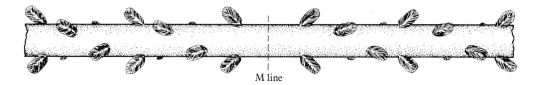

M line

(*b*) Organization of the projecting myosin
heads along the thick filament

Figure 10-2   The myosin molecule and the organization of the thick filament. Myosin (*a*) has a molecular weight of about 470,000 daltons and consists of two long polypeptide chains coiled about each other. Each of the two chains terminates in a globular head. The long rodlike tail is about 1300 Å long and 20 Å in diameter. Both the ATPase activity and the actin-binding capacity of the molecule are localized in the globular head. The thick filaments of the muscle fibril (*b*) are about 1.6 μ long and about 150 Å thick. Individual myosin molecules are integrated into the filament longitudinally, with their ATPase-containing heads oriented away from the M line. The myosin heads project outward from the filament to form a repeating helical pattern, making contact with the six thin filaments thought to surround each thick filament as seen in cross section.

of the sarcomere. This accounts for the length of almost 2 μ noted previously for I bands in extended muscle. The thin filaments contain at least three proteins, of which actin is the most prominent. *Actin* exists in two forms. It is synthesized as a monomer called G-actin (G for globular) with a molecular weight of 47,000 daltons but polymerizes into a long, two-stranded form called F-actin (F for fibrous). The polymerization involves end-to-end joining of G-actin units and is driven by ATP hydrolysis, with the resulting ADP remaining attached to the actin:

$$n(\text{G-actin}) + n\text{ATP} \rightarrow (\text{actin-ADP})_n + n\text{P}_i \qquad (10\text{-}1)$$
$$\text{F-actin}$$

(It should be emphasized that the ATP used in this way for actin polymerization is *not* the same ATP that drives muscle contraction; ATP is required *both* for the polymerization of Equation 10-1 and for the actual contraction process.) It is F-actin that is found as the main structural component of the thin filaments. In addition the thin filaments contain small amounts of *tropomyosin* and *troponin*, two regulatory proteins to which we shall return shortly.

Recognizing that muscle fibrils consist of alternating but interdigitating thick and thin filaments, we come now to examine the molecular mechanism of contraction, which must explain how the A band remains fixed in length while the I band shortens progressively during contraction. This mechanism involves an endergonic sliding of thick and thin filaments relative to each other such that the thin filaments are drawn progressively into the spaces between adjacent thick filaments, overlapping more and more with the thick filaments in the process. The result is a shortening of the individual sarcomeres and fibrils and hence also of the muscle cell and the whole tissue. This in turn causes the movement of the body parts attached to the muscle.

The Mechanism of Muscle Contraction

Since this sliding of thin filaments past thick filaments requires energy, three questions arise. By what mechanism are the thin filaments drawn or pulled progressively into the spaces between thick filaments to cause the actual contraction? How is the energy of ATP used to drive this process? And what keeps the thin filaments of a partially contracted fibril from sliding right back out again?

We can begin by taking the last question first, since it implies a need to link the partially interdigitated thin filaments reversibly but firmly to the adjacent thick filaments. This requirement is met by the heads of the myosin molecules, which project outward at regular intervals along the thick filament and represent binding sites for the

F-actin of the thin filaments. Regions of overlap between thick and thin filaments, whether extensive (in the shortened muscle) or minimal (in the lengthened muscle), are always characterized by the presence of *cross-bridges* between the myosin heads of the thick filaments and the action of the thin filaments. For actual contraction, however, cross-bridges cannot simply be present. They must actually break and re-form in a progressive manner, such that there is a ratchetlike movement as a given myosin head on the thick filament repeatedly breaks its attachment with a specific actin site on the thin filament and forms a new cross-bridge with the next actin site further along the thin filament toward the Z line. In this way the protruding heads of the myosin molecules in the thick filament draw the thin filaments along unidirectionally toward the center of the A band. Overall contraction, then, is the net result of the repeated making and breaking of many such cross-bridges, each cyle of bridge formation causing the translocation of a small length of thin filament of a single fibril in a single cell.

The driving force for cyclic formation and breakage of cross-bridges is clearly the hydrolysis of ATP, catalyzed by the ATPase located strategically in the myosin head. The four-step cycle shown in Figure 10-3 is the best single model currently available for the process whereby ATP hydrolysis drives the sliding of filaments, as mediated by cross-bridge formation. In step 1 a specific myosin head (in an energized configuration it acquired in step 4 of the previous cycle) binds to a specific actin site (designated site s_1) on an adjacent thin filament. Step 2 is the "power stroke"; as the myosin head that forms the cross-bridge reverts from a high-energy to a low-energy configuration, it causes the translocation of the thin filament with respect to the thick filament. Detachment follows in step 3 as the cross-bridge is broken; this is accompanied by the binding of a molecule of ATP in preparation for the next step. (Note that, once detached, the thick and thin filaments would be free to slip back to their previous position except that they are held together at all times by the many other cross-bridges formed along their length at any given moment—just as some legs of a millipede are always in contact with the surface on which it is walking.) Finally, in step 4, the energy of ATP hydrolysis is used to "cock" the myosin head for the next cycle by returning it to the high-energy configuration necessary for the next round of cross-bridge formation and filament sliding.

This brings us back to where we started, for the myosin head is now activated and ready to form a bridge to the actin again. But the new bridge in our example will be formed with site s_2, since the first

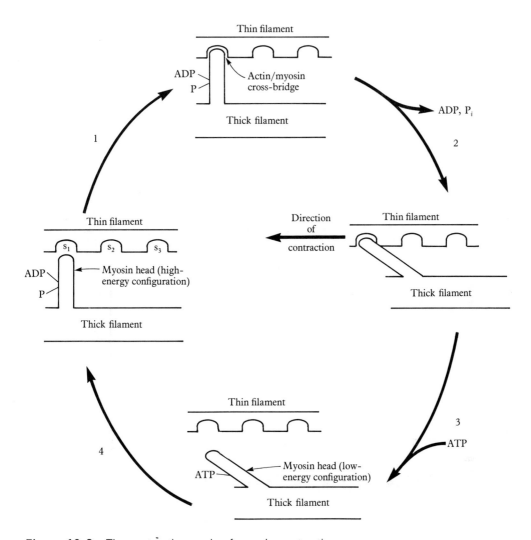

Figure 10-3 The contraction cycle of muscle contraction.

cycle has resulted in a net displacement of the thin filament with respect to the thick filament. In succeeding cycles, the particular myosin head shown in Figure 10-3 will advance progressively to sites s_3, s_4, and so on as the thin filament continues to be displaced in the direction of further contraction.

Of particular interest in this contraction cycle are the separation of the actual hydrolysis of ATP (step 4) from the contraction event (step

2) it drives and also the use of an energized configuration of the myosin head as the "carrier" of the energy. Note also that the products of hydrolysis, ADP and P_i, remain bound to the myosin as long as the protein is in the energized form but are released upon reversion to the low-energy form in the power stroke of step 2.

The Regulation of Muscle Contraction

The sequence of events depicted in Figure 10-3 indicates how muscle contraction is accomplished, but it provides no insight into how the process is regulated. Indeed the description provided so far suggests that muscles ought to contract continuously as long as ATP supplies remain. Yet common experience tells us that muscle contraction is always complemented by relaxation and that the two processes must be regulated to result in the coordinated movements associated with muscular activity. The key to regulation lies in the dependence of the contraction process on calcium ions (Ca^{++}) and in the ability of the muscle cell to raise and lower calcium levels rapidly in the cytoplasm around the fibrils.

The dependence on calcium arises because the thin filaments contain, in addition to actin, the regulatory proteins *tropomyosin* and *troponin*. These molecules act in concert to regulate the availability of the binding sites on the thin filament in a manner that is affected critically by the cytoplasmic calcium concentration. Essentially this is because the binding sites on actin can be occupied by *either* tropomyosin or myosin and because the affinity of tropomyosin for the binding sites is determined by the calcium level. When the cytoplasmic concentration of calcium is relatively low (10^{-7} to 10^{-8} M), tropomyosin binds to the actin sites, effectively preventing their interaction with myosin. As a result, cross-bridge formation is blocked and the muscle becomes or remains relaxed. At higher calcium concentrations (greater than 10^{-6} M), however, calcium binds to troponin. This in turn induces a configurational change in tropomyosin, resulting in its dissociation from the binding sites on actin. The binding sites are therefore free to interact with myosin instead. As a result, cross-bridges can again form, and the sequence of events depicted in Figure 10-3 is set in motion, leading to contraction. When the calcium concentration falls again, the troponin-calcium complex dissociates. The free troponin causes tropomyosin to revert to its high-affinity form, which then binds to the actin, preventing further cross-bridge formation and thereby stopping the contraction cycle.

To understand how the muscle cell regulates its calcium level and how muscle contraction can be triggered by nerve impulses, we need to know that muscle cells contain an extensive *sarcoplasmic reticulum* (SR). The SR is a specialized membranous network throughout the muscle cell that serves to store calcium and release it on signal. There is an active transport protein for calcium (similar to those discussed in Chapter 9) in the membrane of the SR, which pumps calcium ions continuously from the cytoplasm into the chambers or *cisternae* of the SR. (In addition the cisternae contain a calcium-binding protein that takes up the Ca^{++} ions as they are pumped in, thereby increasing the calcium storage capacity of the cisternae.) Normally the SR membrane is only poorly permeable to the free passage of calcium, such that the calcium ions, once pumped in, are effectively sequestered in the cisternae. Upon excitation of the muscle cell by a nerve impulse, however, the SR membrane becomes temporarily very permeable to calcium. The calcium ions therefore flow out of the SR readily because their concentration is much higher in the cisternae than in the surrounding cytoplasm.

The only remaining question is how the nerve signal arriving at the surface of the muscle cell can be transmitted quickly to the SR membranes in the cell so that calcium release occurs simultaneously throughout the cell. This is possible because of a series of *transverse tubules* (the *T system*) that are actually invaginations of the cell membrane that penetrate the cell and allow rapid conduction of electrical impulses into the interior of the cell.

The events that occur when a nerve impulse arrives at the surface of a muscle cell can be summarized as follows:

1. The nerve impulse arrives at the membrane surface of the muscle cell and is transmitted rapidly to the interior of the cell via the T system, which is continuous with the cell membrane.

2. The wave of depolarization passing through the T system causes a transient increase in the permeability of the SR membrane to calcium.

3. Calcium ions flow rapidly out of the SR cisternae into the cytoplasm surrounding the fibrils.

4. With its concentration in the cytoplasm elevated, calcium binds reversibly to troponin. This causes tropomyosin to vacate the myosin binding sites on actin, thereby permitting cross-bridge formation between thick and thin filaments.

5. Actin/myosin cross-bridges form, and the cycle of events depicted in Figure 10-3 is set in motion. The energy of ATP hydrolysis is converted into a configurational change of the cross-bridge protein, which in turn causes filaments to slide and work to be done.

6. Many such cycles of rapid making and breaking of cross-bridges continue to occur at each myosin head along every thick filament, as long as calcium is available. As the thin filaments move further and further into the space between thick filaments, sarcomeres shorten and the muscle contracts.

7. After the T system recovers from the initial excitation process (and assuming no further stimulation), the SR membrane regains its normal impermeability to calcium. The continued pumping of calcium from the cytoplasm back into the SR cisternae rapidly lowers the cytoplasmic calcium level to the point where troponin releases calcium, tropomyosin again binds to the actin sites, and further cross-bridge formation is prevented.

8. With the actin sites occupied by tropomyosin molecules, step 1 of Figure 10-3 cannot occur, and cross-bridges rapidly disappear as actin dissociates from myosin and becomes blocked by tropomyosin.

9. The muscle is now relaxed and can be reextended by contraction of the opposing muscle, since in the absence of cross-bridges the thin filaments are free to slide back out from between the thick ones. (Notice that during contraction some bridges must always be formed at any given moment to avoid backsliding of the filament. But during relaxation most bridges must be broken so that the filaments will slide past each other as they are pulled apart by contraction of the opposing muscle.)

Meeting the Energy Needs of Contraction

Because muscle contraction involves the hydrolysis of an ATP molecule for every cross-bridge cycle, muscle cells obviously need ways of regenerating ATP continuously and also of providing extra energy for temporary exertion in excess of the immediate ATP-regenerating capacity of the cell. Muscle tissue is a fascinating system from this point of view because of the variety of mechanisms the cell has at hand to ensure maximum ATP availability, even during prolonged periods of intense activity.

Ultimately, of course, the ATP needs of the muscle cell are met either by glycolysis or mitochondrial respiration. The extent to which one or the other of these pathways is favored depends upon the kind of muscle involved and the conditions under which it is functioning. Skeletal muscles that are characterized by frequent use and high activity (the flight muscles of birds and your own leg muscles are good examples) usually depend upon complete respiratory metabolism. They draw both glucose and oxygen from the circulatory system, oxidize the glucose completely to carbon dioxide and water, and generate ATP by oxidative phosphorylation in the mitochondrion. Such muscles are characterized by an abundance of mitochondria (usually in close association with the fibrils) and by a red color. The color is due both to the cytochromes in the mitochondria and to the presence of *myoglobin*, a protein related in structure and function to hemoglobin but localized in muscle cells, where it binds and stores oxygen.

During periods of intense muscular activity, the demand for ATP regeneration may exceed the rate at which oxygen can be supplied to the tissue by the circulatory system. After depletion of the reserve oxygen available from myoglobin, the tissue begins functioning anaerobically, converting glucose to lactate. Because the ATP yield per glucose is reduced from 38 (aerobic respiration) to 2 (anaerobic glycolysis), much more glucose is required under these conditions. The extra glucose is supplied by degradation of *glycogen*, the storage carbohydrate of muscle cells. The lactate formed under these anaerobic conditions is released into the blood and is eventually used for the resynthesis of glucose in the liver (the gluconeogenesis of Chapter 8). Intense muscular activity cannot be sustained for long under anaerobic conditions, both because of the rapid depletion of glycogen stores and because of the buildup of lactate. This anaerobic option is, however, useful for short bursts of activity when oxygen cannot be supplied fast enough. Moreover there are some muscles, characterized usually by relative inactivity, that routinely depend upon anaerobic glycolysis to meet their limited energy needs. Such muscles usually contain few mitochondria and little or no myoglobin and are consequently pale or white in appearance. The contrast between the "white meat" of the breast of a chicken and the "dark meat" of the leg illustrates the point well. Such white muscles depend upon the blood both to supply glucose and to carry away lactate. Because the lactate is then used for glucose synthesis in the liver, a cycle is established with glucose moving from the liver via the blood to the anaerobic muscle and lactate returning from the muscle to the liver.

Although ATP serves as the immediate source of energy to drive muscle contraction and is also the form in which energy appears during glycolysis and respiratory metabolism, the muscle cell does not use ATP as its major storage form of energy. A surprising observation made early in muscle research was that the ATP content of working muscle remains almost constant until the muscle is near exhaustion. What decreases instead during prolonged muscular activity is the cellular level of *creatine phosphate*, a high-energy compound with the structure shown in the margin. Understandably, it was thought for a time that creatine phosphate, rather than ATP, was the actual high-energy

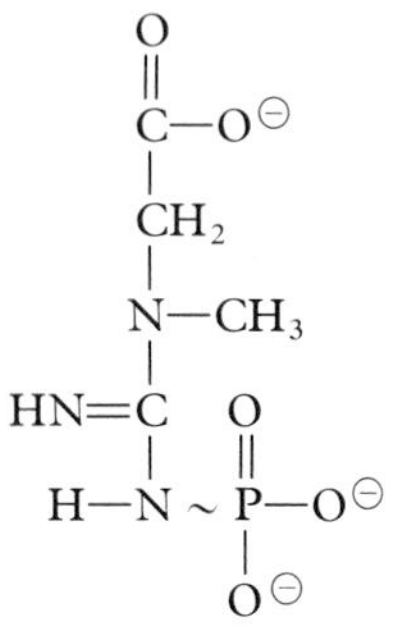

$$
\begin{array}{c}
\text{O} \\
\parallel \\
\text{C}-\text{O}^{\ominus} \\
| \\
\text{CH}_2 \\
| \\
\text{N}-\text{CH}_3 \\
| \\
\text{HN}{=}\text{C} \quad \text{O} \\
| \quad\quad \parallel \\
\text{H}-\text{N}\sim\text{P}-\text{O}^{\ominus} \\
| \\
\text{O}^{\ominus}
\end{array}
$$

Creatine phosphate

compound directly involved in contraction. Now, however, it is known that creatine phosphate represents a reservoir of high-energy phosphate that can be used to recharge ADP (Equation 10-2) and thereby maintain a high ATP level in the cell.

$$\text{Creatine phosphate} + \text{ADP} \xrightarrow[\text{creatine kinase}]{} \text{creatine} + \text{ATP} \quad (10\text{-}2)$$

The highly negative ΔG^0 for the hydrolysis of creatine phosphate (-10.3 kcal/mole) ensures that the equilibrium of Equation 10-2 lies far to the right and that the recharging of ADP is driven effectively.

The variety of mechanisms available to muscle cells to meet their energy needs for continued contraction under a wide variety of conditions can be summarized as follows:

1. With adequate oxygen and glucose available from the blood, complete respiratory metabolism is possible, and most muscle cells are well endowed with mitochondria for this purpose.

2. In addition oxygen can actually be stored within the cell as a complex with myoglobin and can be released to prolong respiratory metabolism for a short time, even when the circulatory system is no longer able to supply sufficient oxygen.

3. When oxygen supplies are inadequate, the muscle cell can revert to anaerobic glycolysis to meet energy needs.

4. Stored glycogen can be used as a source of glucose to continue glycolysis, even when the supply of blood glucose is inadequate.

5. In addition the cell has a reservoir of creatine phosphate (often much more creatine phosphate than ATP, in fact), and the energetics of phosphate transfer to ADP are sufficiently favorable to maintain a high ATP/ADP ratio despite the fall in creatine phosphate levels during muscle activity.

6. Finally, though not mentioned earlier, muscle cells also contain an enzyme (myokinase) capable of phosphorylating one ADP molecule at the expense of another:

$$2\text{ADP} \underset{\text{myokinase}}{\rightleftharpoons} \text{ATP} + \text{AMP} \quad (10\text{-}3)$$

If all else fails and large quantities of ADP accumulate, this mechanism provides a last-ditch means of extracting energy from the remaining acid anhydride bond of ADP ($\Delta G^0 = -7.3$ kcal/mole).

The overall picture of the muscle cell, then, is one of an almost incredible specialization for its role in contraction—specialization in design of the actual contractile elements as well as in the mechanisms available to ensure that the ATP needed for contraction can be supplied under virtually any condition.

Although muscle is clearly the best-understood contractile system, it is becoming increasingly apparent that there are other systems in which actin and myosin are involved in cellular movement, probably also by a contractile mechanism. A common phenomenon in many cell types is that of protoplasmic streaming, in which the cytoplasm of the cell exhibits a regular, directional flow. In some cases the function appears to be one of locomotion and food gathering. Amoebae, for example, depend upon cytoplasmic extensions called *pseudopods* (Greek, "false feet") for both movement and food engulfment (see Figure 10-4a). The progressive extension of a pseudopod is accomplished by a unidirectional streaming of cytoplasm in the direction of extension. A gel-like region of cytoplasm called the *cortex* forms at the periphery of the cell, and the flow of more fluid cytoplasm forward into the pseudopod appears to be vital for further extension.

In slime molds like *Physarum* (Figure 10-4b), cytoplasm streams back and forth in the branched network of protoplasm of which the *Physarum* cell mass consists. The flow of cytoplasm reverses direction with predictable periodicity and is called *shuttle streaming*. Again the purpose seems to be nourishment and locomotion, since the streaming process is correlated with the further extension of the fingerlike projections by which the slime mold reaches out to its environment in search of nutrients. Many plant cells display a circular flow of cell contents around a central vacuole, a streaming process called *cyclosis*. In this case the movement seems designed to circulate and mix cell contents.

The unifying feature in these and other cases of protoplasmic streaming seems to be the presence of actin- and myosinlike filaments in the cytoplasm. These filaments are considered by many to be intimately involved in the streaming process, and the supporting evidence is becoming increasingly convincing.

Similar filamentous structures are seen in cells of higher animals as dense bundles or networks just beneath the cell membranes and also at cleavage furrows during cell division. Such *microfilaments* are about 40–60 Å in diameter and appear to be involved in oriented movements, in shape changes, and in the constriction of the cell surface during cell division in animals. In many cases microfilaments have been identified as *actin* filaments, further strengthening the suggestion that

Other Filament-based Contractile Systems

actin (and perhaps also myosin) plays a much more universal role in cellular movement than might be suggested by its involvement in the classic example of muscle contraction.

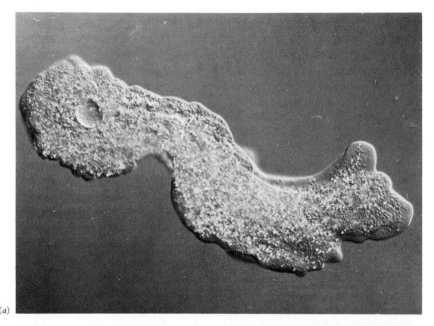

(a)

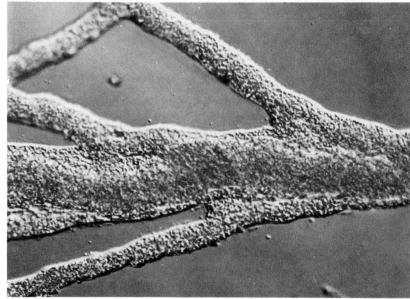

(b)

Thus far our attention has been occupied by motility systems based on the proteins actin and myosin organized into filaments. We come now to the second major category of cellular systems used to accomplish mechanical work: the microtubule-based systems as exemplified by the motile appendages of eukaryotic cells, including cilia, flagella, and sperm tails. These appendages share a common composition and structure, but they differ from each other in relative length and number and in mode of beating. *Cilia* tend to be about 2–10 μ long and are usually numerous if present at all. Ciliated protozoans are illustrated in Figure 10-5a and b. Flagella and sperm tails, on the other hand, are often much longer (10–200 μ) and are usually limited to one or two per cell. Figure 10-5c shows the flagellated protozoan *Trypanosoma equiperdum*, while Figure 10-5d illustrates sperm cells and tails. All three types of appendages shown in Figure 10-5 are used to propel cells and are therefore most characteristic of single cells (protozoa, algae, and sperm of both plants and animals). Cilia also occur in multicellular organisms, however, where they serve to move the environment past the cell rather than vice versa. The cells that line the air passages of your lungs, for example, have several hundred cilia each, and it is the coordinated, wavelike beating of these cilia that both keeps and carries dust and other foreign matter out of the lungs. One of the health hazards of cigarette smoking lies in the inhibitory effect smoke has on normal ciliary beating.

The patterns of movement of cilia, flagella, and sperm tails are intriguing in their variety. Flagella move with an undulating, whip-like lash that may even have a helical or circular pattern to it. The locomotory pattern of most sperm cells and flagellated cells usually involves propulsion of the cell by the trailing flagellum, but examples are also known in which the flagellum actually precedes the cell. Cilia display a wavelike, often coordinated pattern of beating, apparently designed to ensure a steady flow of fluid or air past the cell surface.

Cilia, flagella, and sperm tails have a common structure consisting of an *axoneme* or main cylinder of tubules about 0.20 μ in

Microtubule-based Motility

Figure 10-4   Filament-based motility systems may involve either pseudopod extension or cytoplasmic streaming. (*a*) Cells like those of *Amoeba proteus* (238×) depend upon cytoplasmic extensions called pseudopods for both movement and food engulfment. (*b*) In slime molds such as *Physarum polycephalum* (74.5×), cytoplasm streams back and forth in the branched network of protoplasm, serving both as a means of nourishment and of locomotion. (Both micrographs are by Nomarski differential interference microscopy, courtesy of D. L. Taylor.)

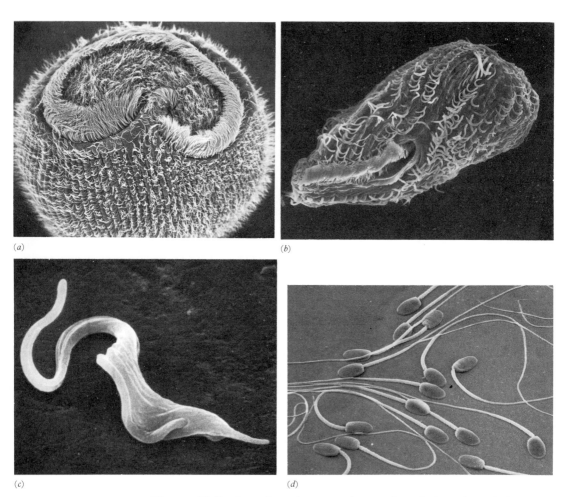

(a)

(b)

(c)

(d)

Figure 10-5 Scanning electron micrographs of the microtubule-based appendages responsible for motility of eukaryotic cells. (a) The protozoan *Stentor coeruleus* (336×) has a zone of cilia around the oral cavity to create the water currents that drive food into the cytostome or "mouth." In addition, meridians of cilia run from the base of the oral zone to the posterior pole of the cell. (b) The cilia of *Urostyla grandis* (483×), on the other hand, are fused into groups that serve for locomotion over the substratum and into rows that sweep food into the "mouth." (c) An example of a flagellated cell is the protozoan *Trypanosoma equiperdum* (7,080×). This is the bloodstream form of the organism, which causes dourine, a venereal disease of horses. (d) Closely related to cilia and flagella are the sperm tails, shown here on spermatazoa from the Rhesus monkey, *Macaca mulatta* (1,920×). (Photo (a) courtesy of J. J. Paulin and A. Steiner, (b) H. Wessenberg, (c) J. J. Paulin, and (d) J. Hren.)

diameter, connected at the base to a *basal body* and surrounded by an extension of the cell membrane (see Figure 10-6a). The basal body (Figure 10-6c) consists of nine sets of tubular structures arranged around its circumference, with three tubules per set. The axoneme, on the other hand, has a characteristic "9 + 2" pattern (Figure 10-6b) with nine outer doublets of two tubules each and an additional pair of tubules in the center. The nine peripheral doublets of the axoneme are thought to be extensions of two of the three tubules in each of the nine sets of the basal body. Radial and circumferential connections are characteristic of the tubules of the axoneme and are thought to be essential to axoneme movement.

The tubular structures of both axoneme and basal body are *microtubules*—straight, hollow cylinders with a diameter of about 240 Å and a core of about 140 Å. The 50-Å wall of the cylinder is made up of longitudinal arrays of *protofilaments* consisting in turn of strings of spherical subunits. Each subunit is thought to be a single molecule of the protein *tubulin*, which is therefore the basic repeating unit of the microtubule. Monomeric tubulin has a molecular weight of about 55,000 daltons and a diameter upon folding equal to that of the wall of the microtubule (50 Å).

Like microfilaments, microtubules occur in eukaryotic cytoplasm with a frequency and ubiquity that cell biologists have only recently begun to appreciate. Both kinds of structures are clearly involved in cellular motility, but they are distinctly different, both in the specific functions they serve and in the protein subunits of which they are composed. Table 10-1 provides a comparison and summary of the chemical, structural, and functional properties of microtubules and microfilaments.

The mechanism of motility in microtubule-based systems is not yet well understood. It is, however, noteworthy that the two tubular structures in each of the nine outer doublets of the axoneme differ from each other in that one has "arms" extending from it. The arms contain a distinctive protein called *dynein* and, significantly, have ATPase activity. Isolated cilia and flagella can be induced to beat if ATP is added to the medium, so it appears likely that the appendages are actively involved in their own motility rather than being passively moved by the action of the cell. In fact it can be demonstrated that energy is expended all along the length of the appendage. A reasonable hypothesis at present envisions a sliding of the outer doublets relative to each other by a "ratchet" mechanism of the dynein arms, analogous, perhaps, to the sliding of filaments in muscle. Presumably the sliding of the outer doublets is resisted by the radial connections between tubules and is thereby converted into bending.

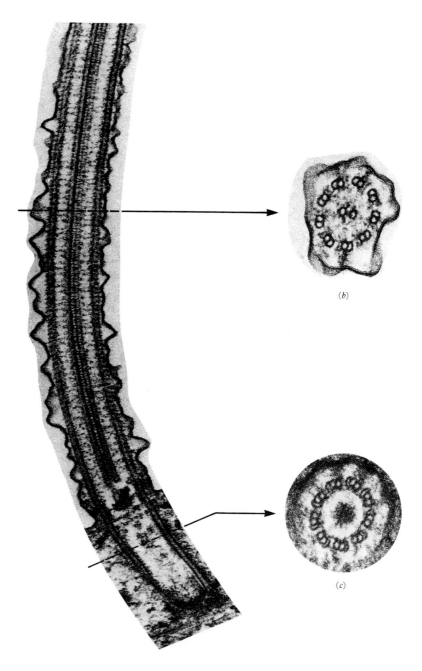

(b)

(c)

(a)

TABLE 10-1   Comparison of the Chemical, Structural, and Functional Properties of Microtubules and Microfilaments

Property	Microtubules	Microfilaments
Structure	Hollow cylinders with 50-Å walls made up of linear strands of globular subunits	Solid rods of polymerized actin
Diameter	240 Å	60 Å
Subunit	Tubulin	Actin
ATPase activity	Dynein arms (characterized only in eukaryotic appendages)	Myosin heads (characterized only in muscle)
Functional roles	Motility of eukaryotic cells (cilia, flagella, sperm tails) Chromosome movement (centrioles and spindle fibers) Movements of materials and particles within cells (synaptic vesicles, pigment granules, cellulose fibers, etc.) Development and maintenance of cell shape (cytoskeletal role)	Muscle contraction (in association with myosin filaments) Protoplasmic streaming Shape changes and oriented movements in cells of higher animals Cell division in animals (constriction of membrane at midpoint of cell to form cleavage furrow)

In addition to their occurrence in the motile appendages of eukaryotic cells, microtubules are also found in the centrioles and spindle fibers involved in chromosome separation during mitosis. *Centrioles* are self-replicating, cylindrical structures about 0.15 μ in diameter and 0.3–0.5 μ long. They occur in pairs in animal (but apparently not in plant) cells and serve as the center of the radiating network of fibers that makes up the *spindle* responsible for ordering chromosome movement. Centrioles have a structure strikingly like that of the basal body of a flagellum or cilium; centrioles and basal bodies may, in fact, be alternative forms of the same organelle. The spindle itself is an array of fibers that radiate out from the poles of the mitotic apparatus. *Continuous fibers* extend from pole to pole, whereas *chromosomal fibers* extend from chromosome to pole. Both kinds of fibers are bundles of microtubules, with 1 to 100 microtubules per

Figure 10-6   The structure of the eukaryotic cilium. The cilium shown in longitudinal section in (*a*) is from the protozoan *Tetrahymena pyriformis*. Note the central microtubules, the outer doublets, the radial spokes, and the basal body structures. Cross-sectional views are shown for both the axoneme (*b*) and the basal body (*c*) to illustrate the characteristic "9 + 2" and "9 + 0" patterns of tubule arrangement, respectively. (All micrographs courtesy of W. L. Dentler.)

fiber in cross section. The mitotic spindle can be shown to assemble spontaneously from the tubulin subunits already present in the cytoplasm. The mechanism whereby the spindle fibers shorten at anaphase to move chromosomes from the metaphase plate to the two poles is ill understood. One explanation envisions a sliding of microtubules such as that thought to occur in cilia and flagella, while an alternative possibility involves a progressive dissembly of the fibers to monomeric tubulin at the pole. Yet a third suggestion currently in vogue is that microtubules provide the structural organization but the motile force is provided by an associated actin/myosin system.

We can summarize cellular motility in eukaryotes by noting that it seems invariably to be either filament- or tubule-based, with the best-understood examples being striated muscle and the motile appendages, respectively. Numerous other phenomena of contraction and motility involving microfilaments or microtubules are now recognized, however, and these are currently the focus of much research effort. Both the filament- and the tubule-based systems are characterized by the presence of an ATP-cleaving activity at intervals along the motile or contractile structure. It is this ATPase that enables the free energy of ATP hydrolysis to be used in the performance of mechanical work, although it is only in the case of muscle contraction that we currently understand the transduction process in any detail at all.

Bacterial Flagella   To complete our discussion of mechanical work in cells, we turn to the bacteria, which accomplish cellular motility by an appendage strikingly unlike those of eukaryotic cells. It is unfortunate, in fact, that the single term "flagellum" is used to refer to motile appendages in both eukaryotes and prokaryotes, when in fact the two kinds of structures have nothing in common except function. The *bacterial flagellum* is usually about 100–200 Å in diameter and about 10 μ long. It is not membrane-bounded and is thus an extracellular structure. The flagellum protrudes through the cell membrane and cell wall and is anchored to an intracellular basal body, but again no structural homology with the eukaryotic basal body is implied. Chemically, bacterial flagella consist of parallel strands of protein coiled around each other. The common subunit is *flagellin*, with a molecular weight of 40,000 daltons in the monomeric form.

The bacterial flagellum has been recognized as a locomotory structure for more than a hundred years, but its mode of propulsion has been elucidated only recently. Initially the characteristic helical motion of the flagellum was thought to result from waves of bending, originating from the base and propagated along the length of the flagellum. It is now quite clear, however, that the individual filaments which make up the flagellar bundle actually *rotate* as rigid, helical structures, driven by a rotary "motor" at the base of each filament as shown diagrammatically in Figure 10-7. Requiring, as it does, the structural equivalents of a rotor, a stator, and rotary bearings, such a mechanism was originally considered highly unlikely—certainly without precedent in the biological world. But a recent series of ingenious experiments in a number of laboratories have provided conclusive evidence in favor of just such a propellerlike rotary drive. Much of this evidence is owed to the availability of antibodies that react specifically with the flagellar filaments. Silverman and Simon, for example, used antibodies to link or "tether" flagellar filaments to a glass surface, thereby preventing normal rotation of the filament. In so doing they made the remarkable observation that if the filament cannot rotate, the whole body of the cell rotates instead!

The same researchers were also able to visualize flagellar rotation directly in an equally clever manner. Using antibodies as "glue," they succeeded in linking microscopic beads of polystyrene-latex to individual flagella filaments as visible markers along the filament surface. Viewed under the microscope, the beads were seen to remain fixed in location with respect to each other but to revolve together around the filament. It is not true, therefore, that Nature failed to invent the wheel—there is one in the rotary motor at the base of each flagellar filament in every motile bacterial cell.

Once it became clear that the driving force for flagellar rotation lies in the basal motor rather than in the flagellum itself, it was no longer surprising that the bacterial flagellum, unlike the motile appendages or contractile filaments of eukaryotic cells, possesses no ATPase activity along its length. (Flagellin has in fact no known enzymatic activity.) Still somewhat puzzling, however, is the evidence that ATP appears not to be involved at all as an energy source for flagellar rotation. Flagellar motility in the bacterium *Escherichia coli* clearly requires oxygen but seems to be independent of the ATP status of the cell, suggesting the involvement of an alternative high-energy intermediate or process instead. The eventual answer to this question promises to be intriguing indeed, for with it is likely to come the explanation of how chemical energy, whether of a high-energy bond or a membrane transport process, can be used to drive the rotary motor that turns the stiff helical propeller responsible for bacterial motility.

In Conclusion

With the bacterial wheels still turning, we come to the close of this discussion of cellular energy metabolism. Throughout these ten

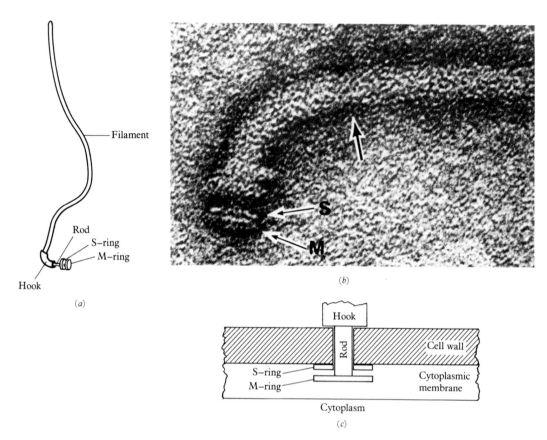

Figure 10-7 Model for the structure of a bacterial flagellum and the "motor" responsible for its rotation. A bacterial flagellum consists of many filaments twisted together yet rotating individually. The helical filament is attached to a hook structure, and this is joined in turn by a rod to two ringlike structures in the base. The S ring is regarded as the nonrotating stator, while the M-ring is thought to be the revolving rotor. The schematic illustration shown in (a) can be compared directly with an electron micrograph (b) of the flagellum from *Escherichia coli*. Structures identified as the S and M rings are indicated on the electron micrograph by arrows; the other arrow indicates the point at which the filament joins the hook. Shown as (c) is a model of the postulated "motor" assembly. The S ring is considered in this model to be mounted rigidly in the cell wall; the M ring is attached directly to the rod and rotates in the cytoplasmic membrane. The torque needed to rotate the filament and so propel the cell is generated between the stationary S ring and the rotatable M ring. The source of the energy needed to develop the torque is not yet known but appears not to be ATP. (Micrograph courtesy of J. Adler.)

chapters we have been exploring the importance of energy to cells, the principles governing its flow in biological systems, the major sources of energy available to cells, and some of the most important uses to which that energy is put. We are about to arrive at the end of the book, but in a sense we have really just begun to understand and appreciate the role of energy flow in living systems. For the basic questions of energy utilization that have provided the themes for these chapters are so integral a part of biology that you are almost certain to encounter them again and again, wherever in the life sciences you may choose to turn. And as you do so, you are very likely to find that the only things more fascinating about bioenergetics than the answers we know today are the questions that still remain for tomorrow.

Practice Problems

10-1. Muscle Structure: Frog skeletal muscle consists of thick filaments with a length of about 1.6 μ and thin filaments with a length of about 1.0 μ.

(a) Indicate the length of the A band and the I band in a muscle with a sarcomere length of 3.2 μ. Describe what happens to the length of both bands as the sarcomere length decreases during contraction from 3.2 to 2.0 μ.

(b) If the H zone of each A band decreases in length from 1.2 to 0.0 μ as the sarcomere length contracts from 3.2 to 2.0 μ, what can you deduce about the physical meaning of the H zone? What can you say about the distance from the Z line to the edge of the H zone during contraction?

10-2. Energy Requirement of Contraction: Mammalian skeletal muscle hydrolyzes 1 mmole ATP per minute per gram of muscle during contraction.

(a) The amount of ATP actually present in the tissue at any given time is about 5 μmoles/g. How long could muscle contraction continue if it depended upon the existing ATP supplies only?

(b) Creatine phosphate is present in muscle cells at about five times the concentration of ATP. How long could contraction continue if it depended solely upon existing supplies of ATP and creatine phosphate? Can you think of circumstances when these reserves of immediate energy would be essential despite the short times over which they could sustain contraction?

(c) Assuming that the need for ATP is in fact being met by aerobic respiration, at what rate must the tissue be supplied with oxygen (in milliliters O_2 per minute per gram of tissue) to sustain activity? (Recall that 1 mole of a gas occupies 22.4 liters at standard temperature and pressure.)

(d) Now assume that the need for ATP is met by anaerobic glycolysis instead, with stored glycogen (polymerized glucose) as the only energy source. If muscle glycogen reserves are equal to 1.0 percent of the muscle by weight and all of this is assumed available for catabolism, how long can anaerobic activity be sustained at the expense of glycogen?

(e) Finally, assume that oxygen is not available, glycogen has been consumed, creatine phosphate stores are depleted, and all ATP has been converted to

ADP, leaving the myokinase reaction as the cell's last resort. How long can contraction go on now? Are the assumptions of the question realistic?

10-3. Rigor Mortis: At death the muscles of an animal body become very stiff and inextensible, and the corpse is said to go into *rigor*.

(a) Explain the basis of rigor. Where in the contraction cycle is the muscle arrested? Why?

(b) Would you be likely to go into rigor faster if you were to die while racing to class or while sitting in lecture? Why?

10-4. Creatine Phosphate: The ΔG^0 of phosphate hydrolysis is -10.3 kcal/mole for creatine phosphate versus -7.3 kcal/mole for ATP. For a long time after its involvement in muscle energetics was first realized in the 1920s, creatine phosphate was thought to be the immediate source of the energy needed for muscle contraction. It was not until inhibitors of creatine kinase were found in the 1960s that the mechanical work of muscle contraction could be correlated directly with ATP consumption and the true nature of creatine phosphate as a "reservoir" of high-energy phosphate for recharging ADP was realized.

(a) What is the ΔG^0 for the creatine kinase reaction (Reaction 10-2) in the direction of ATP regeneration? What is the value of the equilibrium constant K^0 at 25°C?

(b) Assuming conditions under which $\Delta G \simeq \Delta G^0$, calculate the fraction of ADP that is phosphorylated (i.e., the ratio ATP/ATP + ADP) when the following percentages of creatine are phosphorylated: 90; 50; 10; 1. If you were assaying for creatine phosphate and ATP levels during the process of muscle contraction, why might you conclude, as the early workers did, that the creatine phosphate rather than the ATP was the immediate source of the energy needed for contraction?

(c) If, on the other hand, you were to assay for creatine phosphate and ATP levels during contraction as in part (b) but in the presence of an inhibitor (such as 2,4-dinitrofluorobenzene) that inactivates creatine kinase in intact muscle, how would your results differ from those of part (b)? How does this result support the "reservoir" nature of creatine phosphate?

10-5. Some Fowl Biology: Why is the breast meat of flightless birds like the chicken or turkey so white in color while that of other fowl such as the duck or goose is much darker?

10-6. Microtubule Structure: The proteins making up the outer doublets have been isolated from the cilia of the protozoan *Tetrahymena*. These proteins were found to have the following properties:

Protein	Molecular Weight	ATPase Activity
A	55,000	No
B	600,000	Yes
C	Some small multiple of 600,000	Yes

When a solution of protein C was mixed with a suspension of isolated *Tetrahymena* outer doublets from which the arms had previously been removed, protein C bound to the doublets, and electron microscopy established that the arms of the outer doublet were thereby restored. Explain the likely nature and purpose of proteins A, B, and C in outer doublet structure. Suggest additional experiments to test your explanation.

‡10-7. Colchicine and Microtubular Aggregation: According to the *dynamic equilibrium theory* of microtubule assembly, the polymerized tubulin of the microtubules exists in equilibrium with a pool of monomeric subunits. *Colchicine* is a plant alkaloid that disrupts microtubular structure by binding to the tubulin subunit and solubilizing microtubules.

(a) Why is colchicine an effective antimitotic agent?

(b) Suggest a possible mode of action for colchicine in light of the dynamic equilibrium theory.

‡10-8. Tetanus: Single nerve impulses to striated muscle cause separate twitches, but when a string of impulses arrives in rapid succession, it results in a summation of twitches into a steady contraction. This leads to *tetanus*, a condition of rigidity with the muscles temporarily "locked" in the fully contracted state. So defined, tetanus is a normal physiological phenomenon, since this kind of extreme contraction occurs naturally under conditions of intense muscular effort and is followed by relaxation as soon as the nerve stimulation stops. However, the same term is used in a pathological sense to describe a disease caused by the bacterium *Clostridium tetani*, a spore-producing anaerobe. *C. tetani* is normally a nonpathogenic inhabitant of the intestinal tract of animals. But if it gains entry into tissue (usually through a wound), it can produce a powerful exotoxin (a poison liberated from the bacterial cell) capable of inducing an abnormal state of extreme muscle contraction.

(a) Explain the physiological phenomenon of tetanus in terms of stimulation of muscle contraction by nerve impulses.

(b) What is a likely explanation for the ability of *C. tetani* to produce a pathological tetanus condition?

(c) Why is vaccination with *C. tetani* (a "tetanus shot") especially recommended for surface wounds involving skin puncture?

References

Bendall, J. R., *Muscles, Molecules, and Movement: An Essay on the Contraction of Muscles* (New York: American Elsevier, 1968).

Berg, H. C., "How Bacteria Swim," *Sci. Amer.* (Aug. 1975), *233*:36.

Carlson, F. D., and D. R. Wilkie, *Muscle Physiology* (Englewood Cliffs, N.J.: Prentice-Hall, 1973).

Cohen, C., "The Protein Switch of Muscle Contraction," *Sci. Amer.* (Nov. 1975), *233*:36.

Ebashi, S., M. Endo, and I. Ohtsuki, "Control of Muscle Contraction," *Quart. Rev. Biophysics* (1969), *2*:351.

Huxley, H. E., "The Structural Basis of Muscular Contraction," *Proc. Royal Soc.* (1971), *178*:131.

Margaria, R., "The Sources of Muscular Energy," *Sci. Amer.* (Mar. 1972), *226*:84.

Merton, P. A., "How We Control the Contraction of Our Muscles," *Sci. Amer.* (Apr. 1972), *226*:30.

Mommaerts, W. F. H. M., "Energetics of Muscle Contraction," *Physiol. Rev.* (1969), *49*:427.

Satir, P., "How Cilia Move," *Sci. Amer.* (Oct. 1974), *231*:44.

Appendixes

Appendix A:
Useful Symbols, Equations, and Definitions

Specific formulas and definitions are often difficult to locate just when they are needed, so some of the most useful information of that sort is presented here.

Useful Symbols and Abbreviations

A = Helmholtz (extractable) energy

ADP = adenosine diphosphate (see Figure 4-1)

AMP = adenosine monophosphate (see Figure 4-1)

ATP = adenosine triphosphate (see Figure 4-1)

c = velocity of light = 2.997×10^{10} cm/sec

cal = calorie = heat required to raise the temperature of 1 g of water 1°C

CoA—SH = coenzyme A (with free sulfhydryl group; see Figure 5-1)

CoQ = coenzyme Q (see Figure 5-5)

E = internal energy

E^0 = standard internal energy

E_0 = standard reduction potential

F = faraday = 23,040 cal/volt = 96,406 joules/volt

FAD = flavin adenine dinucleotide (see Figure 5-3)

g = gram

G = (Gibbs) free energy

G^0 = standard free energy

h = Planck's constant = 1.57×10^{-34} cal/sec = 6.62×10^{-27} ergs/sec

$[H^+]$ = hydrogen-ion concentration

k = rate constant

K = equilibrium constant

K_m = Michaelis constant (concentration of substrate required for half maximum velocity)

kcal = kilocalorie = 1000 calories ($=1$ Calorie, the unit commonly used to express the energy value of foodstuffs)

$\ln x$ = natural logarithm of x (to the base e) = $2.303 \log_{10} x$

n = number of electrons transferred; number of subunits involved in polymer formation

N = Avogadro's number = 6.023×10^{23} molecules/mole

NAD = nicotinamide adenine dinucleotide (see Figure 4-5)

pH = $-\log[H^+]$

P = pressure

P_i = inorganic phosphate (ionized form of phosphoric acid, H_3PO_4)

PP_i = inorganic pyrophosphate (ionized form of pyrophosphoric acid, $H_4P_2O_7$)

PC = plastocyanin, an electron carrier in photosynthesis (see Figure 7-6)

PQ = plastoquinone, an electron carrier in photosynthesis (see Figure 7-6)

Q = ratio of product concentrations to reactant concentrations

R = gas constant = 1.987 cal/mole-degree

S = entropy

S^0 = standard entropy

$[S]$ = initial substrate concentration

T = temperature (usually in °K: °K = °C + 273.16)

v = velocity of a reaction or process

V = volume

V_{max} = maximum velocity of an enzyme-catalyzed reaction when substrate is saturating

$[X]$ = concentration of species X

Δ = change in

λ = wavelength

milli $= 10^{-3}$ micro $= 10^{-6}$ nano $= 10^{-9}$ pico $= 10^{-12}$ **Useful Hints**

$1\ \mu$ (micron) $= 10^4$ Å $= 10^3$ nm $= 10^{-3}$ mm $= 10^{-4}$ cm $= 10^{-6}$ m

Area of a circle $= \pi r^2$ **Useful Formulas**
and Definitions
Volume of a sphere $= \frac{4}{3}\pi r^3$

Volume of a cylinder $= \pi r^2 h$

$^\circ$K $= {}^\circ$C $+ 273.16$

Density $=$ weight per unit volume

1 dalton $=$ 1 atomic weight unit $=$ weight of hydrogen atom $= 1.67 \times 10^{-24}$ g

1 einstein $=$ 1 mole of photons $= 6.023 \times 10^{23}$ photons

Energy of 1 einstein (mole) of photons $= Nhc/\lambda$

$(10^x)(10^y) = 10^{(x+y)}$ **Useful**
Mathematical
$(10^x)^y = 10^{xy}$ **Relationships**

If $x = a^y$, then $\log_a x = y$

$\log xy = \log x + \log y$

$\log x/y = \log x - \log y$

$\log x^a = a \log x$

$\log_b x = (\ln a/\ln b) \log_a x$

$\ln x = 2.303 \log_{10} x$

Appendix B:
Answers to Problem Sets

Chapter 1

1-1. (a) 1.94 cal/cm^2-min $\times$ 5.26 $\times$ 10^5 min/year $\times$ 1.28 $\times$ 10^{18} cm^2 = 1.3 $\times$ 10^{24} cal/year.

   (b) It is absorbed by components of the earth's atmosphere. (Atmospheric ozone plays an important role in filtering out ultraviolet radiation, while water vapor is responsible for most of the absorption in the infrared range.)

   (c) Much of the radiation falls on areas of the earth's surface where the climate is not favorable (too hot, too cold, too dry) for growth of phototrophic organisms during at least part of the year. In addition, about two-thirds of the earth's surface is covered by oceans which, though quantitatively significant in global photosynthesis, have in general only a very low density of phototrophic organisms and therefore a low efficiency of photosynthetic light utilization. Moreover incident radiation represents a spectrum of wavelengths that can be used with varying degrees of efficiency by the photosynthetic pigments, as is discussed further in Chapter 7.

1-2. (a) Glucose = 40 percent carbon (72/180 = 0.4), so 5 $\times$ 10^{16} g carbon = 12.5 $\times$ 10^{16} g organic matter. (That is about 140 billion tons, if nonmetric units help you imagine the magnitude of the process.)

   (b) 12.5 $\times$ 10^{16} g $\times$ 3.8 kcal/g = 4.75 $\times$ 10^{17} kcal (or 4.75 $\times$ 10^{20} cal).

   (c) 4.75 $\times$ 10^{20} cal/1.3 $\times$ 10^{24} cal = 3.7 $\times$ 10^{-4} = 0.037 percent. (This means that more total solar energy will be received by the earth during the next 12 months than has been trapped photosynthetically since 700 BC!)

   (d) Virtually all of it is consumed by chemotrophs, since the earth is

not undergoing any appreciable annual increase in amount of autotrophic organic matter.

1-3.   (a) Autotroph, phototroph.
      (b) Autotroph, chemotroph.
      (c) Heterotroph, chemotroph.
      (d) Heterotroph, chemotroph.

1-4.   Respiration requires a supply of oxygen (O_2) and oxidizable organic
molecules. Since oxidizable organic molecules can be produced in quantity only by autotrophs, it is generally concluded that the apparatus for
oxidative breakdown of foodstuffs (respiration) is of more recent evolutionary origin than the capability for photosynthesis. Primitive autotrophs
are thought to have evolved from anaerobic heterotrophs that would have
passed unceremoniously out of existence after consuming abiotically
generated organic molecules had they not developed the ability to trap
light energy from the sun and use it in the synthesis of organic compounds from CO_2 and H_2O.

Chapter 2

2-1.   (a) Freezing of water involves a greater ordering of the system and
hence a decrease in the randomness of components (water molecules). Invalid as a measure of spontaneousness; negative under
all conditions.
      (b)' Valid measure; thermodynamically spontaneous reactions always
result in an increase in randomness or entropy of the universe.
      (c) The freezing of water always results in evolution of heat (just the
opposite of melting, which obviously requires heat). Not a valid
criterion; negative under all conditions.

      (d)

$Temp.$ ($°C$)	ΔG_{system} (cal/mole)	$(= \Delta E_{system} - T \Delta S_{system})$
+ 0.1	+ 0.5	
0.0	0.0	
− 0.1	− 0.5	

      (e) Just as valid as ΔS_{total}; all thermodynamically spontaneous reactions
have $\Delta G < 0$.

2-2.   (a) F; delete "the rate at which the reaction proceeds."
      (b) F; ΔG cannot be positive *under any conditions* if reaction is spon-
taneous.
      (c) F; $\Delta G = G_{products} - G_{reactants}$
      (d) T; same magnitude, opposite sign.
      (e) F; can occur spontaneously if concentrations are such that ΔG
(not ΔG^0) is negative.

2-3.   (a) $\Delta G = -750$ cal/mole.

(b) $\quad \Delta G = \Delta G^0 + 2.303RT \log_{10} \dfrac{[\text{malate}]}{[\text{fumarate}]}$

$$0 = -750 + (2.303)(1.987)(298) \log_{10} \dfrac{[\text{malate}]}{1 \times 10^{-3}\ M}$$

$$\log_{10} \dfrac{[\text{malate}]}{10^{-3}} = \dfrac{750}{1364} = 0.55$$

$$\dfrac{[\text{malate}]}{10^{-3}} = \text{antilog}\ (0.55) = 3.5$$

$$[\text{Malate}] = 3.5 \times 10^{-3}\ M$$

At still higher concentrations, ΔG would become positive and the reaction would go to the left instead.

2-4.　(a)　Endergonic under standard conditions.

　　　(b)　Yes because the reaction will proceed to the right until an equilibrium ratio of glucose-6-phosphate and glucose-1-phosphate is attained.

$$\log_{10} K^0 = \dfrac{-1800}{(2.303)(1.987)(298)} = -1.32$$

$$K^0 = \text{antilog}\ (-1.32) = 0.048$$

$$[\text{glucose-1-phosphate}]_{\text{equil}} \simeq 0.0046\ M$$
$$[\text{glucose-6-phosphate}]_{\text{equil}} \simeq 0.095\ M$$

　　　(c)　If glucose-6-phosphate is supplied constantly and glucose-1-phosphate is removed constantly by other reactions so that a high glucose-6-phosphate/glucose-1-phosphate ratio is maintained, then the ΔG value will be negative and the reaction can proceed to the right.

2-5.　(a)　Reaction 2-24: $\Delta G^0 = -5.4$ kcal/mole; $\Delta G = +2.2$ kcal/mole. Reaction goes to the right at standard conditions, to the left in the red cell. (Note that calculations require expression of concentration as M, not μM.) Reaction 2-25: $\Delta G^0 = -1.9$ kcal/mole; $\Delta G = -1.5$ kcal/mole. Reaction goes to the right at standard conditions and in the red cell.

　　　(b)　Reaction 2-26: $K^0 = 2.5 \times 10^5$; $\Delta G^0 = -7.3$ kcal/mole; $\Delta G = +0.6$ kcal/mole. Reaction 2-27: $K^0 = 400$; $\Delta G^0 = -3.5$ kcal/mole; $\Delta G = +3.7$ kcal/mole. Neither reaction will proceed spontaneously to the right in the red cell.

　　　(c)　FDP $\rightarrow$ 2DHAP is feasible in the red cell since both steps have a negative ΔG. FDP $\rightarrow$ 2G3P is not feasible even though it has a negative overall ΔG since ΔG for one of the steps is positive under the concentration conditions that prevail in the red cell.

2-6.　No. The relationship between Helmholtz (extractable) energy A and Gibbs free energy G involves a pressure-volume term

$$G = A + PV$$

and the full expressions for E and ΔE in terms of G and ΔG become

$$E = A + TS = G - PV + TS$$
$$\Delta E = \Delta G - \Delta(PV) + \Delta(TS)$$

The only conditions for which Equation 2-28 is valid are those for which pressure, volume, and temperature are constant during a process such that $\Delta(PV) = 0$ and $\Delta(TS)$ becomes $T\Delta S$.

2-7. For the reaction sequence

$$A \rightarrow B \rightarrow C \rightarrow D$$

ΔG^0 for the overall reaction can be expressed as follows, where K^0 is the ratio of product concentration to reactant concentration at equilibrium under standard conditions of temperature, pressure, and pH:

$$\Delta G^0_{A \rightarrow D} = -2.303RT \log_{10} \left[\frac{(D)}{(A)} \right]_{\text{equil}}$$

This equation can be expanded to show the additive property of ΔG^0 as follows:

$$\Delta G^0_{A \rightarrow D} = -2.303RT \log_{10} \left[\frac{(B)(C)(D)}{(A)(B)(C)} \right]_{\text{equil}}$$

$$= -2.303RT \log_{10} \left[\frac{(B)}{(A)} \right]_{\text{equil}} - 2.303RT \log_{10} \left[\frac{(C)}{(B)} \right]_{\text{equil}}$$

$$- 2.303RT \log_{10} \left[\frac{(D)}{(C)} \right]_{\text{equil}}$$

$$= \Delta G^0_{A \rightarrow B} + \Delta G^0_{B \rightarrow C} + \Delta G^0_{C \rightarrow D}$$

2-8. (a) $\Delta G^0 = 3300 \text{ cal/mole} = -2.303RT \log_{10} K^0$,
       so $\log_{10} K^0 = -3300/1420 = -2.32$ and $K^0 = 4 \times 10^{-3}$.

$$K^0 = \frac{[\text{glucose-6-phosphate}]}{[\text{glucose}][\text{phosphate}]} = 4 \times 10^{-3}$$

$$= \frac{[\text{glucose-6-phosphate}]}{(1 \times 10^{-4})(1 \times 10^{-4})} = 4 \times 10^{-3}$$

$$= [\text{glucose-6-phosphate}] = 4 \times 10^{-11} M = 4 \times 10^{-5} \mu M$$

Most emphatically no!

(b) $K^0 = (0.25 \times 10^{-4})/(1 \times 10^{-4}) [\text{glucose}] = 4 \times 10^{-3}$
     $[\text{glucose}] = 0.25/(4 \times 10^{-3}) = 0.0625 \times 10^3 = 62.5 M$. Not so good either; a *saturated* glucose solution barely reaches 3 M!

(c) $\Delta G^0 = +3.3 \text{ kcal/mole} - 7.3 \text{ kcal/mole} = -4.0 \text{ kcal/mole} = -4000 \text{ cal/mole}$.

$$\log_{10} K^0 = \frac{4000}{RT} = \frac{4000}{1420} = 2.82$$

$$K^0 = \text{antilog } (2.82) = 660$$

$$\frac{[\text{glucose-6-phosphate}][\text{ADP}]}{[\text{glucose}][\text{ATP}]} = 660 \quad \text{or} \quad \frac{25 \times 10^{-6}}{[\text{glucose}](10)} = 660$$

Thus $[\text{glucose}] = 25 \times 10^{-6}/6600 = 3.8 \times 10^{-9} \, M = 3.8 \times 10^{-3} \, \mu M$!

(d) This would require very high local concentrations of phosphate that might have adverse effects on other reactions involving phosphate and would create problems due to high concentration of negative charge.

(e) There is no need for localized elevation in phosphate.

Chapter 3
Question Set

3-1. (a) All molecules of a given enzyme are identical.

(b) Enzymes do not use up or consume substrate except by converting it to product, which is eventually released from the enzyme.

(c) Enzymes do no work on the substrate; they just speed what would occur spontaneously anyway from a thermodynamic point of view.

(d) Most biological reactions are reversible, and the same enzyme catalyzes a reaction in both directions.

3-2. Because concentration measures amount per unit volume and the "concentration" of peanuts is expressed as amount per unit area.

3-3. It should be negligible compared to the time required to shell peanuts.

3-4.

$S =$

Concentration of Peanuts $(peanuts/ft^2)$	Time Required per Peanut			Rate of Shelling	
	To Find $(sec/peanut)$	To Shell $(sec/peanut)$	Total $(sec/peanut)$	Per Monkey $(peanut/sec)$	Total (v) $(peanut/sec)$
1	9	1	10	0.10	1.0
3	3	1	4	0.25	2.5
9	1	1	2	0.50	5.0
18	0.5	1	1.5	0.67	6.7
90	0.1	1	1.1	0.91	9.1
∞	0	1	1	1.00	10.0

The graph of v versus S will have the shape of a hyperbola, with v approaching a finite upper limit (10 peanuts/sec) as S approaches infinity.

3-5. Let t_f represent the time required for a monkey to find a peanut and let t_s be the time required to shell. Note that t_f is inversely proportional to the peanut "concentration" S, while t_s is a constant. Thus, for the general case, we can write

$$t_f = \frac{k_i}{S} \quad \text{and} \quad t_s = k_{ii}$$

where k_i and k_{ii} are constants (equal, respectively, to 9 and to 1 for the particular monkeys and peanuts used in the example). The total time t_t required for one monkey to process one peanut is therefore

$$t_t = t_f + t_s = \frac{k_i}{S} + k_{ii}$$

Since t_t has the units of seconds per peanut, its reciprocal will obviously have the units of peanuts per second, which is the rate of shelling per monkey:

$$\text{Rate (per monkey)} = \frac{1}{k_i/S + k_{ii}}$$

And if this rate per monkey is now multiplied by the total number of monkeys, n, we arrive at a general expression for the overall velocity v:

$$v = \text{rate for } n \text{ monkeys} = \frac{n}{k_i/S + k_{ii}}$$

Multiplying through by S/k_{ii} will convert this into the following form:

$$v = \frac{(n/k_{ii})S}{(k_i/k_{ii}) + S}$$

And if we now define a and b as constants such that

$$a = \frac{n}{k_{ii}} \quad \text{and} \quad b = \frac{k_i}{k_{ii}}$$

we can write the equation above as

$$v = \frac{aS}{b + S}$$

which has the desired form.

3-6. As S gets larger and larger, the time spent finding a peanut gets smaller and smaller. At infinity, with a vast excess of peanuts, the monkeys would be able to find a peanut in "no time at all"; all the monkeys are at all times busy shelling peanuts, which is the step that limits the overall velocity. The constant a is the upper limit to which the velocity of peanut shelling goes as the peanut concentration goes to infinity.

3-7. First-order kinetics means that there is a linear relationship between rate of shelling and peanut concentrations. This is so only at low concentrations, where rate is limited mainly by the time it takes to find a peanut, because only the finding time is affected by changes in peanut concentrations. Zero-order kinetics means that the rate of shelling is insensitive to changes in peanut concentration. This occurs only at high concentrations, where the overall rate is limited mainly by the time it takes to shell a peanut once found, because this shelling time is not affected by changes in concentration.

3-8. V_{max} is the constant a of the derivation in Question 5. In that derivation, a was defined such that $a = n/k_{ii}$ (or $a = n/t_s$ from the definition of k_{ii}).

Thus V_{max} is inversely proportional to shelling time (or reaction time) and directly proportional to the number of monkeys (or enzyme molecules) engaged.

3-9. $V_{max} = 10$ peanuts/sec; $K_m = 9$ peanuts/ft². Not likely at all, since most reactions are carried out at substrate concentrations that are not saturating, and v can only be approximated.

3-10. Because the graph is nonlinear; V_{max} is attained only at infinite substrate concentration, and it is not possible to estimate its asymptotic value accurately just by looking at the low-concentration part of the curve. Similarly, without knowing V_{max} you cannot find the point $V_{max}/2$ and thus cannot determine K_m.

3-11. Inversion of both sides of the Michaelis-Menten equation yields

$$\frac{1}{v} = \frac{K_m + S}{V_{max} S} = \frac{K_m}{V_{max} S} + \frac{S}{V_{max} S} = \frac{K_m}{V_{max}} \left[\frac{1}{S} \right] + \frac{1}{V_{max}}$$

To find the y intercept, set $x = 0$. If $1/S$ is zero, the expression becomes

$$\frac{1}{v} = \frac{1}{V_{max}}$$

so the y intercept is $1/V_{max}$. To find the x intercept, set $y = 0$. If $1/v$ is zero, the expression becomes

$$0 = \frac{K_m}{V_{max}} \left[\frac{1}{S} \right] + \frac{1}{V_{max}}$$

or, upon moving one term across the equal sign and multiplying through by V_{max}/K_m,

$$\frac{1}{S} = -\frac{1}{K_m}$$

3-12.

From Question 4		Reciprocal Values	
S	v	$1/S$	$1/v$
1	1.0	1.000	1.00
3	2.5	0.333	0.40
9	5.0	0.111	0.20
18	6.7	0.056	0.15
90	9.1	0.011	0.11
∞	10.0	0.000	0.10

The plot of $1/v$ (on the y axis; scale: 0 to 1) versus $1/S$ (on the x axis; same scale) should be a straight line intersecting the y axis at $+0.1$ and the x axis at -0.111. From the intercepts, K_m should be $-1/-0.111 = +9$ peanuts/ft² and V_{max} should be $1/0.1 = 10$ peanuts/sec. These values agree with the slope of $+0.9$ sec/ft².

3-13. $1/S$ approaches zero as S approaches infinity. Thus if V_{max} is the value that v approaches as S approaches infinity, its reciprocal must be the value that $1/v$ approaches as $1/S$ goes to zero.

Problem Set

3-1. (a) Thermal activation imparts kinetic energy to molecules, thereby increasing the proportion of molecules that possess adequate energy to collide and react. Once initiated, the reaction is self-sustaining in the sense that reacting molecules release enough energy to energize and activate neighboring molecules for reaction.

(b) Life is basically isothermal; living cells could not tolerate the elevated temperatures required for thermal activation.

(c) Reaction of molecules is facilitated (by positioning on the catalyst surface, for example), thereby requiring less energy to activate and so ensuring that a substantially larger proportion of the molecules possess adequate energy to initiate reaction.

(d) Advantages: specificity, more exacting control. Disadvantages: much more susceptible to inactivation by heat, extremes of pH, and other factors affecting protein stability.

3-2. (a) See Figure B-1. The increase in rate is always proportionately less than the increase in concentration because the curve is hyperbolic rather than linear; it is constantly "bending over" further with increasing S.

(b)

1/Glucose (liters/μmole)	1/Rate (min/unit)
0.100	100.0
0.050	58.8
0.025	37.0
0.020	33.3
0.010	25.0

See Figure B-2 for the double-reciprocal plot.

(c) x intercept $= -0.02$; $K_m = -1/(-0.02) = 50$ μM.

(d) y intercept $= +16.7$; $V_{max} = 1/16.7 = 0.060$ unit/min.

(e) V_{max} is the limiting value to which the rate tends as more and more glucose is added and represents the highest possible velocity (with this amount of enzyme) when the enzyme is completely saturated. K_m is the concentration of substrate that causes the reaction to run at one-half the maximum velocity. From the original data, once we know $V_{max} = 0.060$ unit/min we can see readily that $K_m = 50$, since that is the concentration which gives a velocity of 0.030 unit/min.

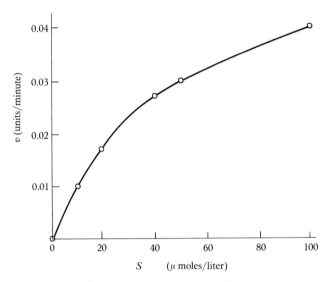

Figure B-1   The plot of v versus S for Problem 3-2.

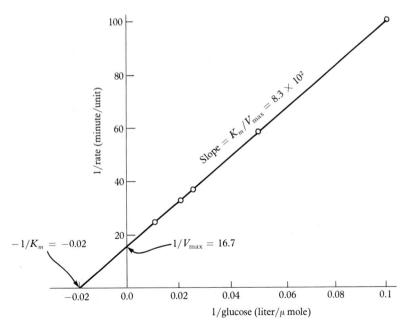

Figure B-2   The double reciprocal plot for Problem 3-2.

3-3.   (a)   Normal enzyme: $V_{\max} = 10$ μmoles/min; $K_m = 1 \times 10^{-5}$ M. Mutant enzyme: $V_{\max} = 10$ μmoles/min; $K_m = 2 \times 10^{-4}$ M.

    (b)   For normal enzyme: $v = 5/(5 + 10)V_{\max} = 33$ percent of maximum velocity. For mutant enzyme: $v = 5/(5 + 200)V_{\max} = 2.4$ percent of maximum velocity. Perhaps impaired glycolysis results from the reduced rate at which the phosphoglyceromutase reaction proceeds under the existing cellular substrate concentration due to the higher K_m of the mutant enzyme.

3-4.   (a) $1/v = 1/V_{\max} + (K_m/V_{\max})(1/S)$. Multiply this by $V_{\max}v$:

$$V_{\max} = v + vK_m\left(\frac{1}{S}\right) = v + K_m\left(\frac{v}{S}\right)$$

$$v = V_{\max} - K_m\left(\frac{v}{S}\right)$$

    (b)   Plot v versus v/S.

    (c)   Intercept $= V_{\max}$; slope $= -K_m$.

3-5.

$$v_b = \frac{0.2V_{\max}}{0.2 + K_m} = 1.5v_a = 1.5\left[\frac{0.1V_{\max}}{0.1 + K_m}\right]$$

$$(0.2V_{\max})(0.1 + K_m) = (0.03 + 0.15K_m)(V_{\max})$$
$$(0.2)(0.1 + K_m) = 0.03 + 0.15K_m$$
$$0.02 + 0.2K_m = 0.03 + 0.15K_m$$
$$0.05K_m = 0.01$$
$$K_m = 0.2 \text{ m}M.$$

3-6.   (a)   If E_1 is effectively regulated by the intracellular level of isoleucine, regulation of the rest of the sequence is automatic in the sense that shutting off E_1 will result in a rapid depletion of α-ketobutyrate and subsequent intermediates, such that enzymes E_2, E_3, E_4, and E_5 will soon be turned off by a lack of substrate. It is an economical means of regulation because only one of the five enzymes requires an allosteric recognition site.

    (b)   They are self-regulating in the sense that the end product of such a pathway turns off its own synthesis as its level rises in the cell and turns it back on again as it falls. Assuming that the cell requires a fairly constant level of a given end product (as it might require a constant level of isoleucine for protein synthesis, for example), the logic of a self-regulating supply is obvious.

    (c)   The enzyme exists in two forms (one active in deaminating threonine, the other inactive), and isoleucine apparently stabilizes the enzyme in the inactive form. How this is actually accomplished is shown diagrammatically in Figure B-3. Isoleucine binds reversibly to the effector site of the enzyme, stabilizing it in the inactive form. As cellular levels of isoleucine drop, the equilibrium shifts, the isoleucine-enzyme complex dissociates, and the enzyme is free to return to the active form.

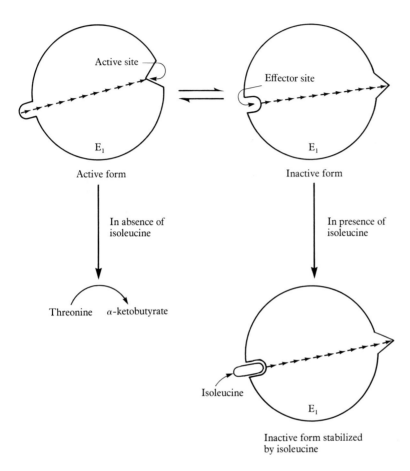

Figure B-3 Allosteric inhibition of threonine deaminase by isoleucine (see Problem 3-6).

(d) A mutation in the active site would be detectable because the cell could not deaminate threonine and therefore could not synthesize isoleucine (though it would be able to carry out the unregulated synthesis of isoleucine if supplied with α-ketobutyrate or any of the subsequent intermediates). A mutation in the effector site would lead to uncontrolled synthesis of isoleucine without regard to the cellular level of isoleucine.

3-7. The point of departure for the derivation is the simple enzyme-catalyzed reaction of Equation B-1 and the Michaelis-Menten model for its mechanism shown in Equation B-2:

$$S \xrightarrow{\text{enzyme}} P \qquad \text{(B-1)}$$

$$E_f + S \underset{k_2}{\overset{k_1}{\rightleftharpoons}} ES \underset{k_4}{\overset{k_3}{\rightleftharpoons}} P + E_f \qquad \text{(B-2)}$$

The velocity v can be expressed as the rate of disappearance of substrate or the rate of appearance of product:

$$v = \frac{-d(S)}{dt} = \frac{+d(P)}{dt} \tag{B-3}$$

To derive the dependence of v upon (S), we begin by writing equations expressing the rates of change in concentrations of S, ES, and P:

$$\frac{d(S)}{dt} = -k_1(E_f)(S) + k_2(ES) \tag{B-4}$$

$$\frac{d(P)}{dt} = k_3(ES) - k_4(P)(E_f) \tag{B-5}$$

$$\frac{d(ES)}{dt} = k_1(E_f)(S) - k_2(ES) - k_3(ES) + k_4(P)(E_f) \tag{B-6}$$

Since we are confining ourselves to the initial stages of the reaction where (P) is essentially zero, Equations B-5 and B-6 simplify to

$$\frac{d(P)}{dt} = k_3(ES) \tag{B-7}$$

$$\begin{aligned}\frac{d(ES)}{dt} &= k_1(E_f)(S) - k_2(ES) - k_3(ES) \\ &= k_1(E_f)(S) - (k_2 + k_3)(ES)\end{aligned} \tag{B-8}$$

To proceed we must now assume the *steady state* at which the enzyme-substrate complex is being broken down at the same rate at which it is being formed, such that the net rate of change in (ES) is zero:

$$\frac{d(ES)}{dt} = k_1(E_f)(S) - (k_2 + k_3)(ES) = 0 \tag{B-9}$$

This can be rewritten as

$$k_1(E_f)(S) = (k_2 + k_3)(ES) \tag{B-10}$$

Obviously the total amount of enzyme present, E_t, is simply the sum of the free form E_f plus the amount of complexed enzyme ES:

$$E_t = E_f + ES \tag{B-11}$$

Hence

$$E_f = E_t - ES \tag{B-12}$$

which, when substituted into Equation B-10, yields

$$k_1(E_t - ES)(S) = (k_2 + k_3)(ES) \tag{B-13}$$

or

$$k_1(E_t)(S) - k_1(ES)(S) = (k_2 + k_3)(ES) \tag{B-14}$$

Upon rearrangement this yields

$$k_1(E_t)(S) = k_1(ES)(S) + (k_2 + k_3)(ES) \tag{B-15}$$

$$= [k_1(S) + k_2 + k_3](ES) \tag{B-16}$$

which can be rewritten as

$$(ES) = \frac{k_1(E_t)(S)}{k_1(S) + k_2 + k_3} \tag{B-17}$$

This is useful since Equations B-3 and B-7 tell us that the velocity v is simply $k_3(ES)$:

$$v = k_3(ES) = \frac{k_1 k_3(E_t)(S)}{k_1(S) + k_2 + k_3} \tag{B-18}$$

$$= \frac{k_3(E_t)(S)}{(S) + (k_2 + k_3)/k_1} \tag{B-19}$$

For a given concentration of enzyme, E_t is a constant, so we can define two kinetic constants as follows:

$$V_{\max} = k_3(E_t) \tag{B-20}$$

$$K_m = \frac{k_2 + k_3}{k_1} \tag{B-21}$$

Rewritten in these terms, Equation B-19 becomes

$$\boxed{v = \frac{V_{\max}(S)}{(S) + K_m}} \tag{B-22}$$

Chapter 4

4-1. (a) T.
   (b) F; glucose is phosphorylated and isomerized to fructose; no oxidation is involved in the initial steps.
   (c) F; the ΔG^{0}'s are almost identical because the resonance stabilization possibilities are very similar (compare phosphate structure with pyrophosphate).
   (d) F; only one is—the oxidation of glyceraldehyde-3-phosphate. The other ATP-generating reaction is an enol-to-keto conversion.
   (e) T.

4-2. Pyruvate is the end product of glycolysis under aerobic conditions in almost all cells. Lactate is expected from most animal and bacterial cells under anaerobic conditions, and ethanol + CO_2 are the end products for some plant and microbial cells (such as yeast). The equations appear in the text. Under anaerobic conditions, $NAD/NADH_2$ does not appear in the net equation because there is no way for NAD to be regenerated from $NADH_2$ in the absence of aerobic electron transport. Aerobically there can be continued reduction of the coenzyme during glycolysis because $NADH_2$ can be reoxidized by O_2 via the electron transport chain.

4-3. No. The step that takes glyceraldehyde-3-phosphate to glycerate-1,3-diphosphate is crucial to the energy (ATP) yield of the overall glycolytic pathway because it couples the oxidation of the aldehyde to the uptake of inorganic phosphate into a high-energy anhydride bond that in the next step is used for the direct synthesis of ATP. If the oxidation of glyceraldehyde-3-phosphate goes directly to glycerate-3-phosphate, no ATP is generated. Since this sequence is repeated twice per glucose (once for each of the two 3-carbon units), the mutation has eliminated two ATPs from the glycolytic pathway such that the net yield is zero. The mutant thus carries out glycolysis with no net ATP synthesis, which would obviously be lethal to the anaerobic cell.

4-4. (a) Galactose + ATP → ADP + glucose-6-phosphate.

(b) Quite similar, since the net result in both cases is an ATP-linked phosphorylation of a six-carbon sugar, and thermodynamic values are independent of route.

(c)

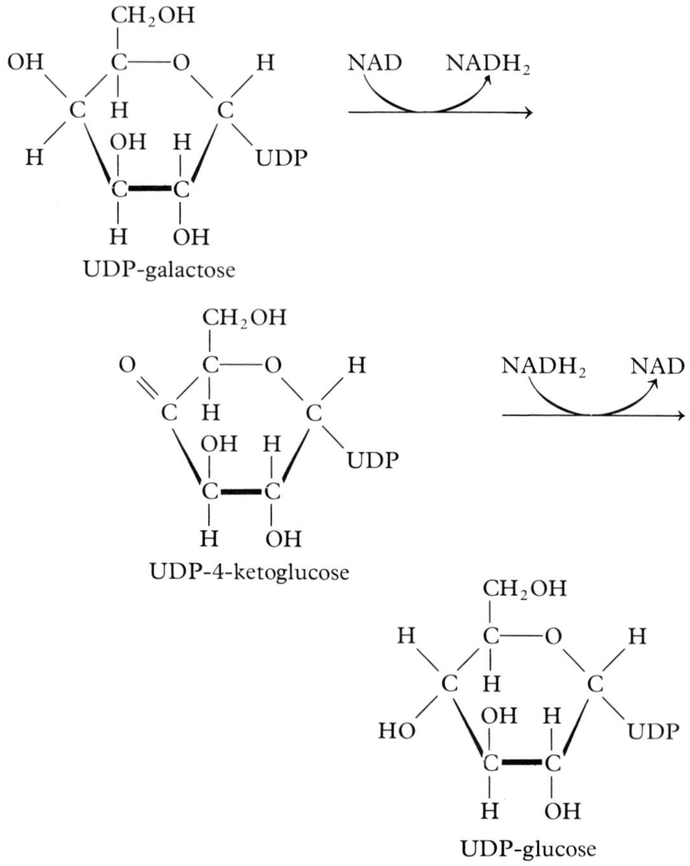

CH_2OH

UDP-galactose

$NAD \quad NADH_2$

CH_2OH

UDP-4-ketoglucose

$NADH_2 \quad NAD$

CH_2OH

UDP-glucose

(d) The inability to metabolize galactose might lead to galactose accumulation in the tissues and blood of an organism that continues to absorb and hydrolyze lactose to get the glucose it needs for energy metabolism. Apparently the elevated level of galactose in the blood is deleterious to brain and lens cells.

4-5. (a)

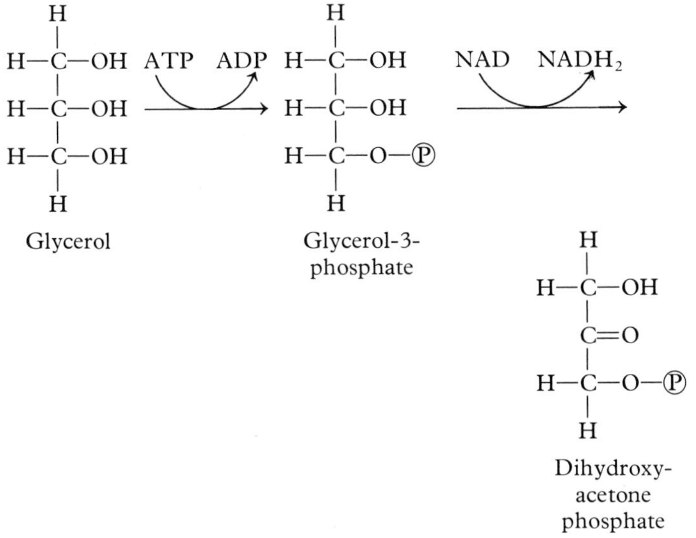

Glycerol

Glycerol-3-phosphate

Dihydroxy-acetone phosphate

(b) $C_3H_8O_3 + 2NAD + 1ADP + 1P_i \rightarrow C_3H_4O_3 + 2NADH_2 + 1ATP$

Glycerol   Pyruvate

4-6. (a) Number 3; (b) number 3; (c) number 2; (d) number 2; (e) number 1; (f) number 3 and/or number 4.

4-7. If arsenate is used instead of phosphate and the resulting intermediate loses the arsenate spontaneously without conserving the energy of the unstable (high-energy) arsenate linkage, the result is exactly the same as in Problem 4-3. The energy yield under anaerobic conditions would be zero, so the arsenate would be fatal.

Chapter 5

5-1. *Glycolysis:* In cytoplasm; to split the 6-carbon starting compound glucose into two 3-carbon units that are oxidized to two pyruvate molecules, with synthesis of two ATPs and reduction of two NADs to $NADH_2$ per glucose. The pyruvate may be reduced to lactate or ethanol + CO_2 under anaerobic conditions or may be passed into the mitochondrion for further oxidative metabolism via the TCA cycle:

$C_6H_{12}O_6 + 2NAD + 2ADP + 2P_i \rightarrow 2C_3H_4O_3 + 2NADH_2 + 2ATP$

TCA (Krebs) Cycle: In matrix of mitochondrion; to oxidize pyruvate (via acetate as acetyl CoA) completely to CO_2, with generation of one high-energy phosphate bond per turn of the cycle (as GTP but convertible to ATP) and transfer of all hydrogen atoms to coenzymes (FAD or NAD):

$$2C_3H_4O_3 + 2ADP + 2P_i + 2FAD + 8NAD + 6H_2O \rightarrow$$
$$6CO_2 + 2ATP + 2FADH_2 + 8NADH_2$$

This summarizes both the gateway step and the TCA cycle itself; it could be written per glucose as above or per pyruvate by dividing all the above coefficients by 2.

Electron Transport (Oxidative Phosphorylation): On inner membrane (and hence cristae) of mitochondrion; to transfer electrons stepwise from the high-energy reduced coenzymes $NADH_2$ and $FADH_2$ through a chain of cytochrome proteins eventually to the ultimate electron acceptor, O_2, coupling the transport of electrons to the phosphorylation of ADP to ATP at three steps where the transport is sufficiently exergonic to drive ATP synthesis:

Per glucose: $10NADH_2 + 2FADH_2 + 6O_2 + 34ADP + 34P_i \rightarrow$
$$10NAD + 2FAD + 12H_2O + 34ATP$$

5-2.   (a) Number 6; (b) number 3; (c) number 3; (d) number 2 (methyl carbon).

(e) $^*CH_2{-}COOH$ (f) $^*CH_2{-}COOH$

         $CH{-}COOH$ $CH_2{-}COOH$

$HO{-}CH{-}COOH$

5-3.

Compound	ATP Equivalents	Free Energy (kcal/mole)	Molecular Weight (g/mole)	Amount per kcal (g/kcal)	Cost ($/g)	($/kcal)
ATP	1	7.3	507	69.5	5.90	410.05
$NADH_2$	3	21.9	663	30.3	25.00	757.50
Acetyl CoA	12	87.6	809	9.2	1910.00	17,572.00
Pyruvate	15	109.5	88	0.80	0.39	0.31
Glucose	38	277.4	180	0.65	0.002	0.0013

Carbohydrates pack much energy into a gram and are therefore efficient storage compounds.

5-4.   (a)   NAD-coupled oxidations involve the removal of one electron from a carbon atom and one from an adjacent oxygen (of an alcohol, aldehyde, or keto group). In FAD-coupled oxidations, both electrons come from carbon atoms, creating a carbon-carbon double bond.

(b)   The oxidation of C—C to C=C is apparently just barely exergonic enough to drive the reduction of FAD, since the ΔG^0 for the coupled reaction is about zero. Oxidations involving NAD must be thermodynamically much more exergonic, since $NADH_2$ is sufficiently more energy-rich than $FADH_2$ to allow the generation of an ATP in the electron transport chain when it is reoxidized by transfer of electrons to FMN. Any attempt to couple succinate oxidation to NAD reduction would be endergonic by at least 10 kcal/mole, and there is no way that changes in product or substrate concentrations could overcome odds like that.

5-5. (a) If electron transport from FMN to CoQ is blocked, reduced coenzymes cannot be reoxidized, food molecules cannot be oxidized, and no ATP can be synthesized except by glycolysis, which apparently cannot meet energy needs of insects or fish adequately (because, in fact, higher organisms have no mechanism to excrete lactate directly).

(b) Rotenone might be expected to inhibit electron transport in other animals as well, because all aerobic cells use very similar electron transport systems, with FMN and CoQ as common features. Therefore rotenone is potentially poisonous to all animals—unless they cannot absorb the compound (at the level of the intestinal tract, the individual cell, or the mitochondrion) or have mechanisms to detoxify it.

(c) Plants, like all other eukaryotes, possess mitochondria with an electron transport chain similar to that in animals. (Plants are, after all, chemotrophs all night every night.) Therefore one might expect rotenone to poison electron transport in plants as well—unless they cannot take up the compound or have a means of detoxifying it.

5-6. (a) Total ATP yield: 12 (one directly, via GTP in Equation TCA-5a; three each from the $NADH_2$ molecules of Equations TCA-4, TCA-8, and 5-21; and two from the $FADH_2$ of Equation TCA-6).

(b) $C_5H_9O_4N + 2O_2 + 12ADP + 12P_i \rightarrow$
   Glutamate
$$C_4H_4O_5 + CO_2 + NH_3 + 12ATP + H_2O$$
   Oxaloacetate

5-7. (a) It is not the initial enzyme in the pathway as we normally represent glycolysis and, more important for the present discussion, it does not obey the general tenet that the allosteric effector should be a substance chemically distinct from both products and reactants.

(b) Because the reaction, though an ATP-consuming step, is actually part of an ATP-generating pathway. ATP is the important end product that can be expected to serve as a feedback inhibitor; it is, in a sense, simply coincidental that the enzyme it inhibits allosterically happens to catalyze one of the "pump-priming" steps of glycolysis that actually consumes ATP.

(c) Apparently the effector site and the active site differ in their affinity for ATP. The active site must have a high affinity, while that of the effector site must be low. At a low ATP level, therefore, binding can occur at the catalytic site but not at the effector site, so the enzyme remains in the active form and functional. At high ATP levels, binding also occurs at the effector site, converting the enzyme to the inactive form and turning off glycolysis.

(d) A given molecule is either turned on or off, but the total enzymatic activity will depend upon the ratio of molecules in the active and inactive forms at a given moment. And this ratio in turn results from the equilibrium between the two forms, which is determined by the level of effector.

5-8. (a) Because the effect of uncoupling is to remove the normal control on electron transport, which will then be free to proceed, in the uncoupled state, at a rapid, uncontrolled rate, oxidizing substrate and transferring the electrons rapidly to oxygen.

(b) Lack of transport-driven ATP synthesis will result in low levels of ATP and high levels of ADP, which in turn will activate both phosphofructokinase and isocitrate dehydrogenase, thereby increasing the rates of both the glycolytic pathway and the TCA cycle.

(c) The free energy released in electron transport is not conserved as ATP but is quantitatively lost as heat, which will tend to raise the temperature of the organism and, in some species, initiate a sweating response.

(d) Dinitrophenol may act to hydrolyze the high-energy phosphate bond of $X \sim P$ before the phosphate group can be transferred to ATP.

(e) Since the chemiosmotic theory of coupling presumes a proton gradient across the membrane, a compound that can facilitate the movement of protons across the membrane unaccompanied by ATP synthesis will tend to "discharge" the proton gradient, leading to the transport of electrons without energy conservation.

(f) Since dinitrophenol results in an enhanced rate of cellular metabolism, it will cause the body tissues to draw upon and utilize food reserves (rather like exercising without having to move). Its use was discontinued after a death had occurred because a "sublethal" dose was not.

5-9. (a) Further oxidation of the $-OH$ group is impossible until the tertiary alcohol (citrate) is converted to a secondary alcohol (isocitrate).

(b) Citrate to aconitate:
$$K^0 = 0.04/0.90 = 0.044$$
$$\Delta G^0 = -(1.987)(298)(2.303)\log_{10}(0.044)$$
$$= -(1364)(0.644 - 2.0)$$
$$= +1.85 \text{ kcal/mole}$$

Aconitate to isocitrate:
$$K^0 = 0.06/0.04 = 1.50$$
$$\Delta G^0 = -1364\log_{10}(1.5)$$
$$= -0.24 \text{ kcal/mole}$$

Citrate to isocitrate:
$$K^0 = 0.044 \times 1.50 = 0.066$$
$$\Delta G^0 = +1.85 - 0.24 = +1.61 \text{ kcal/mole}$$

(c)   No; the conversion of citrate to isocitrate is endergonic under standard conditions; a reaction sequence will not occur unless every step in the sequence is exergonic under the prevailing conditions.

(d)   The conditions in the cell are such that the ΔG is actually negative; this is because citrate is being continuously consumed such that exergonic reactions both before and after the aconitase reaction change concentrations sufficiently to render the aconitase reaction exergonic as well, despite the high positive ΔG^0.

Chapter 6

6-1.   (a)   F; protein catabolism requires many more enzymes because of the diversity of amino acids that must be degraded.

(b)   T.

(c)   F; only $n-1$ cycles are required because the product resulting from the last cycle is itself acetyl CoA.

(d)   T.

(e)   T; the product of transamination is oxaloacetate.

6-2.   (a)   Thirty-four ATP molecules per molecule of palmitate (seven cycles of β-oxidation minus one ATP for initial activation). Note that to split a 16-carbon chain into eight 2-carbon pieces requires only seven cleavage cycles.

(b)   Ninety-six ATP molecules per molecule of palmitate (8 acetyl CoAs; 12 ATPs per acetyl CoA).

(c) $34 + 96 = 130$ ATP molecules;

$$C_{16}H_{32}O_2 + 23O_2 + 130ADP + 130P_i \rightarrow$$
$$16CO_2 + 16H_2O + 130ATP.$$

6-3.   (a)

$$1 \text{ g glucose} \times \frac{1 \text{ mole glucose}}{180 \text{ g glucose}} \times \frac{38 \text{ moles ATP}}{1 \text{ mole glucose}} = 0.21 \text{ mole ATP/g}$$

$$0.21 \text{ mole ATP} \times 12 \text{ kcal/mole} = 2.5 \text{ kcal per gram of glucose}$$

(b)

$$1 \text{ g palmitate} \times \frac{1 \text{ mole palmitate}}{256 \text{ g palmitate}} \times \frac{130 \text{ moles ATP}}{1 \text{ mole palmitate}} = 0.51 \text{ mole ATP/g}$$

$$0.51 \text{ mole ATP} \times 12 \text{ kcal/mole} = 6.1 \text{ kcal per gram of palmitate}$$

(c)   Fat is $6.1/2.5 = 2.4$ times better a source of energy than carbohydrate on a per-gram basis. Organisms store energy as fat rather than as carbohydrate because it is more efficient; more calories can be packed into a gram of fat than into a gram of glucose. You may despair putting on 5 pounds of weight ("getting fat"), but you would have gained $5 \times 2.4 = 12$ pounds if you had stored the same amount of energy by "getting carbohydrate" instead!

(d)   The difference arises because fat is more highly reduced than is carbohydrate and hence must be oxidized more to get to the terminal level of CO_2. More oxidation means more energy, since oxidative reactions are invariably exergonic.

6-4. (a) It is a maximally conservative scheme; instead of providing a separate deaminating enzyme for each of the 20 amino acids, deamination occurs from a single carbon skeleton.

(b)

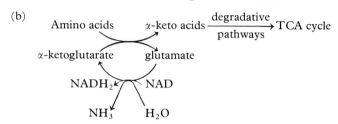

(c) Terrestrial animals retain their nitrogenous wastes in a concentrated form (urine) that is voided only periodically and therefore must ensure that nitrogen is not present as the toxic ammonia form. Aquatic organisms live in an environment in which water is plentiful and can be used for rapid dilution and removal of ammonia.

(d) The tadpole is an aquatic form and can afford the luxury of excreting nitrogen directly as ammonia, which though toxic is highly soluble and readily diluted by the aqueous environment. In preparation for transition to a terrestrial life-style, the frog must acquire the ability to convert ammonia to urea for holding and excreting as urine.

(e) Plants take up nitrogen in inorganic form (as nitrate or ammonium ions, usually) and apparently adjust intake to meet needs. They are also apparently good at recycling nitrogen, such that nitrogen from catabolic pathways is reutilized efficiently in anabolic sequences.

6-5. (a) The complete sequence is as follows:

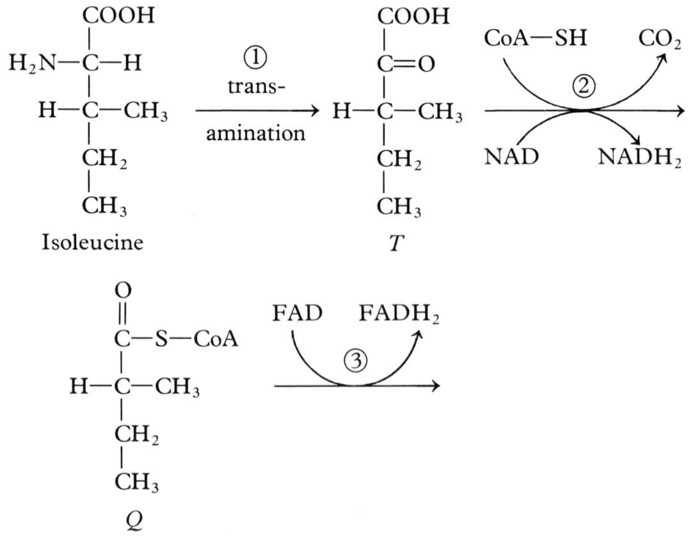

$$
\begin{array}{c}
COOH \\
| \\
H_2N-C-H \\
| \\
H-C-CH_3 \\
| \\
CH_2 \\
| \\
CH_3 \\
\text{Isoleucine}
\end{array}
\quad
\xrightarrow[\text{amination}]{\overset{①}{\text{trans-}}}
\quad
\begin{array}{c}
COOH \\
| \\
C=O \\
| \\
H-C-CH_3 \\
| \\
CH_2 \\
| \\
CH_3 \\
T
\end{array}
$$

with CoA—SH, CO₂, NAD, NADH₂ at step ②

$$
\begin{array}{c}
O \\
\| \\
C-S-CoA \\
| \\
H-C-CH_3 \\
| \\
CH_2 \\
| \\
CH_3 \\
Q
\end{array}
$$

with FAD, FADH₂ at step ③

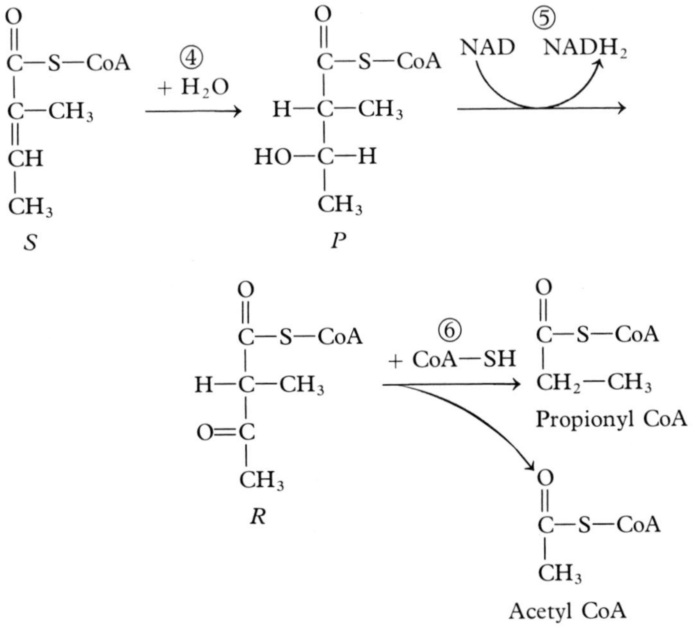

$$
\begin{array}{c}
\overset{\displaystyle O}{\overset{\displaystyle \|}{C}}-S-CoA \\
| \\
\overset{}{C}-CH_3 \\
\| \\
CH \\
| \\
CH_3 \\
S
\end{array}
\quad\xrightarrow[+\,H_2O]{\textcircled{4}}\quad
\begin{array}{c}
\overset{\displaystyle O}{\overset{\displaystyle \|}{C}}-S-CoA \\
| \\
H-\overset{}{C}-CH_3 \\
| \\
HO-\overset{}{C}-H \\
| \\
CH_3 \\
P
\end{array}
\quad\xrightarrow[\text{NAD}\quad\text{NADH}_2]{\textcircled{5}}\quad
$$

$$
\begin{array}{c}
\overset{\displaystyle O}{\overset{\displaystyle \|}{C}}-S-CoA \\
| \\
H-\overset{}{C}-CH_3 \\
| \\
O=\overset{}{C} \\
| \\
CH_3 \\
R
\end{array}
\quad\xrightarrow[+\,CoA-SH]{\textcircled{6}}\quad
\begin{array}{c}
\overset{\displaystyle O}{\overset{\displaystyle \|}{C}}-S-CoA \\
| \\
CH_2-CH_3 \\
\text{Propionyl CoA}
\end{array}
$$

$$
\begin{array}{c}
\overset{\displaystyle O}{\overset{\displaystyle \|}{C}}-S-CoA \\
| \\
CH_3 \\
\text{Acetyl CoA}
\end{array}
$$

(b)

Step	Prototype Reaction
1	Transamination (Reactions 6-4 and 6-5).
2	Oxidative decarboxylation of pyruvate to acetyl CoA or of α-ketoglutarate to succinyl CoA (Reactions 5-1 and TCA-4).
3 4 5 6	See Reactions Fat-2 to Fat-5. (You can, in fact, just regard Q as the CoA derivative of a fatty acid that happens to be methylated on the α-carbon.)

(c)

Position in Isoleucine	Eventual Position
2	Carboxyl carbon of succinyl CoA bearing the thio-ester link.
5	Methyl carbon of acetyl CoA.
1	Does not appear in either end product; lost as CO_2 in step 2 (though a trace might appear in the non–CoA-bearing carboxyl group of succinyl CoA if the $^{14}CO_2$ evolved in step 2 is reused in the carboxylation of propionyl CoA to form succinyl CoA).

All positions	All carbons of both end products except the non–CoA-bearing carboxyl group of succinyl CoA (and that could be labeled to a limited extent due to reutilization of liberated $^{14}CO_2$, as explained above).

6-6. (a) Both are genetic defects of amino acid catabolism, both result in the appearance in the urine (and therefore presumably in the blood) of keto acids, and both are characterized by brain damage and mental retardation.

   (b) The defect apparently involves the loss of the enzyme activity that catalyzes reaction 2 in the degradative pathway for isoleucine (Problem 6-5); if the keto acid cannot be oxidatively decarboxylated, it accumulates in the blood and spills over into the urine. The high blood levels apparently lead to the brain damage, as for phenylketonuria. (But note that here the keto acid which accumulates is part of the normal metabolic pathway, whereas for phenylketonuria the phenylketo acids are produced in significant quantities only in the absence of a normal functional pathway.)

   (c) Apparently there are other amino acids that yield keto acids B and C upon transamination, and these keto acids are also unable to undergo further metabolism in the presence of the defect, perhaps because they also require oxidative decarboxylation as the next step. (The amino acids are in fact valine and leucine.)

   (d) A reasonable conclusion is that further metabolism of all three keto acids depends upon the presence of some common enzyme. The α-keto acids corresponding to the amino acids valine, leucine, and isoleucine are in fact oxidatively decarboxylated by a single enzyme that, when genetically absent or deficient, leads to the simultaneous accumulation of all three keto acids.

   (e) Assuming prompt detection, the most reasonable course of action would be to restrict the dietary intake of isoleucine, valine, and leucine. These three amino acids cannot be eliminated from the diet since they are still needed for protein synthesis, but they must not be present in excesses requiring significant catabolism. This is likely to be considerably more difficult to achieve than in the case of phenylketonuria, since intake of three amino acids instead of one must be restricted. Further problems would arise if any of the intermediates in the normal metabolism of isoleucine, valine, or leucine are required as starting points for synthetic pathways in the cell. In the case of phenylketonuria, such needs can be provided by dietary tyrosine, which avoids the phenylalanine hydroxylase block; but in maple syrup urine disease the lack of the decarboxylating enzyme common to the catabolism of three amino acids means that none of the intermediates in these three degradative pathways can be formed in the cell.

Chapter 7

7-1.   Here are three major reasons. (1) ATP would be an inefficient way to transport energy since one molecule of ATP (MW = 500) has 2 high-energy phosphate bonds whereas one molecule of glucose (MW = 180) is equivalent to 38 high-energy bonds. (2) ATP is highly charged (four negative charges per molecule) and requires special mechanisms to move across biological membranes. (3) The supply of ATP would disappear shortly after the lights were turned out, so the plant would not live through the night; ATP is too bulky and unstable a form in which to store energy for very long periods of time.

7-2.   (a)   3 tons × 2000 lb/ton × 454 g/lb × 1 mole/342 g × 12CO_2/sucrose × 10 einsteins/mole = 9.6×10^5 einsteins required.

      (b)   1.4 cal/cm^2-sec × 43,560 ft^2 × 144 in.^{2}/ft^2 × 6.45 cm^2/in.^{2} × 105 days × 15 hr/day × 3600 sec/hr = 3.21×10^{14} cal or 3.21×10^{11} kcal or 5.8×10^9 einsteins available.

      (c) $9.6 \times 10^5/5.8 \times 10^9 = 1.7 \times 10^{-4}$ or 0.017 percent.

      (d)   Only a small portion of the total spectrum of incident light can be used by the photosynthetic cells, since chlorophyll absorbs only red and blue light; not all of the area is covered by photosynthetic tissue for all the growing season; not all days have 15 hr of actual sunshine; and not all of the light absorbed by the plant is stored as sucrose; most of it is used to carry out the necessary metabolism of the plant during the growing season. (Note, for example, that only 20 percent of the mass of the beets actually harvested is sugar.)

7-3.   Carbons 3 and 4 of glucose will be labeled first since they are derived from the carboxyl carbon of glycerate-3-phosphate. (The mathematically inclined will note that if every incoming CO_2 is labeled with ^{14}C, then 50 percent of the 3-PGA will be labeled in carbon 1 and 50 percent will be unlabeled initially. Of the glucose molecules formed initially, 25 percent will be unlabeled, 25 percent will be labeled in carbon 3 only, 25 percent in carbon 4 only, and 25 percent in carbons 3 and 4. Shortly, however, label will begin appearing in all carbon atoms as ribulose-1,5-diphosphate becomes regenerated from labeled carbons.)

7-4.   Since our cells can already fix CO_2 and make ATP and NADPH$_2$, we could presumably carry out the complete Calvin cycle, given the needful enzymes. But the only way we have of making the ATP and NADPH$_2$ is by oxidizing organic molecules, so we would in effect be oxidizing glucose to obtain ATP and NADPH$_2$ to make glucose. Since no system is 100 percent efficient, we could not make as much glucose by our hypothetical Calvin cycle as we would have to degrade to get the ATP and NADPH$_2$ and could therefore not effect net synthesis. The critical missing feature, of course, is the ability to use sunlight as the energy source.

7-5.   (a)   The difference in ATP yield per electron pair reflects the difference in reduction potentials that the transport systems span. In the transport system of respiratory metabolism, electrons move from NADPH$_2$ ($E_0 = -0.32$ volt) to O$_2$ ($E_0 = +0.82$ volt). This is a total potential difference of 1.14 volt, which corresponds to about 52 kcal/mole for a two-electron transfer, clearly enough energy to drive three phosphorylation events (assuming the ΔG for ATP hydrolysis to be in the usual cellular range of -12 to -14 kcal/mole). In noncyclic electron flow, the transport chain between photosystems I and II spans a potential gradient of about 0.5 volt, representing about 23 kcal/mole. This is apparently enough energy to drive one, but not two, phosphorylations.

(b)   Cyclic electron flow around photosystem I spans a gradient of about 1.0 volt (46 kcal/mole), about twice that of the noncyclic chain. This might reasonably be expected to couple to two phosphorylation events rather than to one, if the same general efficiency of energy conservation is to be realized.

(c)   The quantum requirement would be decreased from ten to nine because cyclic electron transport (which accounts for two of the ten photons per CO$_2$) would require only half as many photons.

7-6.   (a)   At 720 nm, because photosystem I will be active at this wavelength, pumping electrons *from* the transport chain to NADP.

(b)   Because at 650 nm, photosystem II will be active, pumping electrons *to* the intermediates of the transport chain from H$_2$O.

(c)   Because it is only upon illumination with both wavelengths that net, sustained flow of electrons from H$_2$O to NADP is possible. Without such a flow, NADPH$_2$ is not available to allow continued functioning of the Calvin cycle.

7-7.   (a) $\Delta G = \Delta G^0 + RT \ln \dfrac{[\text{ATP}]}{[\text{ADP}][\text{P}_i]}$

$= 9.25 \text{ kcal/mole} +$

$(1.987)(273 + 25)(2.303) \log_{10} \dfrac{1.38 \times 10^{-4}}{(5 \times 10^{-6})(6.7 \times 10^{-4})}$

$= 9250 \text{ cal/mole} + 1364 \log_{10} (4.1 \times 10^4)$

$= 9250 \text{ cal/mole} + 1364(4.6)$

$= 15{,}500 \text{ cal/mole} = 15.5 \text{ kcal/mole}$

(b) $\Delta G = -nF\Delta E$, so $15.5 \text{ kcal/mole} = -2(23.04)\Delta E$; $\Delta E = 15.5/46.08$
$= 0.34 \text{ volt}$.

(c)   No. One would need to know the value of E under the actual conditions existing in the chloroplast; one would also have to know whether a mechanism exists for coupling phosphorylation of ADP with electron transport at that step.

7-8.   Carbons 4 and 5 of the regenerated ribulose-1,5-diphosphate will be unlabeled.

7-9.   (a)   Each cytochrome has a characteristic absorption spectrum; that for cytochrome *c*-553 is apparently absent from the absorption profile for the cytochromes of mutant *ac*-206.

(b) II $\rightarrow$ plastoquinone $\rightarrow$ cytochrome b-559 $\rightarrow$ M component $\rightarrow$ cytochrome c-553 $\rightarrow$ (plastocyanin $\rightarrow$ chlorophyll P-700) $\rightarrow$ I. (Note that you cannot tell the order of plastocyanin and chlorophyll P-700 from these data. In point of fact, however, chlorophyll P-700 is a part of photosystem I, so it is the final component in the chain.)

(c) They use acetate as both an energy and a carbon source and therefore have no need for photosynthesis.

(d) Sure.

Chapter 8

8-1. (a) Glucose + 2ADP + 2P$_i$ + 2NAD $\rightarrow$
$$2 \text{ pyruvate} + 2\text{ATP} + 2\text{NADH}_2.$$

(b) 2 pyruvate + 6ATP + 2NADH$_2$ $\rightarrow$
$$\text{glucose} + 6\text{ADP} + 6\text{P}_i + 2\text{NAD}.$$

(c) $\Delta G^0 = -686 + 2(273.5) + 2(52) + 2(7.3)$
$$= -686 + 665.6 = -20.4 \text{ kcal/mole of glucose}$$

(d) $\Delta G^0 = +686 - 2(273.5) - 2(52) - 6(7.3)$
$$= +686 - 694.8 = -8.8 \text{ kcal/mole of glucose}$$

(e) Because if the pathway is thermodynamically feasible in one direction, it will not be feasible in the other direction without considerable (often enormous) changes in concentrations of key intermediates; moreover regulation would be difficult if the same enzymes were used at all steps in both directions.

(f) Activate; if acetyl CoA levels are high, the cell obviously has adequate fuel for energy metabolism and should therefore convert carbon to glucose for storage rather than burning yet more of it for energy. Activation of the pyruvate-carboxylating enzyme will initiate gluconeogenesis.

8-2. (a)

Glycerate-3-phosphate + NAD + glutamate $\rightarrow$
$$\text{serine} + \text{NADH}_2 + \text{P}_i + \alpha\text{-ketoglutarate}$$

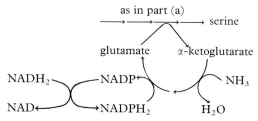

(b)   Pyruvate $\xrightarrow{\text{ATP} \quad \text{ADP} + \text{P}_i}$ oxaloacetate $\xrightarrow{\text{ATP} \quad \text{ADP}}$ phosphoenol pyruvate

as in part (a) $\leftarrow$ glycerate-3-phosphate $\leftarrow$ glycerate-2-phosphate

Pyruvate + NAD + glutamate + 2ATP + $H_2O \rightarrow$

serine + $NADH_2$ + $2P_i$ + 2ADP + α-ketoglutarate

(c)   Pyruvate $\xrightarrow{\text{as in part (b)}}$ glycerate-3-phosphate

$\xrightarrow{\text{as in part (a)}}$ serine

Pyruvate + NH_3 + 2ATP $\rightarrow$ serine + 2ADP + $2P_i$

Note that the sequence consumes one $NADPH_2$ (in the reductive amination of α-ketoglutarate) but generates one $NADH_2$ (in the oxidation of the hydroxyl group of glycerate-3-phosphate), so there is no net requirement for reducing power in the overall equation.

8-3.   (a)   Since the reduction of nitrate to ammonia is highly endergonic, the reverse process of nitrification must be highly exergonic, suggesting that it may be an important source of energy for the nitrifying bacteria. (In fact ammonia is the main electron donor in *Nitrosomonas*; the electrons flow along a cytochrome transport chain from NH_3 to O_2, and the transport of electrons is coupled to the phosphorylation of ADP to ATP. Similarly *Nitrobacter* obtains most of its energy from the oxidation of nitrite.)

(b)   They are *autotrophs* since they can obtain their carbon from CO_2 and do not need it in organic form. However, reducing power and ATP are generated by oxidation of ammonia or nitrite rather than by trapping light, making them *chemotrophic* rather than phototrophic with respect to energy requirements. See Table 1-1.

8-4.   (a)   The labeled compounds are:

ATP: added in labeled form

Glucose-6-phosphate: labeled by phosphorylation with ATP

Glucose-1-phosphate: same labeled phosphate as in glucose-6-phosphate

UDP-glucose: the phosphate derived from glucose-1-phosphate is labeled, the other is not

UTP: labeled by transfer of γ-phosphate from ATP; β-phosphate also labeled after first cycle; see below

UDP: labeled after the first cycle because the second (outer) phosphate comes from the phosphate contributed to UDP-glucose by glucose-1-phosphate

PP_i: both phosphates labeled after the first cycle because they represent the β and γ phosphates of UTP

P_i: as for PP_i

(b) Only ATP and ADP are labeled.

(c) Yes because ATP can be generated as follows (the enzyme that interconverts ATP and UTP is already present in the system): ADP + UTP → ATP + UDP.

(d) Middle phosphate of UTP appears as pyrophosphate upon UDP-glucose formation, so the only labeled compounds would be UTP, UDP, PP_i, and P_i.

8-5. (a) 4.5×10^{-15} g/cell $\times$ 1 mole/320 g $\times$ 6.02×10^{23} molecules/1 mole $= 8.5 \times 10^6$ nucleotides per cell.

(b) 8.5×10^6 nucleotides/cell $\times$ 2ATP/nucleotide polymerized $= 17 \times 10^6$ ATP per cell.

(c) $\dfrac{17 \times 10^6 \text{ ATP/cell}}{20 \text{ min}} \times \dfrac{1 \text{ glucose}}{38 \text{ ATP}} \times \dfrac{1 \text{ min}}{60 \text{ sec}} = 373$ glucose/sec.

(d) Bacterial cells can synthesize nucleotides from other cellular precursors (amino acids, CO_2, and so on; see Figure 8-1). Thus DNA synthesis and cell division could still proceed, but much more energy would be required if the cell had to carry out both the synthesis and the polymerization of the nucleotides.

8-6. The oxaloacetate that serves as the carboxylated intermediate is part of the TCA cycle and can be rapidly converted into malate and fumarate. The latter is a symmetric molecule, so label initially present in one of the terminal carbons of oxaloacetate and malate becomes scrambled upon equilibration with fumarate. Thus if any of the oxaloacetate produced by carboxylation of pyruvate equilibrates with malate and fumarate, the labeled carbon can be randomized between carbons 1 and 4 of all three compounds. And since carbon 1 is retained in the pyruvate and gives rise eventually to carbons 3 and 4 of glucose, the subsequent appearance of label in those positions of glucose attests to the occurrence of such equilibration.

Chapter 9

9-1. (a) P, A; (b) D; (c) A; (d) Ḋ, P, A; (e) D, P.

9-2. (a) The plot for ethanol is linear; that for acetate is hyperbolic, approaching a rate of 100 μmoles/min as the concentration approaches infinity. Ethanol therefore crosses the membrane by simple diffusion, whereas acetate requires a carrier. (Note that a distinction between passive and active transport cannot be made for acetate based on the data available.)

(b) Ethanol is not charged and is known to be miscible with organic solvents, whereas acetate is negatively charged at neutral pH and is water-soluble.

(c) Yes for ethanol (we would not get a linear plot over a thousandfold concentration range for a carrier-mediated process); no for acetate (we do not know the internal concentration).

(d) The solutes under examination are themselves substrates for respiratory metabolism; if the respiratory process is poisoned, substrate will not be metabolized and may accumulate in the cells at levels that affect uptake kinetics.

9-3. (a) Transcellular; the H^+ ions are being moved from the blood, across the epithelial cells, and into the stomach.

(b) Concentration gradient is $(10)^1/(10)^{-7} = 10^6$ (millionfold).

(c) $\Delta G = 2.303RT \log(10)^6 = (2.303)(1.987)(273 + 37)(6) = 8.51$ kcal/mole.

(d) 8.51 kcal/mole required; ΔG^0 for ATP hydrolysis is only -7.3 kcal/mole, but concentrations of ATP, ADP, and P_i are usually such that ΔG for ATP hydrolysis in the cell is in the range of -10 to -14 kcal/mole, so this would probably be adequate to drive H^+ secretion—except that we are dealing here with *transcellular* movement, so it will take at least one ATP to move H^+ into the parietal cell on the blood side and another ATP to move H^+ out on the stomach side. (Notice that by maintaining an intracellular pH somewhere between 1 and 7 the epithelial cell can reduce the amount of energy required to achieve transport across either membrane.)

9-4. (a) The red blood cell lacks mitochondria, so its only source of energy is glycolysis, of which the enolase reaction is a component part. Since both K^+ and Na^+ are kept at levels within the red blood cell that are not in equilibrium with the levels of those ions in the plasma in which the cell is bathed, active transport (and hence energy derived from glycolysis) is clearly necessary to maintain such high levels of intracellular K^+ and low levels of intracellular Na^+.

(b) If fluoride is added to the blood (and provided it can penetrate the red blood cell membrane to gain entrance to the cell), it will inhibit the enolase reaction, thereby stopping or greatly slowing glycolysis and reducing or eliminating ATP generation. Without adequate ATP, the Na^+/K^+ pump in the membrane cannot function and intracellular levels of K^+ and Na^+ will approach those of the blood plasma as K^+ leaks out and Na^+ leaks in.

(c) Probably not, since the algal cells may well be able to depend for transport of inorganic salts upon ATP produced by photophosphorylation and may therefore be relatively insensitive to an inhibitor of glycolysis, at least in the short term.

9-5. (a) No; Na^+ cotransport is not required for passive uptake.

(b) Yes; cotransport of Na^+ drives the inward movement of amino acid and can occur only if Na^+ is actively pumped back out again.

(c) Yes; K^+ must be actively pumped into red blood cells (see the data of Problem 9-4) and occurs via a pump that couples to outward pumping of Na^+.

(d) No; active uptake of sugars and amino acids in bacteria does not appear to be coupled to sodium cotransport as it is in animal cells.

(e) Yes; since about two-thirds of the total ATP requirement of kidney epithelial cells is used to transport sodium from the blood into the urine (see text), inhibition of sodium transport should reduce ATP requirements and hence respiratory rates.

9-6. (a) Aerobic respiration results in the net generation of $NADH_2$ in the cytoplasm due to glycolysis, but all $NADH_2$ reoxidation occurs in the electron transport chain located only in the mitochondria. Prokaryotes do not face this problem because electron transport occurs on the cellular membrane and is not compartmentalized from glycolysis.

(b) Both are doubly charged dicarboxylic acids that are very unlikely to cross the membrane by simple diffusion.

(c)

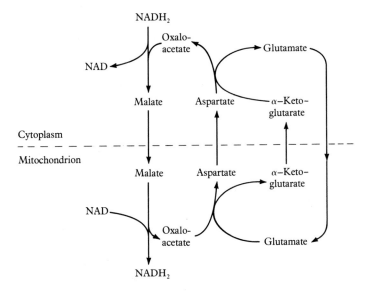

The scheme requires that α-ketoglutarate and glutamate move across the membrane also, in equimolar amounts but opposite directions.

(d) Yes because all the reactions are apparently reversible, each proceeding in one direction in the cytoplasm and in the other direction in the mitochondrion. A malate dehydrogenase that coupled to FAD instead of NAD would make the scheme irreversible, since $FADH_2$ and $NADH_2$ differ in energy by more than 12 kcal/mole (see Figure 5-4). If the NAD-linked dehydrogenase were reversible, then the FAD-linked reaction would be highly exergonic in the direction of malate oxidation.

9-7. (a) Roots were probably excised from young barley seedlings and suspended in a solution of KCl at the desired concentration, labeled with radioactive potassium ($^{42}K^+$). At the end of a uniform exposure period, roots were rinsed in nonradioactive KCl solution to remove $^{42}K^+$ not actually inside cells (loose association with the cell wall, for example) and then assayed for radioactivity. The data were expressed as counts per minute taken up per gram of root tissue, converted to micromoles (from known specific activity of KCl solution) and corrected for exposure time, if less than or greater than 1 hr.

(b) Component 1: y intercept is about 0.08, so $V_{max(1)}$ is about 12.5 μmoles/g-hr; x intercept is about -50, so $K_{m(1)}$ is about 0.02 mM. Component 2: y intercept is about 0.07, so $V_{max(2)}$ is about 14.3 μmoles/g-hr; x intercept is about -0.08, so $K_{m(2)}$ is about 12.5 mM. (Actual values as calculated from original data by Epstein et al.: $V_{max(1)} = 11.9$ μmoles/g-hr; $K_{m(1)} = 0.021$ mM; $V_{max(2)} = 13.2$ μmoles/g-hr; $K_{m(2)} = 11.4$ mM. Differences between their values and yours are probably due to the difficulty of reading intercepts accurately.)

(c) $$v_T = v_1 + v_2 = \frac{V_{max(1)}S}{K_{m(1)} + S} + \frac{V_{max(2)}S}{K_{m(2)} + S}$$

or, with values calculated by original investigators,

$$v_T = \frac{11.9S}{0.021 + S} + \frac{13.2S}{11.4 + S}$$

Choose a series of values for S, plug into the above equation, calculate a series of corresponding v_T values, and plot v_T versus S on a scale with a discontinuity between 0.2 and 0.5 mM on the x axis. Agreement with Figure 9-11 is amazingly good.

(d) Na^+ appears to inhibit the transport mechanism responsible for component 1 of the uptake curve only minimally, and the inhibition is competitive. Na^+ is a very potent inhibitor of the component 2 mechanism and probably inhibits competitively.

Chapter 10

10-1.　(a)　At a sarcomere length of 3.2 μ, length of the A band is 1.6 μ and length of the I band is 1.6 μ (0.8 μ on either side of the Z line). During contraction of the sarcomere from 3.2 to 2.0 μ, length of the A band remains fixed at 1.6 μ and length of the I band decreases linearly with sarcomere length, from 1.6 to 0.4 μ.

　　　(b)　The H zone corresponds to that portion of the thick filament length not overlapped by thin filaments. (It is, in fact, the lack of interdigitated thin filaments that gives the H zone the lighter density responsible for its name.) The distance from the Z line to the edge of the H zone represents the length of the thin filament and remains constant during contraction.

10-2.　(a)　1 mmole/1 min = 0.005 mmole/x min; x = 0.005 min or 0.3 sec.

　　　(b)　Contribution from ATP: 0.3 sec; contribution from creatine phosphate: $5 \times 0.3 = 1.5$ sec (because there is five times as much creatine phosphate); total: 1.8 sec. There are many examples of muscular movement where immediate movement rather than sustained contraction is essential. Try touching a hot stove if you are not convinced that contractions lasting less than 2 sec are important.

　　　(c)　
$$\frac{1 \text{ mmole ATP}}{\text{min} \times \text{g}} \times \frac{1 \text{ mmole glucose}}{38 \text{ mmoles ATP}} \times \frac{6 \text{ mmoles } O_2}{1 \text{ mmole glucose}} \times$$
$$\frac{22.4 \text{ ml}}{\text{mmole } O_2} = 3.5 \text{ ml } O_2 \text{ per minute per gram}$$

(Note that a gram of muscle occupies about 1 cm^3, so the contracting muscle uses $3\frac{1}{2}$ times its own volume in oxygen every minute!)

　　　(d)　
$$\frac{10 \text{ mg glycogen}}{1 \text{ g muscle}} \times \frac{1 \text{ mmole glucose}}{162 \text{ mg glycogen}} \times \frac{2 \text{ mmole ATP}}{1 \text{ mmole glucose}} =$$
$$0.12 \text{ mmole ATP per gram of muscle}$$

Thus 0.12 mmole ATP per gram can support contraction for 0.12 min or 7.2 sec. (The molecular weight used for glucose is 162 instead of 180 because glucose is polymerized into glycogen with the loss of water; see Reaction 8-12.)

　　　(e)　By repeated cycling of ADP through the myokinase reaction, it would in theory be possible to convert it all into AMP. In this way the high-energy anhydride bond of ADP can be used to generate an equimolar amount of ATP, which ought therefore to sustain contraction for another 0.3 sec; see part (a). The question is not realistic, however, inasmuch as the cell would be dead if all its ATP were converted to ADP, and the myokinase reaction would not "wait" until that stage but would become functional as ATP levels dropped and ADP levels climbed. Either way, however, it could not sustain contraction for more than an additional 0.3 sec.

10-3.   (a)   Rigor results from a failure to break the cross-bridges that link thick to thin filaments in the contraction cycle (Figure 10-3). In the living cell this detachment occurs as step 3, upon binding of the next molecule of ATP. After death, however, cellular ATP is hydrolyzed and cannot be replaced. This has two effects: (1) ATP is not available to cause cross-bridge detachment at step 3; and (2) in the absence of ATP, calcium can no longer be pumped into the sarcoplasmic reticulum. It therefore accumulates in the cytoplasm and promotes attachment of cross-bridges to actin (step 1). The net result is an accumulation of cross-bridges that lock the muscle filaments together and give the corpse its characteristic stiffness.

(b)   While running, since ATP levels would be already lower and cross-bridge formation would occur more quickly upon death.

10-4.   (a)
$$\text{Creatine phosphate} \rightarrow \text{creatine} + P_i \qquad \Delta G^0 = -10.3 \text{ kcal/mole}$$
$$\text{ADP} + P_i \rightarrow \text{ATP} \qquad \Delta G^0 = +7.3 \text{ kcal/mole}$$

$$\text{Creatine phosphate} + \text{ADP} \rightarrow$$
$$\text{creatine} + \text{ATP} \qquad \Delta G^0 = -3.0 \text{ kcal/mole}$$

$$\Delta G^0 = -2.303 RT \log_{10} K^0 \qquad \log_{10} K^0 = \frac{-\Delta G^0}{2.303 RT}$$

$$K^0 = \text{antilog}\left(\frac{-\Delta G^0}{2.303 RT}\right) = \text{antilog}\left(\frac{3000}{1364}\right)$$

$$= \text{antilog } (2.20) = 158.5$$

(b)   From part (a):

$$K^0 = \frac{[\text{creatine}][\text{ATP}]}{[\text{creatine phosphate}][\text{ADP}]} = 158.5$$

or $$\frac{\text{ATP}}{\text{ADP}} = 158.5 \, [\text{creatine phosphate/creatine}]$$

| *Relative Amounts* | | | | *Relative Amounts* | | |
Cr-P	Cr	$\frac{Cr\text{-}P}{Cr}$	$\frac{ATP}{ADP}$	ATP	ADP	$\frac{ATP}{ATP + ADP}$
0.9	0.1	9	1427	1427	1	0.999
0.5	0.5	1	158.5	158.5	1	0.994
0.1	0.9	0.11	17.4	17.4	1	0.945
0.01	0.99	0.01	1.6	1.6	1	0.615

As contraction proceeds, creatine phosphate levels fall very substantially before any perceptible decrease in ATP levels occurs. (When only 10 percent of creatine is in the phosphorylated form, more than 94 percent of the ADP still is.) With ATP level so unaffected by consumption of high-energy phosphate during contraction, the conclusion seemed logical that creatine phosphate rather than ATP was the immediate energy source.

(c) With the creatine kinase reaction blocked, creatine phosphate levels would remain high during contraction and the ATP level would fall rapidly and precipitously. It was in fact just such findings that led muscle biochemists to the conclusion that ATP is the immediate source of energy in contraction and that creatine phosphate is a reserve source used for rephosphorylation of the ADP.

10-5. Birds that do not routinely fly place little or no demand upon the breast muscle, and glycolysis will therefore be adequate to meet the limited energy needs. Accordingly such tissue contains few mitochondria and little or no myoglobin and is pale in color. Birds that do (or at least can) fly, on the other hand, must depend upon aerobic respiration to sustain the intense muscular activity of flight and therefore have breast muscle characterized by numerous mitochondria and much myoglobin, which give it a darker color.

10-6. Protein A is most likely the structural subunit of the microtubule wall, since it has no ATPase activity and has the right molecular weight. Protein C is probably a polymerized form of dynein, the protein of the arms. It has ATPase activity, associates with "de-armed" doublets, and restores the arms. Protein B may well be the monomeric form of dynein that polymerizes to form the function protein of the arms. It possesses the requisite ATPase activity and seems from the molecular weight to be a likely subunit of protein C.

    Since de-armed doublets are apparently available, it would be interesting to see if they consist of protein A only and whether they can be reconstituted from a solution of protein A. Protein B should also be tested for its ability to form a polymer and could also be tested with the de-armed doublets to see whether arms can be reconstituted from B directly.

10-7. (a) It disrupts the spindle fiber (or prevents its assembly) by solubilizing the microtubules; without the spindle fibers, chromosomes cannot move in an orderly manner and mitosis cannot occur.

    (b) If monomeric and polymerized form of tubulin subunits exist in equilibrium, anything that removes, inactivates, or complexes with the monomeric form will shift the equilibrium. Most likely colchicine binds to the monomeric, but not to the polymerized, form of tubulin:

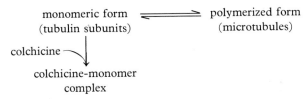

10-8. (a) As a physiological phenomenon, tetanus results from the arrival of successive impulses at the muscle cell membrane at such a rapid rate that calcium ions cannot be pumped back into the SR between pulses. These ions thus remain continuously in the cytoplasm, causing continuous contraction and leading to a "freezing" of the muscle in the fully contracted form that persists as long as the rapid nerve impulses do.

(b) The exotoxin produced by *C. tetani* might be a powerful stimulant of the central nervous system, such that nerve impulses are sent to the muscle at a rapid, unremitting rate. Alternatively it could act directly on the SR, causing release of (or failure to reaccumulate) calcium. (The former explanation is in fact the actual case.)

(c) *C. tetani* is only a danger if it gains entry to tissue and establishes itself. Any skin wound is therefore a possible site of entry, but punctures by sharp objects are a special threat because (1) the sharp object, if contaminated with *C. tetani* spores, can "inject" the spores into the underlying tissue very efficiently, (2) the puncture, unlike an open wound, is unlikely to bleed, such that the spores or cells will probably not be flushed out, and (3) the puncture is likely to be anaerobic, the environment required by *C. tetani*.

Index

341